111 Advances in Polymer Science

Polymer Synthesis

With contributions by
R. Arshady, A. Guyot, J. Lin, D.B. Priddy,
A.L. Rusanov, D.C. Sherrington, K. Tauer

With 48 Figures and 55 Tables

Springer-Verlag
Berlin Heidelberg GmbH

ISBN 978-3-662-14895-2   ISBN 978-3-540-47949-9 (eBook)
DOI 10.1007/978-3-540-47949-9

© Springer-Verlag Berlin Heidelberg 1994
Originally published by Springer-Verlag Berlin Heidelberg New York in 1994
Softcover reprint of the hardcover 1st edition 1994

Library of Congress Catalog Card Number 61-642

Typesetting: Macmillan India Ltd., Bangalore-25
02/3020 5 4 3 2 1 0 Printed on acid-free paper

Editors

Table of Contents

Polymer Synthesis via Activated Esters: A New Dimension of Creativity in Macromolecular Chemistry

Reza Arshady
Department of Chemistry, Imperial College of Science, Technology and Medicine, University of London, London SW7 2AY, UK

The polymerization and chemistry of activated acrylates have been elaborated in recent years to provide a uniquely versatile method of polymer synthesis. The new method is simple, generally applicable, and is ideally suitable for the synthesis of specialty polymers of interest in the emerging technologies in chemistry, engineering, biotechnology and medicine. This article discusses the polymerization and copolymerization of activated acrylates by solution and suspension techniques, and reviews polymer properties such as comonomer distribution, molecular weights, ^{13}C-NMR spectra and gel morphology. The use of copolymers of activated acrylates for the synthesis of a variety of specialty polymers, including amphiphilic gels, graft copolymers, and side chain reactive and liquid crystalline polymers is outlined. Potential applications of these polymers are also highlighted, and the versatility of active ester synthesis as a new dimension of creativity in macromolecular chemistry is emphasized.

Advances in Polymer Science, Vol. 111
© Springer-Verlag Berlin Heidelberg 1994

Abbreviations

AOBt	1-Acryloyloxybenzotriazol
AOSu	*N*-Acryloyloxysuccinimide
AOPcp	Pentachlorophenyl acrylate
AOTcp	2,4,5-Trichlorophenyl acrylate
AONp	4-Nitrophenyl acrylate
AOQu	Acryloyloxy-8-quinoline
AOPy	3-Pyridyl acrylate
AOCp	2-Carboxyphenyl acrylate
AOMcp	2-Methoxycarbonylphenyl acrylate

1 Introduction

The synthesis of specifically tailored macromolecular structures is a fascinating scientific endeavor in its own right, as well as a useful art for the creation of speciality polymeric materials for technological applications. The term "speciality" is employed here to describe a broad range of polymers embodying a "specific" chemical, physicochemical, optoelectronic or biophysical function, as compared with the mechanical function of commodity polymeric materials. In such a broad sense, speciality (or functional) polymers include many types and varieties of chemical structure and morphology, and are of potential interest in numerous emerging technologies in the fields of chemistry, engineering, biotechnology and medicine [1–10].

The author's own interest in this area includes new functional polymers for solid phase synthesis [11–13], polymers with "molecularly imprinted" substrate selectivity [14], polymer-supported transition metal catalysts [15], novel polymers of potential interest for electrocatalysis [16], targeting of colloidal drug carriers [17, 18], molecular composites [19], and biocompatible surfaces [20]. These studies have led to, among other things, a uniquely versatile method of polymer synthesis based on the chemistry of activated acrylates, i.e. polymer synthesis via activated esters. Various aspects of polymers and copolymers of activated (meth)acrylates have also been investigated in this and several other laboratories.

This review introduces the method of active ester synthesis, and discusses its application to the preparation of a variety of specialty polymers, including amphiphilic gels, graft copolymers, and side chain reactive and liquid crystalline polymers. The polymerization and copolymerization of activated acrylates by solution and suspension techniques are discussed, and polymer properties such as comonomer distribution, molecular weights, ^{13}C-NMR spectra and gel morphology are reviewed. Potential applications of these polymers are also highlighted, and the versatility of active ester synthesis as a new dimension of creativity in macromolecular chemistry is emphasized.

2 Synthesis Methodology

Functional polymers are conventionally produced by two alternative routes: (Co)polymerization of specifically functionalized monomers (e.g. 1–5 in Fig. 1 [21]) with suitably chosen structural monomers, or functionalization [1, 7, 22] of preformed non-functional polymers.

For most scientific studies, both of these two approaches are practicable, and the choice depends on experience and availability of starting materials. For the development of technologically viable functional polymers, however, an ideally

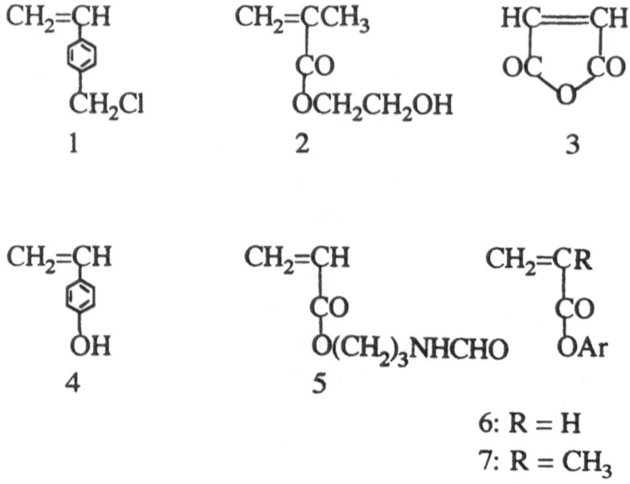

$$CH_2=CH$$

$$CH_2=CCH_3$$

$$HC=CH$$

1

2

3

$$CH_2=CH$$

$$CH_2=CH$$

$$CH_2=CR$$

4

5

6: R = H

7: R = CH$_3$

Fig. 1. Examples of functional monomers carrying specific functional groups (1–5), or general purpose activating (OAr) groups (6–7)

desirable route for the synthesis of functional polymers should embrace the following three criteria.

* A general purpose polymer intermediate, which can be produced readily and in a variety of forms (i.e. low or high molecular weights, soluble or crosslinked, colloid or suspension, surface graft or coating, etc.).
* A versatile chemistry of derivatization, which provides a simple route for functional and/or structural modification of the polymer without side reactions.
* A tuning device, whereby the overall structure of the polymer can be readily tailored for optimum performance in a given application.

This appraisal, and the author's interest in the synthesis of naturally occurring macromolecules, led to the idea that the chemistry of "activated (meth)acrylates" (6–7 in Fig. 1) may serve as the centerpiece of such an ideal synthetic route to functional polymers[1]. According to this idea (Fig. 2) [23–27], copolymerization of activated (meth)acrylates (6–7) with suitably chosen structural monomers produces a new class of general purpose activated polymer intermediates (8). This copolymerization process is technically efficient as will be discussed later. The prepositioned activating (or leaving) groups on the polymer are then displaced by the desired functional residues via a uniquely verssatile single-step reaction pathway. And finally, the functionalizing residues can be

---

[1] It is noteworthy that the formation of naturally occurring macromolecules (including polysaccharides, nucleic acids, proteins and natural rubber) is based on the chemistry of activated esters.

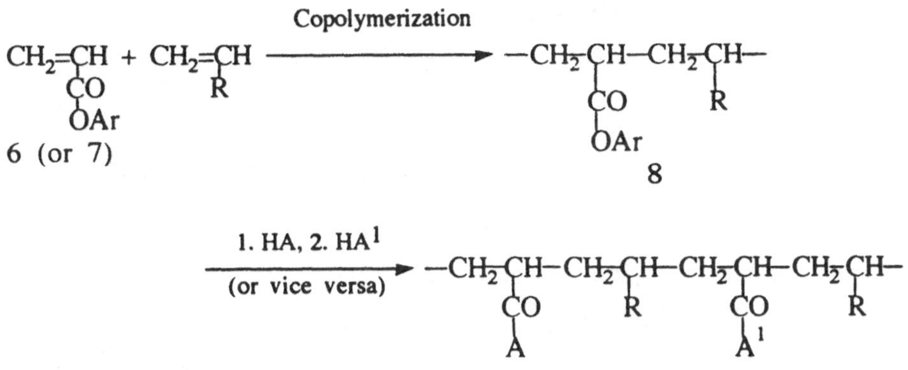

Copolymerization

$$CH_2\!=\!CH + CH_2\!=\!CH \longrightarrow -CH_2\!-\!CH\!-\!CH_2\!-\!CH-$$

6 (or 7)

8

$$\xrightarrow[\text{(or vice versa)}]{1.\ HA,\ 2.\ HA^1} -CH_2\!-\!CH\!-\!CH_2\!-\!CH\!-\!CH_2\!-\!CH\!-\!CH_2\!-\!CH-$$

OAr = Phenoxy or N-oxy residue (see Table 2)

$R = -\langle \rangle$, $CONH_2$, $COOCH_3$, etc.

A = Functional Residue, A^1 = Structural residue

Fig. 2. Polymer synthesis via activated esters (active ester synthesis)

chosen in such a way as to "tailor" or "tune" the overall structure of the resulting polymer in terms of chemical structure, crosslinking, frequency of the functional residues along the main chain, and their distance from the backbone.

The methodology of active ester synthesis, as shown in Fig. 2, is generally applicable and covers a wide range of nucleophiles, including primary, secondary and aromatic amines, primary alcohols and phenols. Thus, chemical modification of polymeric active esters (i.e. active ester synthesis) provides a "single-step" route for the preparation of functional polymers in general. The synthesis of various polymer types by the active ester method is advanced in Sects. 5–7. Here, an example of a relatively simple functional group (OH) is discussed to illustrate the versatility of the active ester method, as compared with conventional methods of polymer functionalization.

The hydroxy group (OH) is ubiquitous for the study and development of functional polymers. Typical examples of hydroxy-bearing polymers which can be produced in a single step by active ester synthesis are shown in Fig. 3. The substitution reaction proceeds quantitatively on the polymer, and the resulting hydroxy-bearing polymers are characterized by well-defined macromolecular structures. In contrast, the introduction of OH groups into, for example, polystyrene by functionalization requires a minimum of 2–4 steps for different derivatives [28–34].

Activated acrylates (**6**) are obtained readily by acylation of the corresponding phenols or N-hydroxy compounds with acryloyl chloride, acrylic anhydride or acrylic acid [21]. In the latter instance, a carbodiimide or other condensing

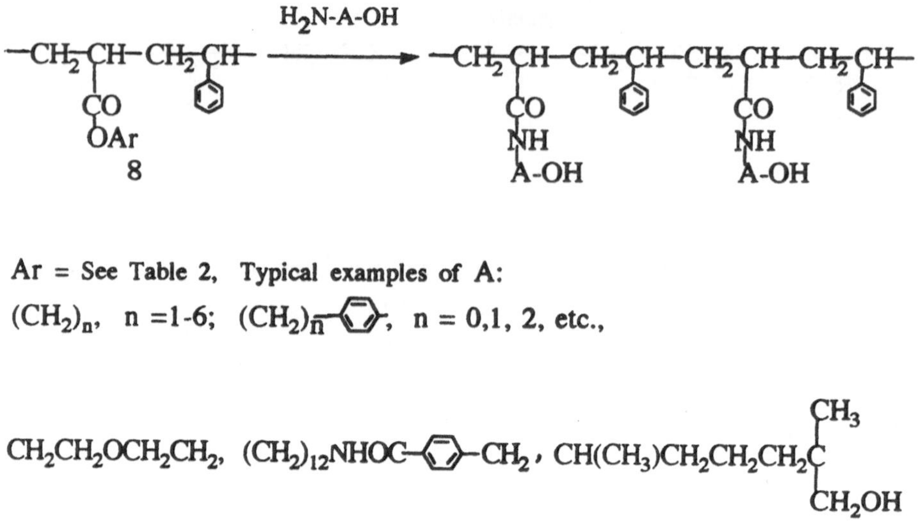

$$H_2N-A-OH$$

$$-CH_2CH-CH_2CH- \longrightarrow -CH_2CH-CH_2CH-CH_2CH-CH_2CH-$$

                CO   ⬡   CO   ⬡   CO   ⬡
                OAr   NH   NH
                 8   A-OH   A-OH

Ar = See Table 2, Typical examples of A:

$(CH_2)_n$, n = 1-6; $(CH_2)_{\overline{n}}$⬡, n = 0,1, 2, etc.,

$CH_2CH_2OCH_2CH_2$, $(CH_2)_{12}NHOC$⬡$-CH_2$, $CH(CH_3)CH_2CH_2CH_2\underset{CH_2OH}{\overset{CH_3}{C}}$

Fig. 3. Preparation of hydroxy-bearing polymers via active ester synthesis

Table 1. Main characteristics of activated acrylates (CH_2=CHCO–OAr, AOAr) reported for the synthesis of carboxyl-activated polymer intermediates

Monomer	Ar	Melting point (°C)	Characteristic IR (γ_{CO}, cm^{-1})	Ref.
AOBt	1-Benzotriazol	67–68		36
AOSu	N-Succinimide	68–69	1735, 1775	25, 36
AOPcp	Pentachlorophenyl			37
AOTcp	2,4,5-Trichlorophenyl	74	1760	25, 36
AONp	4-Nitrophenyl	64–65	1760	23
AOQu	8-Quinoline	53.5		25
AOPy	3-Pyridyl	b.p. 112/200 Pa	1752	25
AOCp	2-Carboxyphenyl	134–136	1720, 1760	25
AOMcp	2-Methoxycarbonylphenyl	b.p. 110/40 Pa	1725, 1760	25

agents used in peptide synthesis [35, 36] are needed. Essential details of a series of activated acrylates employed for the synthesis of activated polymer inter-mediates are summarized in Table 1. A list of abbreviations for the monomers is also given in Table 1 for ease of reference. Activated methacrylates (7 in Fig. 1) are obtained in the same fashion. The synthesis and polymerization of a wide range of activated (meth)acrylates have been reported in the literature [37–51].

3 (Co)polymerization of Activated acrylates

3.1 Reactivity Ratios

Reactivity ratios of a number of activated acrylates and methacrylates with different structural monomers are given in Table 2. Polymer compositions produced at low conversions from equimolar monomer feed compositions are also indicated in the Table. These values are back calculated from the reactivity ratios, and are hence equivalent to experimental data. The actual copolymerizability patterns of three comonomer pairs are also shown in Fig. 4. These results show that, among the various monomer pairs, AOTcp with styrene, and AOTcp with N-vinylpyrrolidone, produce azotropic copolymers at about equimolar monomer feed compositions. The relatively small reactivity ratios of these two monomer pairs indicate that their equimolar copolymerization produces approximately alternating, rather than random or block, copolymers.

Copolymerization of AOCp with N-vinylpyrrolidone was found to be complicated because of the low solubility of the resulting copolymers. This problem is thought to arise from strong interchain H-bonding between the two comonomer units. However, the copolymerization experiments could be carried out satisfactorily by using H-bond breaking solvents such as dimethylformamide (DMF) or dimethylsulfoxide (DMSO).

Table 2. Copolymerization reactivity ratios (r_f and r_s) of activated (meth)acrylates (M_f) and structural monomers (M_s)

Activated monomer	Comonomer	r_f	r_s	M_f incorporation into polymer[a]		Ref.
				10% M_f in feed	50% M_f in feed	
AOPcp	Acrylonitrile	1.44	0.40	20.1	61.6	37
AOPcp	Alkyl acrylate	0.21	0.88	10.3	39.2	37
AOTcp	Styrene	0.29	0.25	23.9	50.8	43
AOTcp	N-vinylpyrrolidone	0.17	0.02	46.0	53.4	44
AOCp	Syrene	0.62	0.50	16.3	51.9	45
AOcp	N-vinylpyrrolidone	0.61	0.01	49.5	61.4	45
MAOSu	Styrene	0.53	0.20	27.5	56.0	36
MAOTcp	Methacrylamide	1.04	0.54	16.4	57.0	36
MAOTcp	Acrylamide	0.49	1.23	8.0	40.0	36

[a] The data show theoretical polymer composition produced at low conversion, as calculated from the copolymerization equation: mole % of M_f in copolymer $= [100(r_f\alpha + 1)]/[r_f\alpha + (r_s/\alpha) + 2]$, $\alpha =$ mole fraction of $[M_f]/[M_s]$

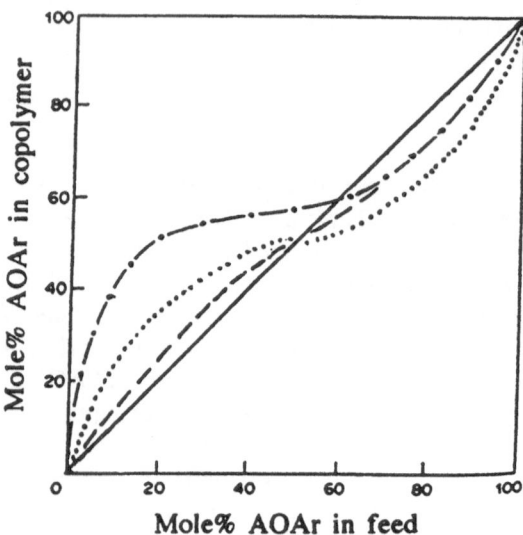

Fig. 4. Copolymerization graphs for 2,4,5-trichlorophenyl acrylate (AOTcp) with styrene (....) [43], AOTcp with N-vinylpyrrolidone (–·–) [44], and 2-carboxyphenyl acrylate (AOCp) with styrene (– – –) [45]. The *solid line* represents an ideally azotropic copolymerization system

3.2 Molecular Weights

Molecular weights data for two series of copolymers of AOTcp with styrene and with N-vinylpyrrolidone obtained at low conversions are given in Table 3 [43, 44]. In the case of the AOTcp-styrene system, M_w of polyAOTcp is about half that of polystyrene obtained under similar conditions. The dispersity indexes (M_w/M_n) for the two homopolymers are 2.0 and 1.69, respectively. The molecular weights of the copolymers increase with increasing styrene content, and a corresponding decrease in the value of the dispersity index is also observed. These results are thought to reflect the pattern of chain termination reactions in the two homopolymers and the intermediate copolymers, as shown in Fig. 5. The indicated termination mechanism is proposed on the basis of the established knowledge for the homopolymerization of acrylates and styrene [52]. These two systems involve termination by disproportionation (acrylates) and combination (styrene), with theoretical dispersity indexes of 2.0 and 1.5, respectively [52]. A similar molecular weight pattern is evident for the AOTcp-vinylpyrrolidone copolymers [44], although the latter data cover a relatively limited range of copolymer compositions.

Molecular weights of copolymers of AOCp with styrene, and AOCp with N-vinylpyrrolidone, are recorded in Table 4 [45]. It is interesting that the dispersity index for polyAOCp is 1.47, i.e. very close to the theoretical value for chain termination by combination. More notable is, however, that N-vinylpyrrolidone copolymers show very broad molecular weight distributions. These are thought to be largely due to poor polymer solubility and H-bonding between the comonomer units [53]. H-bonding between the AOCp units, and within a single AOCp unit, is also possible (cf. Ref. [54]). All of these molecular weights

Table 3. Molecular weight analysis of copolymers of AOTcp with styrene (STY)[a] and with N-vinylpyrrolidone (NVP)[b]. All copolymers obtained at low conversions (about 3–12 wt %). Data compiled from Refs. [43, 44]

Mole fraction of AOTcp in copolymer	Comonomer	Molecular weights[c]	
		$M_w \times 10^{-3}$	M_w/M_n
1.00	STY	20.357	2.00
0.83	STY	21.573	2.14
0.67	STY	22.700	1.99
0.48	STY	25.940	1.83
0.35	STY	30.470	1.83
0.30	STY	40.990	1.66
0.19	STY	46.430	1.74
0.00	STY	43.686	1.69
0.62	NVP	327	4.25
0.56	NVP	791	3.73
0.53	NVP	948	3.37
0.50	NVP	826	3.28

[a] Polymerization at $[AOTcp] + [STY] = 0.8 \ mol \cdot l^{-1}$, $[AIBN] = 9.08 \times \ mol \cdot l^{-1}$, and at 60 °C. [b] Polymerization at $[AOTcp] + [NVP] = 0.695 \ mol \cdot l^{-1}$, $[AIBN] = 2.8 \times 10^{-5} \ mol \cdot l^{-1}$, and 50 °C. [c] Determined by gel permeation chromatography, using polystyrene gels and polystyrene standards

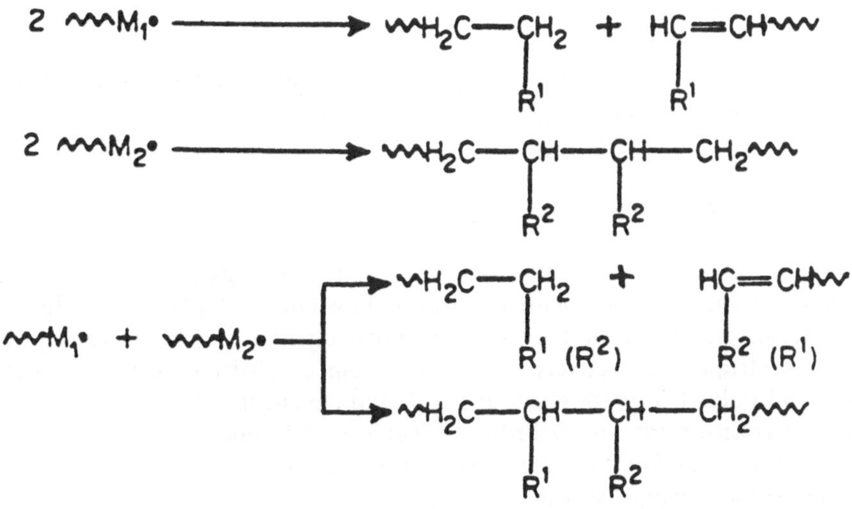

Fig. 5. Proposed chain termination reactions in copolymerization of AOTcp with styrene

were determined by gpc based on polystyrene gels and polystyrene standards. These conditions are considered to be satisfactory for polyAOTcp and copoly(AOTcp-styrene), but not for homo- and copolymers of AOCp and N-vinylpyrrolidone. The last-named polymers differ widely from polystyrene in

Table 4. Molecular weight analysis of copolymers of AOCp with styrene (STY)[a] and with *N*-vinylpyrrolidone (NVP)[b]. All copolymers obtained at low conversions (about 3–12 wt %). Data compiled from Ref. [45]

Mole fraction of AOCp in copolymer	Comonomer	Molecular weights[c]	
		$M_w \times 10^{-3}$	M_w/M_n
1.00	STY	8.82	1.47
0.76	STY	34.40	1.80
0.65	STY	45.69	1.77
0.52	STY	21.36	1.47
0.39	STY	28.36	1.60
0.27	STY	28.82	1.56
0.00	STY	23.02	1.60
1.00	NVP	8.82	1.47
0.69	NVP	7267	40.22
0.63	NVP	1970	21.88
0.58	NVP	3785	83.12
0.54	NVP	895	29.71
0.53	NVP	1765	Double peak
0.00	NVP	5.52	2.94

[a] Polymerization at [AOCp] + [STY] = 0.882 mol·l^{-1}, [AIBN] = 0.3060×10^{-2} mol·l^{-1}, and at 50 °C. [b] Polymerization at [AOTcp] + [NVP] = 0.886 mol·l^{-1}, [AIBN] = 3.049×10^{-2} mol·l^{-1}, and 50 °C. [c] Determined by gel permeation chromatography, using polystyrene gels and polystyrene standards

terms of chemical structure and solubility, and no firm conclusions can thus be drawn from the molecular weights data in Table 4.

3.3 NMR Spectroscopy

^{13}C NMR spectra of 2,4,5-trichlorophenyl acrylate (AOTcp), its homopolymer, and its equimolar copolymer with styrene are shown in Fig. 6 [43]. Peaks due to carbon atoms attached to none, one, or two hydrogen atoms appear as singlet, doublet, or triplet, respectively. The off-resonance [1]H-decoupled spectra of AOTcp and polyAOTcp are given in Fig. 7, and chemical shift assignments for different carbon atoms are recorded in Table 5. It is interesting that all six aromatic carbons in AOTcp are clearly resolved owing to the substituent effects [55] of the chlorine and oxygen atoms.

In polystyrene, C1 is split into three peaks (upfield intensities with ratio of 1:1.3:1.8), reflecting the presence of isotactic, heterotactic and syndiotactic triads in the chain [56]. In polyAOTcp, the aromatic carbon C1 is not affected by tacticity, and is not split. In the copolymer, C1 of styrene units is split into four peaks (1:1.2:3.3:1.7) whereas C1 of AOTcp units is split into two peaks (upfield intensities of 1.5:1). Similarly, the carbonyl carbon is split into two peaks in polyAOTcp (relative upfield ratios 1:1.6), and into three peaks in

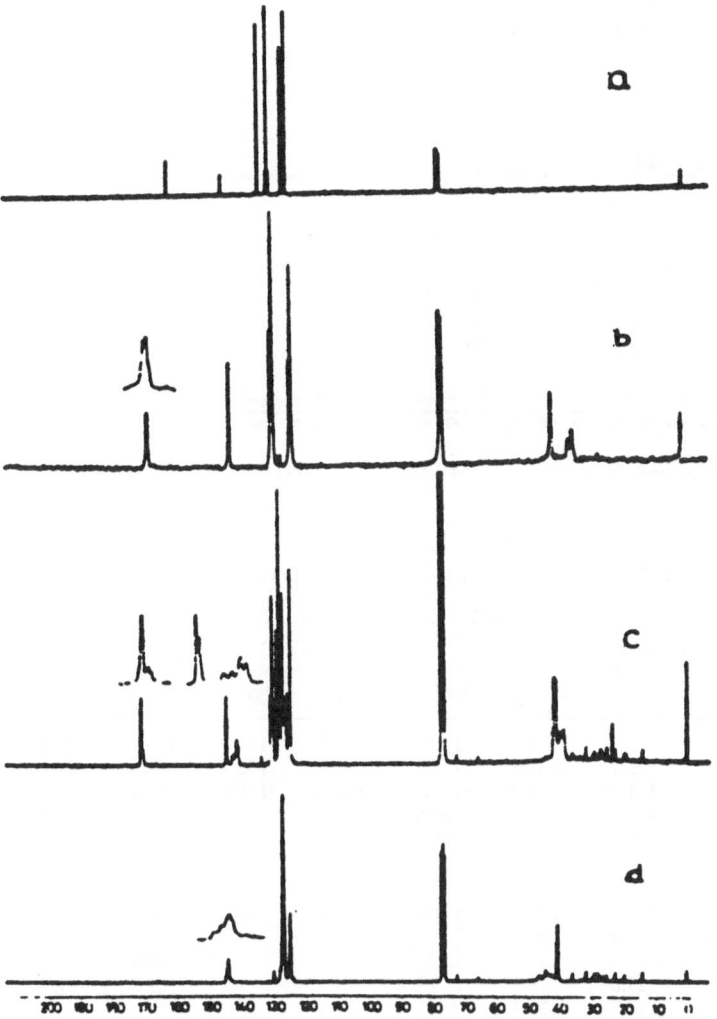

Fig. 6. ^{13}C NMR spectra of AOTcp (*a*), polyAOTcp (*b*), equimolar copoly(AOTcp-styrene) (*c*), and polystyrene (*d*)

copoly(AOTcp-styrene) (10.2:2.3:1.0). These additional splitings in copoly-(AOTcp-styrene), over those of the respective homopolymers, may result from the presence of unsymmetrical triads in the copolymer.

A series of low intensity peaks at 15–35, 66, 72 and 131 ppm in the spectra of polyAOTcp and copoly(AOTcp-styrene) are thought to result from the end groups of relatively low molecular weight chains. For the relatively higher molecular weight polystyrene (see Table 2) these peaks are not observed. However, it is difficult to ascertain the chemical nature of such low concentrations of end groups purely by ^{13}C NMR spectroscopy. A similar ^{13}C NMR

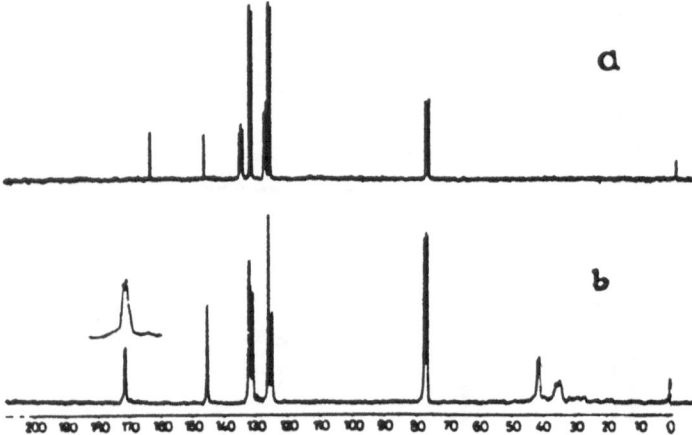

Fig. 7. Off-resonance proton decoupled ^{13}C NMR spectra of AOTcp (*a*) and polyAOTcp (*b*)

Table 5. C-13 chemical shifts of 2,4,5-trichlorophenyl acrylate (AOTcp), polyAOTcp and equimolar copoly(AOTcp-styrene)

Monomer or polymer	Chemical shifts (ppm)[a,b]								
	C1	C2	C3	C4	C5	C6	Cα	Cβ	Cγ
AOTcp	145.8	126.2	126.6	130.6	130.5	131.1	134.0	131.1	162.8
PolyAOTcp	145.2	125.7	131.0	131.7	131.9	125.2	34.8	41.4	171.1
	(155.3)	(121.4)	(131.9)	(129.1)	(134.5)	(118.8)	35.9		171.4
Polystyrene	145.3	127.7	128.3	125.5			44.0	40.5	
	145.6	127.9		125.6			44.3		
Copoly-(AOTcp-styrene)									
AOTcp units	145.4	125.8	130.6	131.4	131.4	125.2	38.5	41.4	171.5
	145.5	125.4	130.8				38.7		171.7
							38.9		172.2
Styrene units	141.8	127.0	128.6	125.2			42.0	41.0	
	142.1	127.7	129.7						
	142.8								
	143.5								

[a] Values given in parentheses are calculated on the basis of substitutional parameters in phenol (cf. Levy CG, Nelson GL (1972) Carbon-13 NMR for organic chemists. Wiley Interscience, New York, p 81)
[b] Carbon numbering in Copoly(AOTcp-styrene)

spectroscopy of homo- and copolymers of AOCp and *N*-vinylpyrrolidone has also been reported [45].

3.4 Suspension Copolymerization

In suspension copolymerization [57] of activated acrylates with styrene [23, 27], the monomer solution (including initiator) is dispersed in an aqueous medium to form a microdroplet suspension. Polymerization is then effected at the desired temperature (ca. 60–80 °C), to convert the monomer microdroplets to the corresponding polymer microspheres (beads or pearls). Essential details of a series of beaded polymer intermediates (**8**) produced by suspension copolymerization of AOTcp with styrene are given in Table 6. The beaded polymers were prepared by using 2–6% of a crosslinking monomer, namely divinylbenzene (DVB), ethylene dimethacrylate (EDM) or *N,N'*-dimethyl-1,6-hexanediacrylamide (HDA). A monomer diluent[2] [chlorobenzene or chlorobenzene-octane (1:1)] was also used to produce gel type or porous beads. Figure 8 shows scanning electron micrographs (SEM) of a typical example of the activated resins (that of **8c**).

Preparation of copoly(AOTcp-styrene) (**8**) by suspension polymerization is basically the same as that of commodity polymers such as polystyrene [57]. In fact, equimolar copolymerization of styrene with AOTcp proceeds at a considerably faster rate than homopolymerization of styrene (see above). As a result, complete monomer conversion is accomplished during a correspondingly shorter period of time (ca. 5 h vs. 15 h).

SEM micrographs of the surfaces and cross-sections of the activated resin samples **8a** and **8c** are shown in Fig. 9. These resin samples were obtained in the

Table 6. Examples of activated resin intermediates (**8**) produced by suspension copolymerization of styrene with 2,4,5-trichlorophenyl acrylate and a crosslinking monomer [46]

Activated resin	Matrix type	Crosslinking monomer (mol %)[a]	Monomer diluent (ml/g)[b]	Active ester content	
				mmol/g	Chlorine (%)[c]
8a	Gel type	DVB (3.0)	A (0.7)	2.77	29.56 (29.02)
8b	Porous	DVB (6.1)	B (1.33)	2.73	29.07 (28.81)
8c	Porous	DVB (5.5)	B (1.33)	2.74	29.18 (29.30)
8d	Porous	DVB (1.33)	B (1.33)	2.85	30.35 —
8e	Gel type	EDM (3.0)	A (0.7)	2.61	27.79 (28.10)
8f	Gel type	HBA (2.1)	A (0.7)	2.84	30.16 (29.12)
8g	Gel type	HBA (4.5)	A (4.0)	2.69	28.67 —

[a] DVB, *m,p*-divinylbenzene; EDM, 1,2-ethanedimethacrylate; HBA, *N,N'*-dimethyl-1,6-hexanediacrylamide. [b] A, chlorobenzene; B, chlorobenzene-*n*-octane (1:1). [c] Values found experimentally are given in parentheses

---

[2] The term "monomer diluent", rather than solvent, is used here to avoid confusion between the polymerization solvent (i.e. the monomer diluent) and water used as the suspension medium.

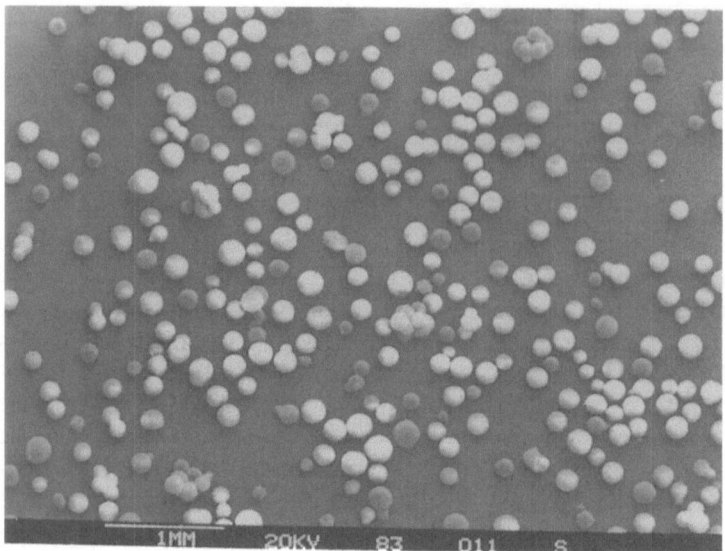

Fig. 8. Scanning electron micrograph of activated resin sample **8c** (see Table 6)

presence of a good solvent (chlorobenzene, **8a**), or a poor solvent (chloro-benzene-octane, **8c**), and represent typical examples of gel type and macro-porous resins, respectively. It is particularly interesting that, in both resin samples, the bead surface is substantially more porous than the cross-section. This is probably a reflection of the different environments (i.e. water or a nonpolar organic solvent) in which the polymers is formed.

The featureless cross-section of the resin sample **8a** observed at a magnific-ation of 12 000 is consistent with the expected gel type (relatively homogeneous) matrix. At a higher magnification of 40 000, however, both samples appear heterogeneous to differing degrees. In sample **8c**, the pores (in the region of 20–200 nm) are rather well defined, whereas in **8a** the pores are less numerous and less well defined.

4 Chemistry of Active Ester Synthesis

Aminolysis of alkyl esters [58, 59] and activated esters [60, 61] is generally known to be catalyzed by tertiary amines and basic solvents such as dimethyl-formamide (DMF). In addition, aminolysis of relatively less activated esters is catalyzed by the hydroxy components of the more reactive ones. This is especially true in the case of N-hydroxysuccinimide (HOSu) and 1-hydroxy-benzotriazol (HOBt), and the latter compound is routinely used as acylation catalyst [61–63] in peptide synthesis. In addition, the susceptibility of activated esters to general base catalysis, intramolecular and intra-polymeric catalysis, and solvent effects presents a fascinating area of organic polymer chemistry.

An interesting example of intra-polymeric catalysis is provided by the effect of polymer side chains on the aminolysis of polymer-bound nitrophenyl ester [41a], as illustrated in Fig. 10. Thus, apparent reactivity of the polymer-bound carbonyl groups is substantially increased by changing the polymer side chains from phenyl to methoxycarbonyl, and to dimethylamide. This type of intra-polymeric catalysis (shown schematically by species 9 in Fig. 11) assumes special significance in crosslinked polymers and solid phase synthesis. An important implication of this catalytic effect for polymer synthesis is that when an activated polymer intermediate (8) is not sufficiently reactive towards a given nucleophile, polymer reactivity can be enhanced by partial aminolysis with dimethyl-amine [25].

A related catalytic effect, namely intramolecular catalysis involving electron-rich residues on the substrate, has also been established [25]. Thus, trans-esterification of the ester groups on copoly(AOTcp-styrene) with $HO(CH_2)_nX$ (n = 2, 3 or 6), proceeds satisfactorily at above $50\,°C$, when X is NHCHO, $N(CH_3)_2$ or PPh_2, but not when X is Cl or Br. These observations clearly indicate that the former species enhance the nucleophilicity of the hydroxyl group via intramolecular catalysis, as represented schematically by the species 10 in Fig. 11. A similar intra-molecular catalysis appears to operate in the case of amines carrying catalytic residues at the ε-carbon. For example, aminolysis of the above-named polymer with $HN(CH_3)CH_2CH_2OH$ proceeds readily at (or below) room temperature; but reaction with $HN(CH_3)CH_2CH_2CN$ is effective only at temperatures higher than $50\,°C$, under otherwise similar conditions.

Chemical modification of polymer-bound active ester groups is also subject to strong solvent effects. In copoly(AOTcp-styrene), both aminolysis and trans-esterification with primary alcohols are positively influenced by solvents in the order of dimethylformamide (DMF) > dioxan > chloroform > chloro-benzene > dimethylsulfoxide (DMSO). However, trans-esterification with phenols proceeds in dioxan, but not in DMF. The last-named solvent effect is probably related to inactivation of the phenolate ion in DMF, as observed also for the acylation of polymer-bound phenolic groups by soluble trichlorophenyl esters [64].

A series of quantitative data for solvent effect on the aminolysis of nitro-phenyl esters attached to polyacrylamides have also been reported [41b]. These data are in broad agreement with the above-mentioned observations. However, the apparent solvent effects in chemical transformation of polymers must be interpreted in terms of a dual function, i.e. "polymer solvation" and "solvent catalysis". For example, DMSO is a poor solvent for copoly(AOTcp-styrene), but a good solvent for polymers carrying amide residues. It should also be noted that alcohols and water are not usually suitable as solvent for chemical transformation of activated esters, because they may themselves enter the reaction as nucleophiles.

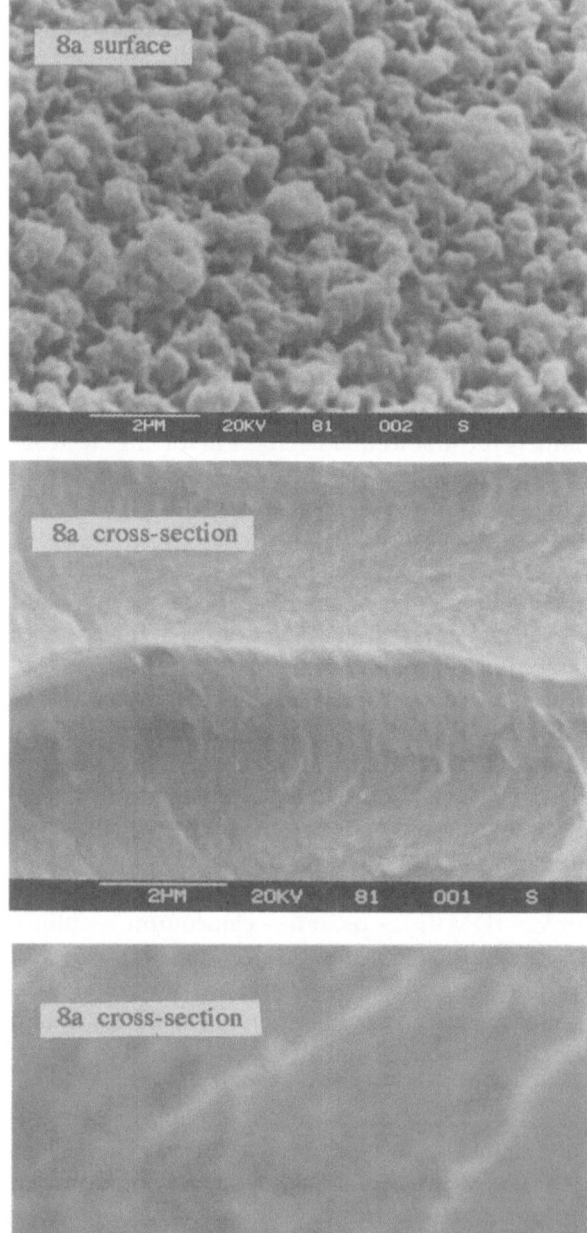

Fig. 9. Scanning electron micrographs of surface and cross-sections of beaded resins obtained by suspension copolymerization of AOTcp with styrene and divinylbenzene (see Table 6)

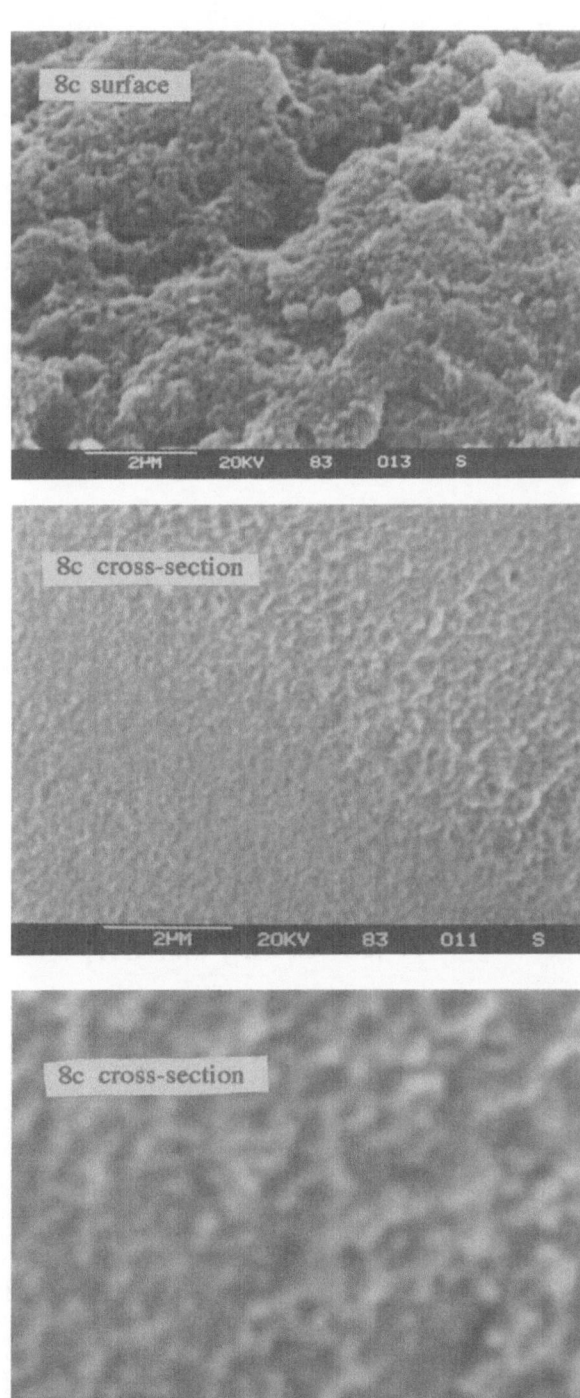

Fig. 9 (contd).

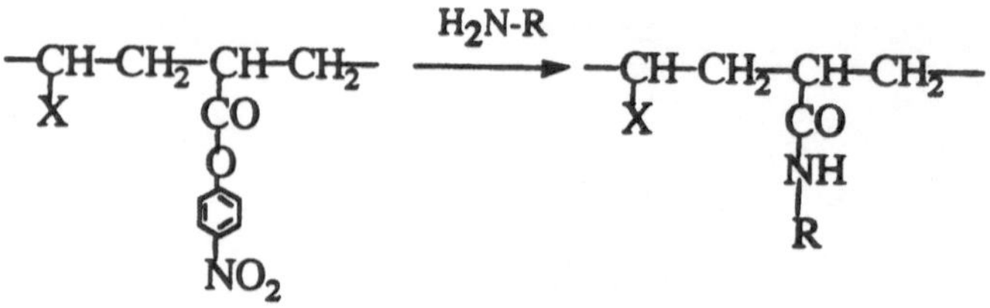

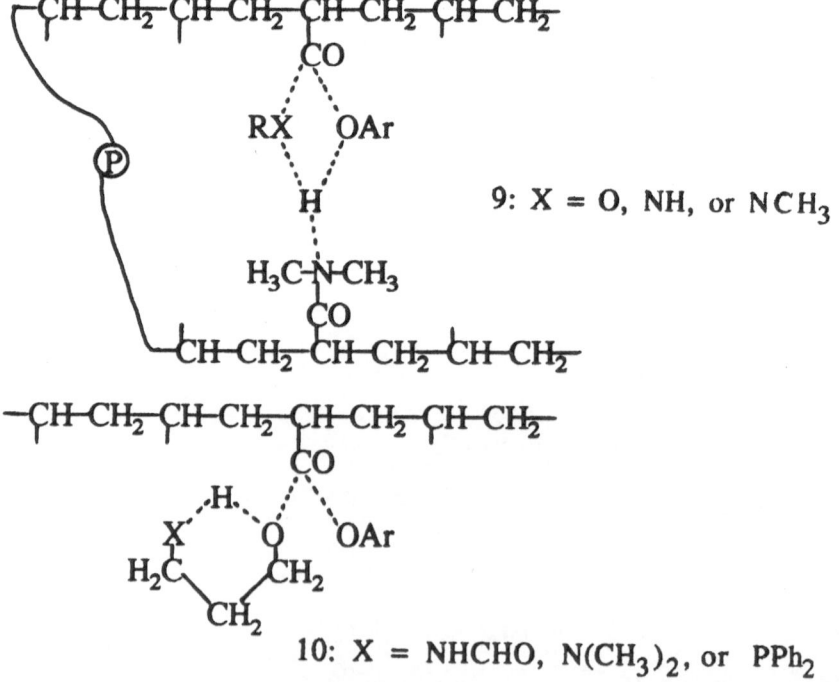

Rate of polymer aminolysis in dioxane at 50 °C
(Relative to that of 4-nitrophenyl 2-methylpropanoate)

X \ R	$(CH_2)_3CH_3$	CH_2Ph	$(CH_2)_9CH_3$	CH_2CH_2OH
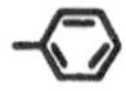 (phenyl)	0.03	0.02	0.03	0.02
COOMe	0.29	0.26	0.32	0.20
$CONMe_2$	5.68	4.67	4.92	2.93

Fig. 10. Effect of polymer side chains on the rate of aminolysis of polymer-bound nitrophenyl ester. The data compiled from Ref. [41a]

9: X = O, NH, or NCH_3

10: X = $NHCHO$, $N(CH_3)_2$, or PPh_2

Fig. 11. Schematic presentation of intra-polymeric (**9**) and intra-molecular (**10**) catalysis in polymer synthesis via activated esters

An interesting side reaction has also been observed in alcohol transesterification of polymer-bound trichlorophenyl esters. Thus, transesterification with secondary alcohols at 50–100 °C is accompanied by the formation of free carboxyl groups on the polymer. Free carboxyl groups may form via elimination of the resulting alkyl esters, or more probably from the corresponding transition states. In the case of tertiary alcohols no transesterification takes place, and only free carboxyl groups are formed.

5 Amphiphilic Gels for Solid Phase Synthesis

The methodology of solid phase peptide synthesis (SPPS) [65, 66] has been credited with the award of 1984 Nobel Prize in chemistry [67] to its inventor, Bruce R. Merrifield of the Rockefeller University. At the heart of the SPPS lies an insoluble polymer support or "gel", which renders the synthetic peptide intermediates insoluble, and hence readily separable from excess reagents and by-products. In addition to peptide synthesis, beaded polymer gels are also being studied for a number of other synthetic and catalytic reactions [2]. Ideally, the polymer support should be chemically inert and not interfere with the chemistry under investigation. The provision of chemical inertness presents no difficulty, but the backbone structure of the polymer may profoundly influence the course of the reaction on the polymer support. This topic has attracted considerable interest, particularly in relation to the properties of polystyrene (nonpolar, hydrophobic), polydimethylacrylamide (polar, hydrophilic), and copoly(styrene-dimethylacrylamide) (polar-nonpolar, amphiphilic) (see later).

5.1 Synthesis of Amphiphilic Gels

Amphiphilic gels (11) suitable for solid phase synthesis are produced by the active ester method according to Fig. 12. The activated polymer intermediates

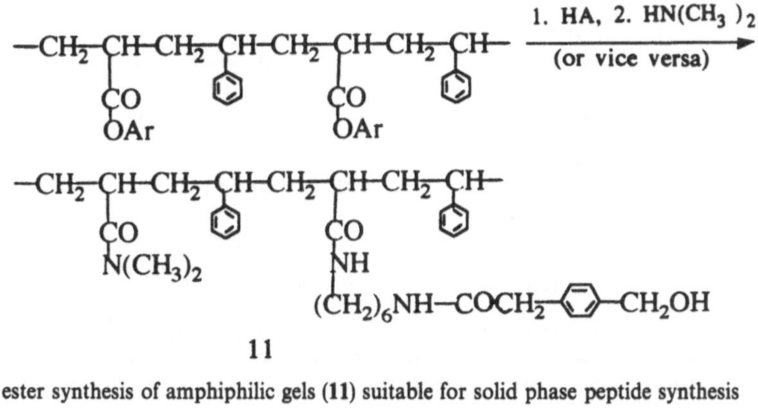

Fig. 12. Active ester synthesis of amphiphilic gels (11) suitable for solid phase peptide synthesis

(8) are treated with the calculated quantity of the desired functionalizing nucleophile (HA), followed by reaction with an excess of the structural amine (HA^1), or vice versa. A wide range of amphiphilic gels with relatively low degrees of functionality (< 1 mmol/g) obtained according to Fig. 12 are listed in Table 7.

More highly functionalized gels are obtained via complete displacement of the activating groups by the functionalizing reagent (see Table 8). Interesting examples of highly functionalized gels are polymeric phenols and amines (12–14 and 17, Fig. 13), which are of potential interest as acylating and deprotecting

Table 7. Amphiphilic polymers (11) with relatively low degrees of functionality obtained via active ester synthesis (see Fig. 12)

Activated resin precursor	A^1	A	Capacity mmol/g[a]
8e	N(CH$_3$)$_2$	NH(CH$_2$)$_6$NHCOOC(CH$_3$)$_3$	1.19
8e	N(CH$_3$)$_2$	NH(CH$_2$)$_6$NH$_2$	1.15
8e	N(CH$_3$)$_2$	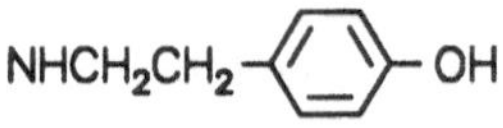 NHCH$_2$CH$_2$—⟨⟩—OH	1.16
8e	N(CH$_3$)$_2$	NH(CH$_2$)$_6$OH	1.15
8c	N(CH$_3$)$_2$	O(CH$_2$CH$_2$O)$_5$OH	0.35
8e	N(C$_2$H$_5$)$_2$	NH(CH$_2$)$_6$NH$_2$	0.48
8e	N(C$_2$H$_5$)$_2$	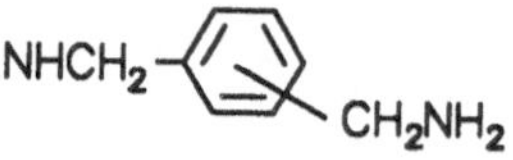 NHCH$_2$—⟨⟩—CH$_2$NH$_2$	0.28
8a	N(C$_2$H$_5$)$_2$	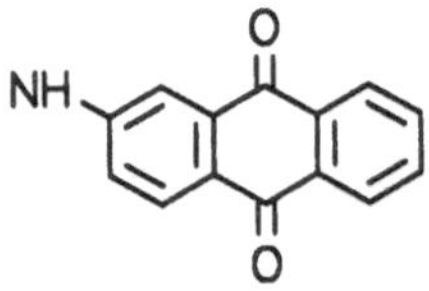 NH—(anthraquinone)	1.15
8e	N(C$_2$H$_5$)$_2$	NH(CH$_2$)$_5$COOH	0.48
8e	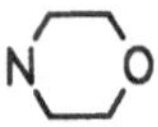 N◯O (morpholino)	OCH$_2$C(CH$_3$)$_2$NHCHO	0.73
8f	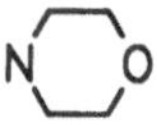 N◯O (morpholino)	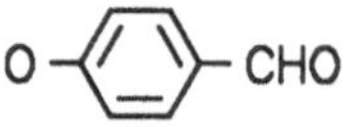 O—⟨⟩—CHO	1.72
8c	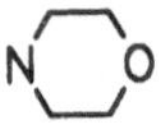 N◯O (morpholino)	O(CH$_2$)$_3$PPh$_2$	0.52
8c	N(CH$_2$)$_3$OH	O(CH$_2$)$_3$PPh$_2$	0.65
8c	N(CH$_2$)$_3$OH	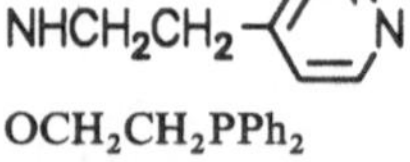 NHCH$_2$CH$_2$—⟨⟩N	0.63
8c	O(CH$_2$CH$_2$O)$_{10}$CH$_3$	OCH$_2$CH$_2$PPh$_2$	0.56

[a] Theoretical values calculated from the quantity of HA^1 or HA used (see Ref. 25)

Table 8. Examples of highly functionalized amphiphilic polymers obtained via active ester synthesis.

Activated resin precursor	Functional residue	Capacity (mmol/g)[a]	Characteristic IR (cm^{-1})
8e	NHCH$_2$CH$_2$OH	4.36	1660–80
8e	NH(CH$_2$)$_5$COOH	3.51	1670, 1700
8e	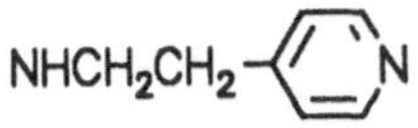 NHCH$_2$CH$_2$-⟨pyrazine⟩N	3.61	1520–60, 1605
8c	NH(CH$_2$)$_3$NHCH$_3$	3.91	Not recorded
8c	NHCH$_2$CH=CH$_2$	4.45	Not recorded
8a	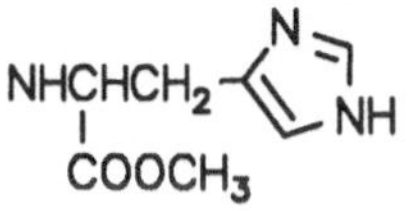 NHCHCH$_2$-⟨imidazole⟩NH COOCH$_3$	3.04	1665, 1730
8e	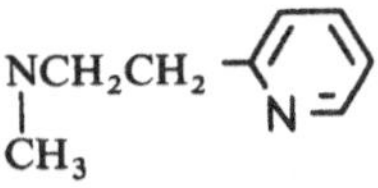 NCH$_2$CH$_2$-⟨pyridine⟩N CH$_3$	3.61	1520–50, 1605
8e	NCH$_2$CH$_2$CN CH$_3$	4.19	1680, 2255
8c	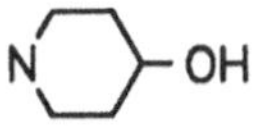 N⟨cyclohexyl⟩-OH	3.72	1625
8e	O(CH$_2$)$_6$NHCHO	3.34	1670, 1725
8c	O(CH$_2$CH$_2$O)$_{16}$CH$_3$	1.09	1630–40, 1725
8d	O(CH$_2$)$_3$N(CH$_3$)$_2$	3.90	1720
8e	 HN-⟨phenyl⟩-OH	3.97	1700–1710
8e	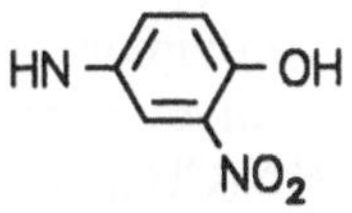 HN-⟨phenyl⟩-OH NO$_2$	3.24	1520, 1700–1710
8c	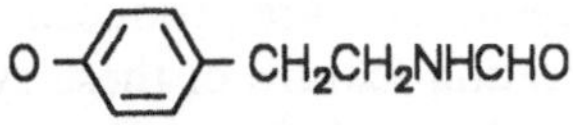 O-⟨phenyl⟩-CH$_2$CH$_2$NHCHO	3.01	1670, 1760

[a] Calculated on the basis of chlorine content of the activated polymer precursors. Values obtained by analysis are, within experimental error, indistinguishable from the calculated data

reagents for peptide synthesis by inverse solid phase (see later). Polymeric acylating reagents (12–14) are obtained from the activated resins (8) via two alternative routes [46]. The polymer derivative (12) was prepared by aminolysis of the activated polymer with N-methyl-1,3-propanediamine, followed by reaction with 4-hydroxy-3-nitrobenzoic acid. Esterification of the resulting phenolic

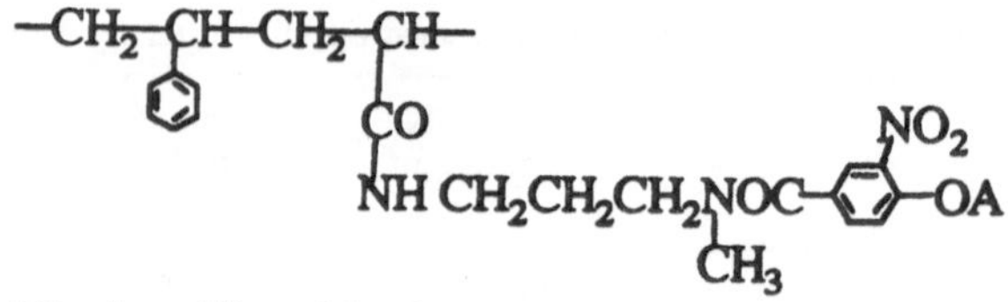

$$-CH_2-CH-CH_2-CH-$$

12: A = H,   13: A = N-Protectected amino acid

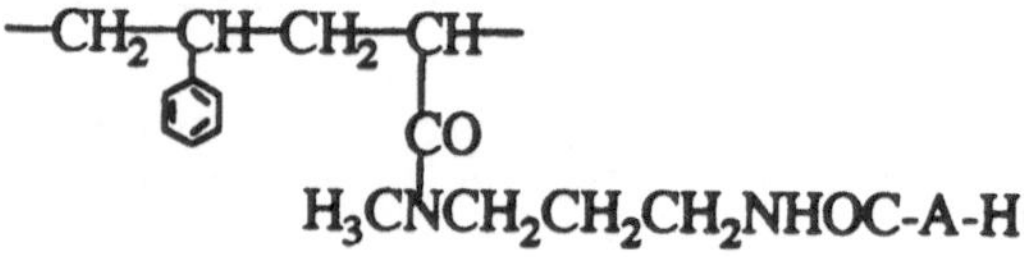

$$-CH_2-CH-CH_2-CH-$$

$$H_3CNCH_2CH_2CH_2NHOC-A-H$$

14: A =

$$-CH_2-CH-CH_2-CH-$$

15: R^1 = , R^2 = -CH$_2$

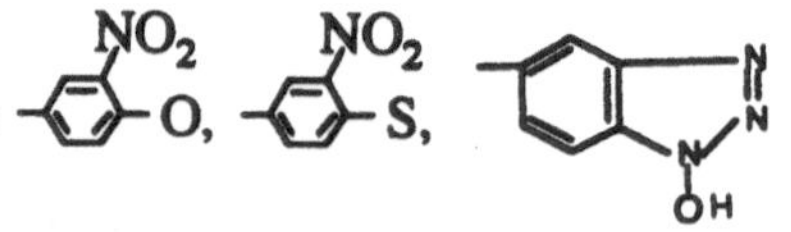

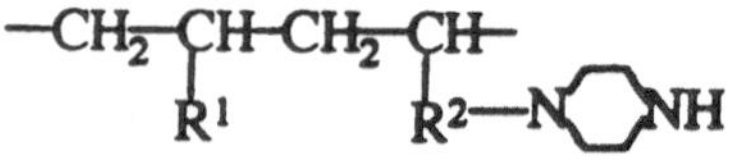

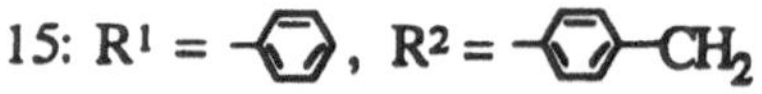

16: R^1 = CON(CH$_3$)$_2$, R^2 = CONHCH$_2$CH$_2$

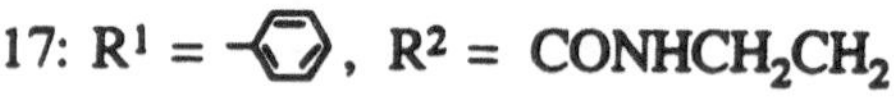

17: R^1 = , R^2 = CONHCH$_2$CH$_2$

Fig. 13. Examples of polymeric reagents useful for peptide synthesis by inverse solid phase method (see Fig. 16)

polymer with N-Fmoc-amino acids (glycine, alanine, or leucine) produced the corresponding polymeric acylating reagents (13) containing about 1 mmol/g amino acid. Preparation of the acylating reagents 14 involves direct aminolysis of the activated polymer by the amine precursor, $HN(CH_3)CH_2CH_2-A-OH$.

The amphiphilic polymeric amine reagents (17) are produced via active ester synthesis, according to Fig. 14. An interesting feature of these syntheses is the use of unsymmetrical diamines for aminolysis of the activated resins. When carried out at relatively low temperatures (usually in an ice bath), these reactions proceed mainly via the primary amino function for most of the unsymmetrical diamines examined (Fig. 14). In the case of 3-(N-methylamino)propylamine and 1-(2-aminoethyl)piperazine, for example, only about 3–4% of the diamine units take part in intra-resin crosslinking. This is in contrast with extensive intra-polymeric crosslinking generally observed when functional polymers react with symmetrical difunctional reagents (cf. 17b, see also Sect. 8.2). Interestingly, crosslinking is also observed in the case of 4-aminomethylpiperidine, leading to the formation of a highly rigid polymer with substantially reduced amine content.

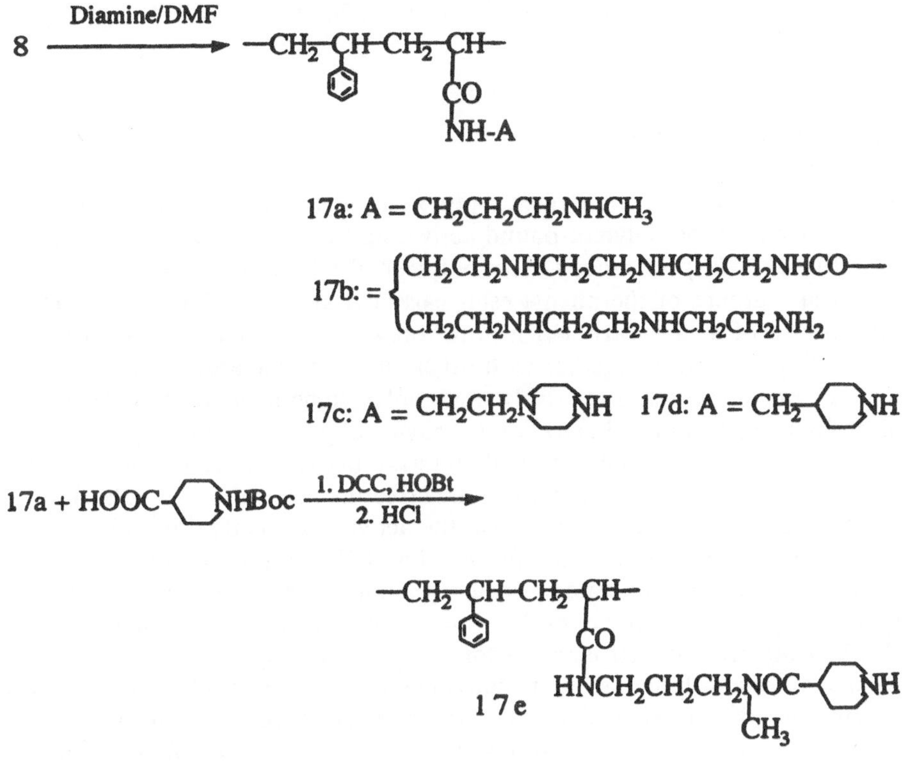

Diamine/DMF

8 $\longrightarrow$ $-CH_2-CH-CH_2-CH-$
                          | |
                       (phenyl)   CO
                                        |
                                      NH-A

17a: A = $CH_2CH_2CH_2NHCH_3$

17b: = $\begin{cases} CH_2CH_2NHCH_2CH_2NHCH_2CH_2NHCO- \\ CH_2CH_2NHCH_2CH_2NHCH_2CH_2NH_2 \end{cases}$

17c: A = CH_2CH_2N(ring)NH 17d: A = CH_2(ring)NH

17a + HOOC-(ring)-NHBoc $\xrightarrow[\text{2. HCl}]{\text{1. DCC, HOBt}}$

$-CH_2-CH-CH_2-CH-$
        | |
     (phenyl)   CO
                       |
17e $HNCH_2CH_2CH_2NOC-$(ring)$-NH$
                        |
                       CH_3

DCC, Dicyclohexylcarbodiimide
HOBt, 1-Hydroxybenzotriazol

Fig. 14. Preparation of basic polymeric deprotecting reagents via active ester synthesis

5.2 Introduction of Spacer Arm

It is recognized that polymer-bound functional groups generally show lower reactivities than the corresponding small molecular species for steric reasons. This reduced reactivity is more pronounced in crosslinked polymers as a result of both "reduced mobility" of the polymer chain segments and "spatial" hindrance within the crosslinked matrix. Reduced accessibility of functional groups caused by low mobility of the polymer chains can, to some extent, be remedied by introducing a spacer arm between the polymer backbone and the functional groups.

In the case of conventionally produced polymers, the positioning of functional groups at the end of a spacer arm involves a multi-step synthesis, in addition to those needed for the introduction of the functional groups themselves. For the copolymer resins produced by the active ester method, the concept of spacer arm is an integral element of the polymer design. Thus, the desired functionality, already positioned at the end of a 1- to 9-bond spacer arm (see Tables 7–8), is introduced into the polymer without any additional synthetic steps.

5.3 Polymer Analysis

An important feature of polymer synthesis by the active ester method is that displacement of the polymer-bound activating (leaving) groups can be readily monitored by IR spectroscopy of the polymer. The IR spectrum generally shows the disappearance of the phenyl ester carbonyl at about 1760 cm^{-1}, and the appearance of a new carbonyl absorption at $1620–80 \text{ cm}^{-1}$ (amide) and/or $1720–30 \text{ cm}^{-1}$ (ester), together with other characteristic absorptions due to A and A^1. In some cases, a relatively weak polystyrene band at ca. 1607 cm^{-1} is also observed. A typical illustration is provided by Fig. 15, showing the aminolysis of the activated polymer with 6-*tert.*-butoxycarbonylaminohexylamine, followed by reaction with dimethylamine (see the first entry in Table 7).

In addition, polymer synthesis via the active ester method usually involves substantial changes in the heteroatom (Cl and N) composition of the polymer. When necessary, the liberated trichlorophenol can also be quantified by UV spectroscopy to confirm the results of polymer analysis. Thus, a combination of IR spectroscopy and microanalysis for Cl and N, usually provides a convenient means of establishing polymer functionality and chemical structure. It is noteworthy that precise chemical characterization of crosslinked polymers obtained by conventional methods of functionalization is often difficult or impracticable.

5.4 Polymer–Solvent Compatibility

Swelling behavior of typical amphiphilic gels obtained by active ester synthesis are compared with those of conventionally available polymers in Table 9. The solvents examined include toluene (TOL), ethyl acetate (EAC), tetrahydrofuran (THF), dichloromethane (DCM), dimethylformamide (DMF), dimethylsulfoxide (DMSO), methanol (MeOH), acetic acid (AcOH) and water. The data in Table 9 clearly demonstrate that amphiphilic copolymer resins obtained by active ester synthesis have general solvent compatibility, ranging from toluene and ethyl acetate on the one hand, to acetic acid and water on the other. This swelling behavior compares favorably with those of conventionally available polymers. Styrene based resins are compatible only with the first five solvents listed in Table 9, whereas dimethylacrylamide resins are permeated by the last six solvents.

It is noteworthy that highly porous resins may also swell in certain non-solvents (e.g. polystyrene in methanol). Swelling of highly porous resins by nonsolvents is the result of liquid storage within the pore structure of the resin matrix; whereas general solvent compatibility of copoly(styrene-

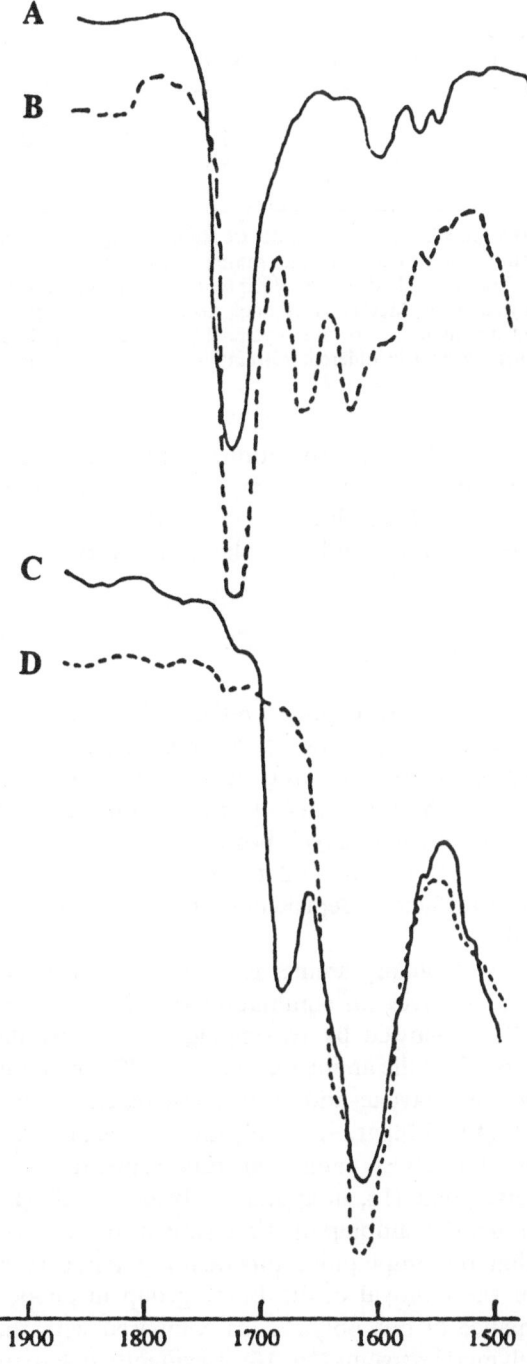

Fig. 15. Infra-red (IR) spectra of (A), beaded copoly(AOTcp-styrene) (sample **8f** in Table 6); (B), **8f** after partial aminolysis by N-(6-aminohexyl)*tert.*-butyl carbamate (see entry 1 in Table 7); (C), after complete aminolysis of above by dimethylamine; and (D), after removal of the *tert.*-butylcarbonyl group

Table 9. Swelling behavior of different polymer types

Polymer type[a]	Expanded volume (ml/g) in different solvents[b, c]								
	TOL	EAC	THF	DCM	DMF	DMSO	MeOH	AcOH	Water
Polystyrene	5.1	4.8	5.0	5.2	4.2	–	–	–	–
PolyDMA	–	–	–	9.5	9.1	10	12	12	9.0
Copoly-(DMA-styrene)	4.7	4.0	5.3	5.8	5.2	4.6	5.5	5.5	3.7

[a] For details of polystyrene and polyDMA (polydimethylacrylamide) see Refs. [7, 12] respectively. Copoly(DMA-styrene) was obtained from the activated copolymer sample **8c** (see Table 6)
[b] TOL = toluene, EAC = ethyl acetate, THF = tetrahydrofuran, DCM = dichloromethane, DMF = dimethylformamide, DMSO = dimethylsulfoxide, MeOH = methanol, AcOH = acetic acid
[c] Polymers with higher, or lower, expanded volumes can be obtained readily for all three polymer types [7, 12], but the pattern of polymer swelling in different solvents is a consequence of the chemical structure of the polymer

dimethylacrylamide) is primarily related to the "solvation" of the amphiphilic structure of the copolymer backbone. Note, for example, that linear (non-crosslinked) polystyrene is soluble in toluene but not in methanol, whereas copoly(styrene-dimethylacrylamide) is soluble in both toluene and methanol.

5.5 Structure-Performance Relationship

In an interesting modification of solid phase peptide synthesis (known as the inverse solid phase method), the sequential peptide assembly is carried out in solution by using polymeric acylation and deprotecting reagents. Among different strategies of inverse solid phase synthesis (see Ref. 46 and citations therein), the method based on the use of fluorenylmethoxycarbonyl (Fmoc) protecting group (Fig. 16) is of particular interest. In this method, the acylation reaction is relatively straightforward, but no satisfactory deprotecting polymeric reagent had until recently been developed.

Removal of the Fmoc group by secondary amines is known to proceed in two steps [68, 69]. The first step involves an elimination reaction and the formation of dibenzofulvene (DBF), followed by scavenging of DBF by the polymeric amine in the second step. All of the amine reagents **17a–17e** shown in Fig. 14 are effective for both of the cleavage and scavenging reactions, but piperazine and piperidine reagents (**17c–17e**) are relatively more efficient. Typical results for the deprotection of relatively nonpolar Fmoc-substrates by piperazine reagents based on polystyrene (**15**, nonpolar), polyacrylamide (**16**, polar), or copoly(styrene-acrylamide) (**17**, amphiphilic) are shown in Table 10. These results clearly illustrate that the amphiphilic piperazine polymer (**17c**) provides an efficient reagents for the removal of the Fmoc group in peptide synthesis. The improved performance of this polymer, as compared with the more conventionally available polymeric reagents (**15–16**), is evidently due to its amphiphilic structure and favorable substrate compatibility [46].

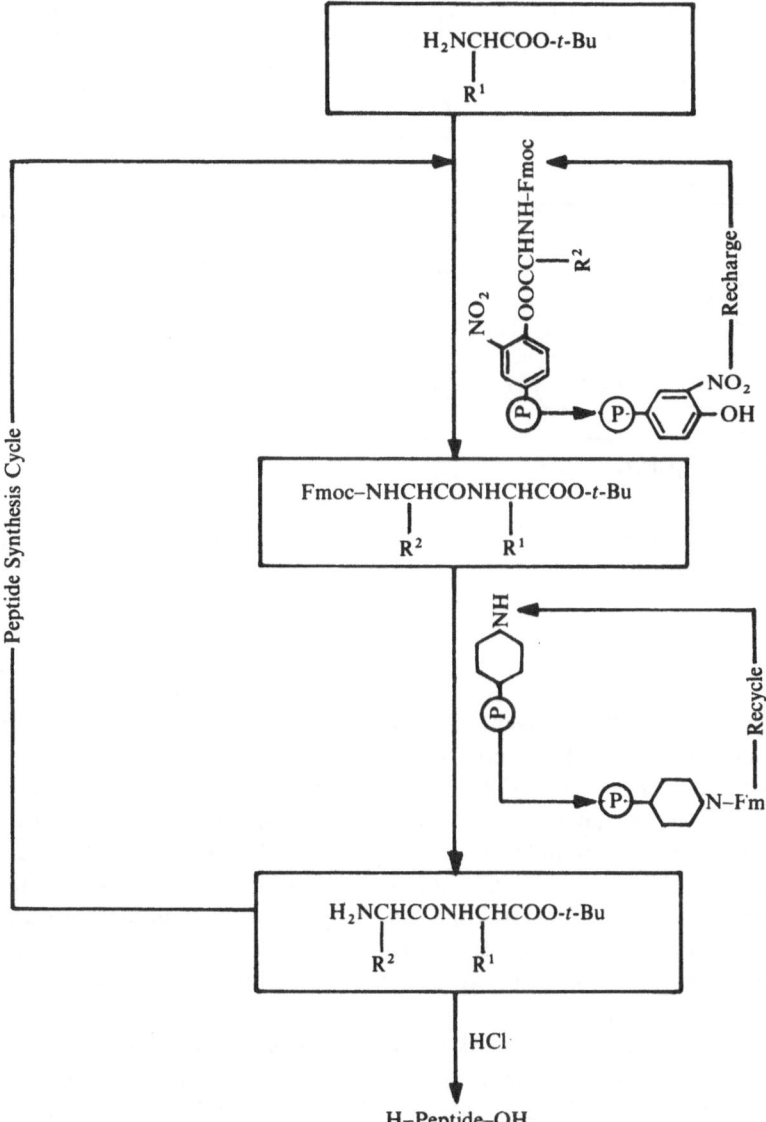

Fig. 16. Schematic presentation of peptide synthesis by inverse solid phase method based on the use of polymer-bound amino acid active esters (coupling reagent) and polymer-bound piperazine (deprotecting reagent)

Table 10. Removal of fluorenylmethoxycarbonyl (Fmoc) group from relatively hydrophobic substrates and scavenging of the by-product dibenzofulvene (DBF) by different polymeric piperazine reagents

Piperazine reagent	Capacity (mmol/g)	Deprotection[a]	
		100% Fmoc cleavage	DBF scavenging
Piperazine solution	10%	10 min	+
⬡–Ⓟ–⬡–CH₂N◯NH	1.7–3.5	1–24 h	−
Me₂NOC–Ⓟ–CONHCH₂CH₂N◯NH	3.0	1–24 h	−
⬡–Ⓟ–CONHCH₂CH₂N◯NH	3.5	> 1 h	+

[a] The exact time for complete Fmoc cleavage depends on the excess reagent used, type of solvent and polymer porosity (see Ref. 46).

Current practice of the conventional solid phase peptide synthesis (the Merrifield method) is based largely on the use of polystyrene and polydimethylacrylamide supports (see Fig. 17). The latter polymer was introduced in the 1970s [12, 13] to provide a relatively more polar support, as compared with polystyrene. However, accumulation of experimental evidence since then (cf. Ref. 70), indicates that an "ideal" polymer support for SPPS should be compatible with both polar (H-bonding) and nonpolar (hydrophobic) residues on the peptide grafts (Fig. 17). When the polymer support is not compatible with the growing peptide grafts, phase separation occurs, and the synthesis becomes inefficient or impracticable.

In general, the synthesis of relatively nonpolar sequences (cf. **18–19** in Fig. 17), proceeds efficiently on the nonpolar polymer matrix, polystyrene, but the assembly of strongly polar sequences (cf. **20–21** in Fig. 17) is particularly difficult on this polymer [70a]. This arises because the hydrophilic grafts (**20–21**) are not compatible with the nonpolar polymer backbone. As a result, the polymer-bound peptide chains interact within themselves, and become inaccessible as a result of *intra-resin H-bonding* [71]. Interestingly, an opposite problem of polymer–peptide incompatibility is observed in the case of the polar polymer, dimethylacrylamide. In this case, peptide synthesis proceeds favorably for polar sequences (cf. **20–21**), but the synthesis of strongly hydrophobic sequences (e.g. **18–19**, in Fig. 17) is not practicable because of *intra-resin hydrophobic aggregation* [70b]. For a recent study of peptide–peptide and peptide–polymer interactions and solvation in solid phase synthesis see Ref. [72].

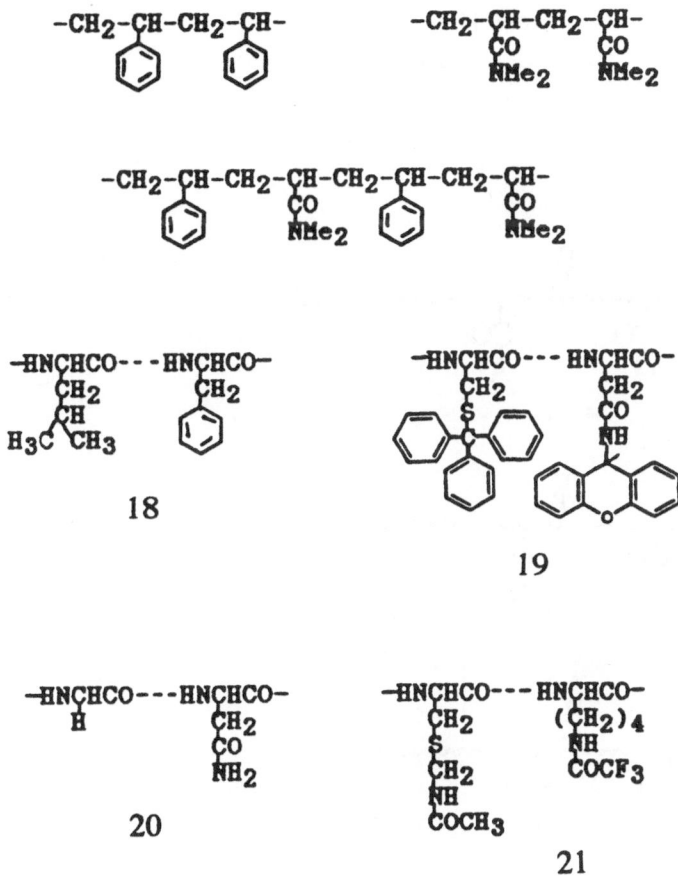

Fig. 17. Chemical structures of polymer supports based on styrene (nonpolar, hydrophobic), polydimethylacrylamide (polar, hydrophilic) and styrene-dimethylacrylamide (amphiphilic). Structures of nonpolar (18–19) and polar (20–21) peptide residues are also shown to illustrate the basis of polymer-peptide incompatibility during peptide synthesis on polystyrene and polydimethylacrylamide. Amphiphilic polymer supports are expected to be compatible with both nonpolar and polar peptide residues

6 Novel Graft Copolymers

Preparation of graft copolymers by the active ester method follows logically from the chemistry of activated acrylates discussed in the preceding two Sections. It involves the displacement of pre-positioned leaving groups on the polymer backbone by grafting chains carrying nucleophilic end groups (amines and alcohols), as illustrated in Fig. 18. The method is generally applicable, and the main intermediates (activated main-chain polymers and amino- and hydroxy-terminated oligomers) are readily accessible.

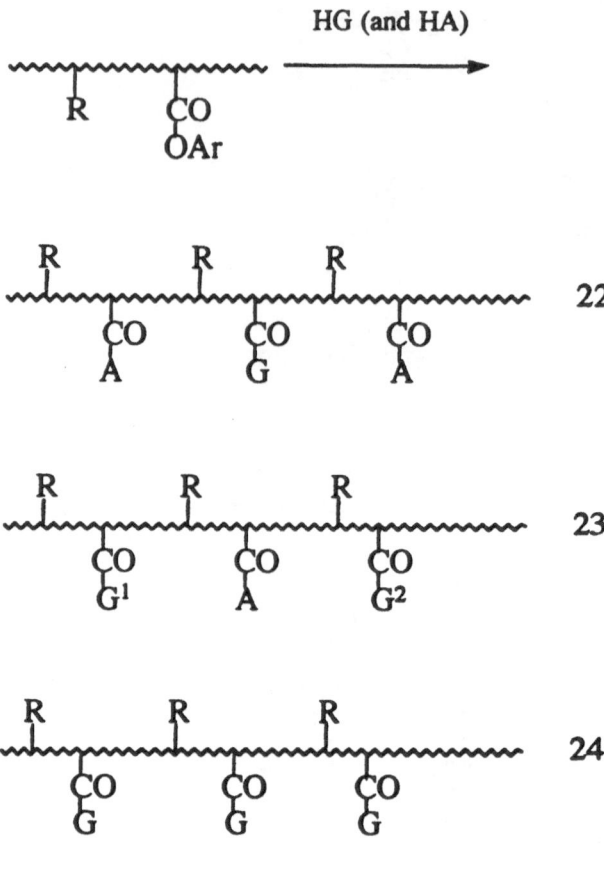

Fig. 18. General scheme for the formation of different types of graft copolymers (22–24) via active ester synthesis (see text for explanation of different structures)

In the case of copoly(AOTcp-styrene), for example, the following three synthetic strategies have been recently communicated [47].

1. Synthesis of single type graft copolymers, in which additional active ester groups on the polymer backbone enable the introduction of other structural units (A) for tailoring the overall structure of the graft copolymer (22 in Fig. 18).

2. Synthesis of double (or multi-) graft structures (G^1, G^2) with high degrees of precision (structure 23).

3. Synthesis of highly crowded graft copolymers with graft frequencies of up to 50% of the total monomeric units in approximately alternate positions on the polymer backbone (structure 24).

Molecular characterization and the study of solid state and solution behavior of such high frequency graft copolymers provide an exciting field of macromolecular research.

A particularly interesting example of graft copolymers produced via active ester synthesis is that of the comb structure (**25**, Fig. 19) with a hydrophobic backbone and hydrophilic grafts. A series of these amphiphilic graft copolymers have been produced by partial displacement of the trichlorophenoxyl groups

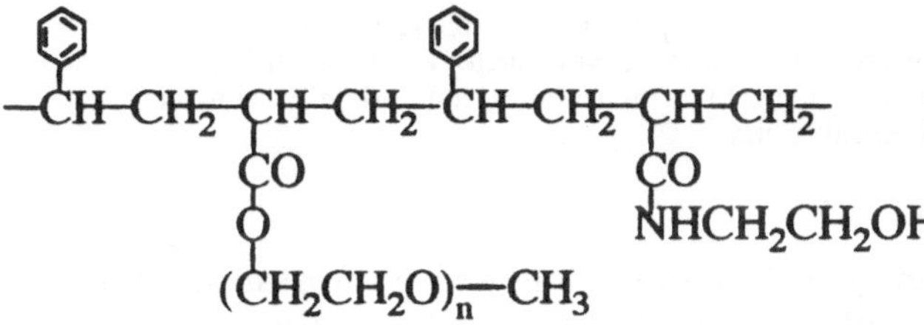

25

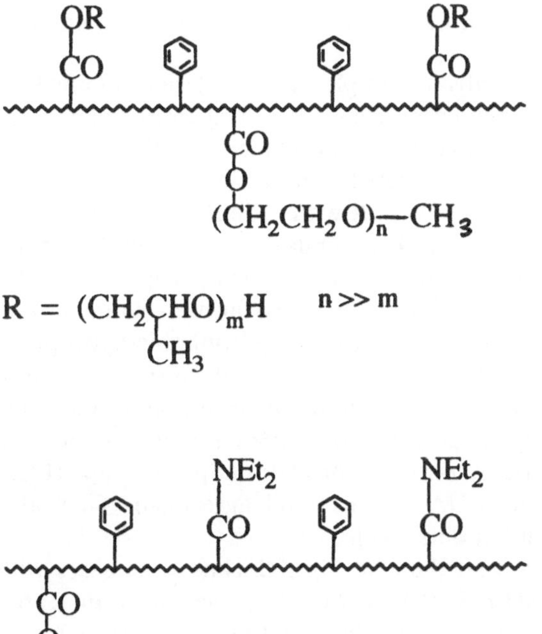

26

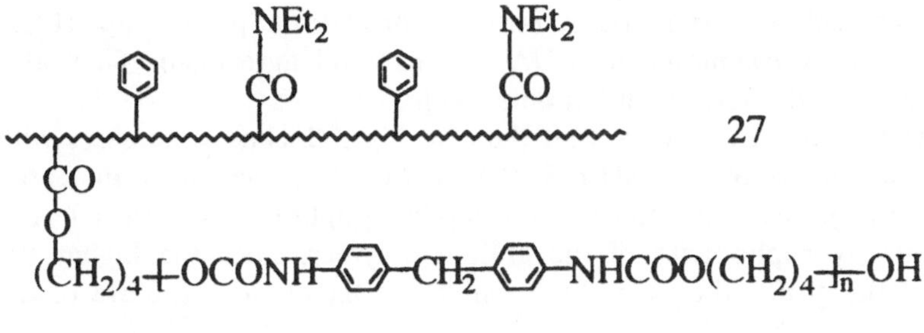

27

$Et = CH_2CH_3$

Fig. 19. Examples of graft copolymers (**25–27**) so far prepared via active ester synthesis

Table 11. Main characterstics of new graft copolymers (25a–25d) prepared according to Fig. 18 [47]

Graft copolymer	Nominal mole ratio of MPEG:HEAM	Molecular weights of MPEG grafts[a]		Estimated[b] M_z of graft coopolymer ($\times 10^{-6}$)
		M_n	M_z	
25a	55:45	8,080	10,800	0.98
25b	70:30	7,410	10,000	1.15
25c	80:20	6,742	9,600	1.26
25d	40:60	7,410	10,000	0.77

[a] Determined by gpc, using polyethyleneglycol/polyethyleneoxide standards.
[b] Calculated on the basis of the molecular weights of the activated copolymer ($M_n = 8,730$, $M_z = 57,000$). Preliminary M_z data obtained by sedimentation equilibrium are consistent with, but lower than, the theoretical values

with methoxypolyethyleneglycol (MPEG), followed by treatment with excess ethanolamine (2-hydroxyethylamine, HEAM). Both the length and the frequency of the polyethyleneglycol (PEG) grafts are readily controlled by the appropriate choice of the quantity of the desired MPEG fraction. All of the reactions are monitored by IR spectroscopy and microanalysis for chlorine and/or nitrogen.

Essential details of four graft copolymer samples (25a–25d) thus obtained are given in Table 11. The samples have been briefly examined by sedimentation equilibrium analysis. The molecular weights (M_z) obtained by this method are consistent with, but lower than, the expected values recorded in Table 11. Another series of amphiphilic graft copolymers with a double comb structure has also been synthesized (26, in Fig. 19) [47]. Similarly, type 27 graft copolymers have been prepared by the reaction of the activated copolymer with hydroxy-terminated polyurethanes, but these have not so far been characterized.

The graft copolymers (25–26) have well-defined hydrophobic-hydrophilic regions, and are therefore ideally suitable for surface modification (coating) in aqueous media. Table 12 shows the result of coating of clean polystyrene latex particles (nanospheres) by the graft copolymers (25a–25d) in aqueous solution. The coat layer thickness measured by Photon Correlation Spectroscopy (PCS) and Laser Doppler Anemometry (LDA) are in good agreement. They also correlate well with the zeta potential data [47].

The data in Table 12 indicate that the coat layer thickness increases with increasing the length of the MPEG chains, but decreases with increased frequency of the grafts. This latter relationship is thought to reflect the effect of the backbone hydrophobicity of the graft copolymer, since the higher the frequency of the grafts, the more hydrophilic the backbone structure of the copolymer.

The coating behaviour of the amphiphilic graft copolymers (25) presented in Table 12, suggests that the nonpolar (hydrophobic) polymer backbone interacts relatively strongly with the hydrophobic polystyrene particles, whereas the

Table 12. Results of surface coating of clean polystyrene nanospheres (188.5 nm) by graft copolymers (**25a–25d**) [47]

Graft copolymer	Surface layer thickness (nm)		Zeta potential (mV)
	By photon correlation spectroscopy	By laser doppler anemometry	
None	–	–	– 67.0
25a	18.6	23.5	– 5.8
25b	18.1	18.0	– 1.3
25c	7.5	5.5	– 57.0
25d	19.7	21.0	– 4.4

hydrophilic grafts remain "floated" in the aqueous medium. This coating mechanism has numerous potential applications in colloid and interface technologies, including suspension, emulsion, solubilization, micellization, and the design of biomedical surfaces. These graft copolymers also show potential as "macromolecular homing devices" for possible targeting of colloidal drug carriers to the bone marrow [17].

7 Side Chain and Surface-Bound Reactive Polymers

A variety of soluble and surface-bound polymers with side chain functional groups are available via active ester synthesis according to the general reaction schemes discussed in preceding Sections. Typical examples of such polymers are outlined below.

Functional polymers which can take part in electron transfer reactions on the electrode surface (i.e. electroactive polymers) are of potential interest for the study of electrochemical phenomena and electrocatalytic applications [74–76]. Main chain aromatic structures (e.g. polypyroles) are the most obvious candidates for the development of electroactive polymers, but a variety of side chain reactive polymers have also been studied for this purpose. Examples of such polymer obtained by active ester synthesis are illustrated in Fig. 20 [16].

Among the polymers represented in Fig. 20, the derivative **28a** carries quinone groups, and is directly electroactive, whereas **28b–28d** become electroactive following metal complexation. The polymers **28e–28f** carry nitrogen ligands and halogen groups, and are precursor to multi-dentate amine/phosphine ligands. Displacement of the halogen with phosphine is accomplished by reaction with diphenylphosphine in the presence of potassium *tert.*-butoxide. Structures of a number of metal complexes (chelates, **29–33**) obtained by reaction of these polymers with various metal substrates are shown in Fig. 21. Quantitative analytical data recorded in Tables 17 and 18 are in good agreement with the suggested structures [16].

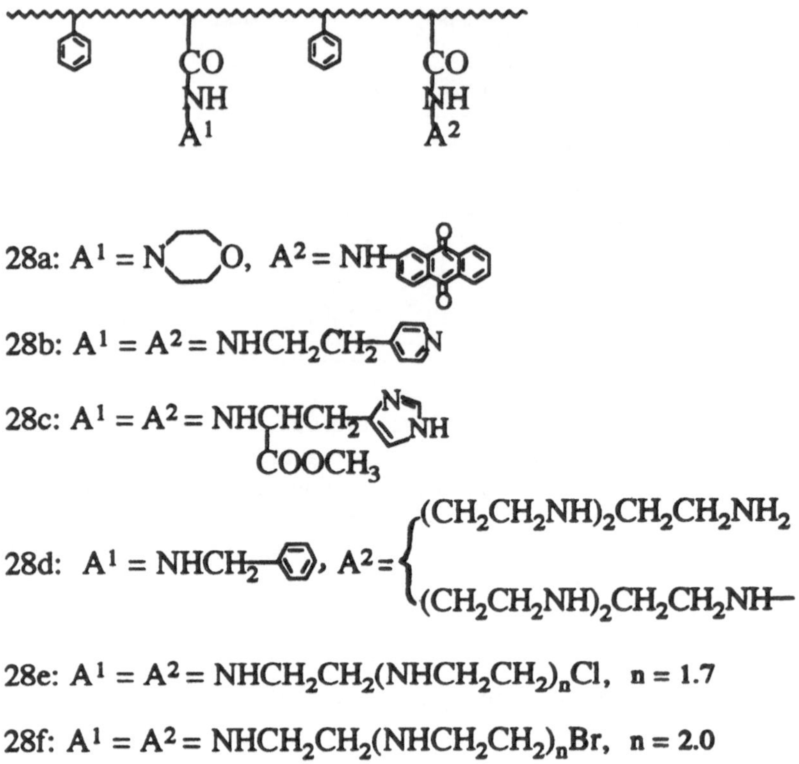

28a: $A^1 = N\bigcirc O$, $A^2 = NH\bigcirc$

28b: $A^1 = A^2 = NHCH_2CH_2\bigcirc N$

28c: $A^1 = A^2 = NHCHCH_2\bigcirc NH$
$\qquad\qquad\qquad\ \ COOCH_3$

28d: $A^1 = NHCH_2\bigcirc$, $A^2 = \begin{cases} (CH_2CH_2NH)_2CH_2CH_2NH_2 \\ (CH_2CH_2NH)_2CH_2CH_2NH- \end{cases}$

28e: $A^1 = A^2 = NHCH_2CH_2(NHCH_2CH_2)_nCl$, $n = 1.7$

28f: $A^1 = A^2 = NHCH_2CH_2(NHCH_2CH_2)_nBr$, $n = 2.0$

Fig. 20. Side chain electroactive polymers obtained via active ester synthesis for the preparation of surface modified electrodes

Electroactive polymers may be coated on, or covalently attached to, the electrode surface. In the latter case, a facile pathway involving the acylation of amino groups on the electrode surface with partially derivatized activated polymer intermediates is shown in Fig. 22 [16]. Similar reactions based on the use of partially functionalized polymer intermediates provide access to a variety of modified surfaces with biocompatibility, substrate selectivity, fluorescence or ferromagnetic properties (proprietary work, unpublished).

Another important property of polymers and copolymers of activated acrylates is their reaction with difunctional nucleophiles under a wide range of experimental conditions. This provides an ideal system for the study of polymer crosslinking (curing) and related applications (e.g. coating, composite formation, encapsulation, bioactive gels, etc.). For example, crosslinking of copoly(AOTcp-styrene) can be effected at room temperature within 1–2 min by primary amines, or within 2–10 min by secondary amines. Crosslinking with diols and aromatic diamines takes place at above 40 °C. Alternatively, the curing reaction can be programmed to take place in two steps by using unsymmetrical crosslinkers, as exemplified in Fig. 23 (proprietary work, unpublished). Crosslinking of

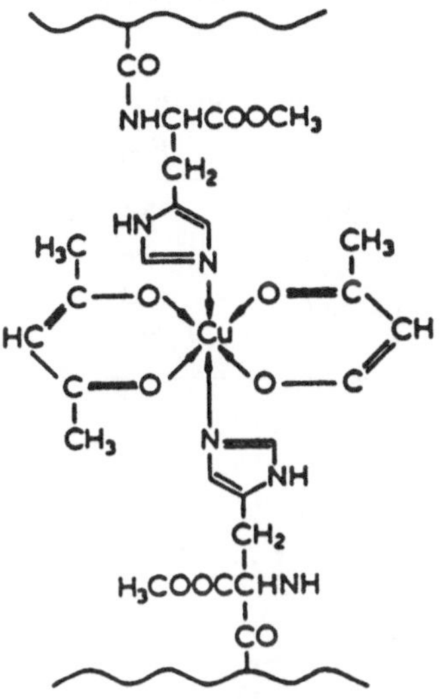

Fig. 21. Tentative chemical structures of polymeric transition metal complexes (chelates, 29–33) of potential interest for electrocatalysis

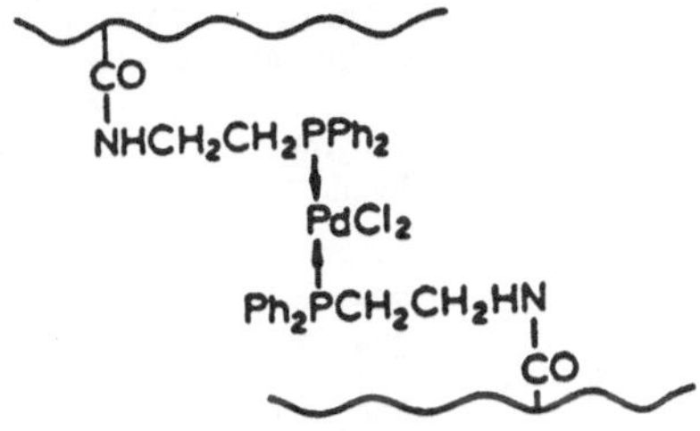

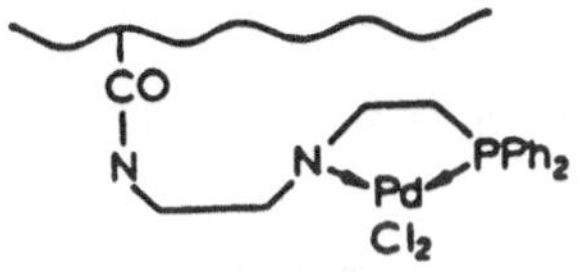

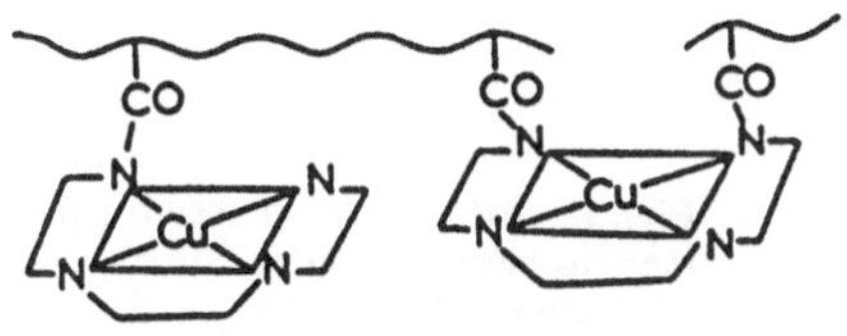

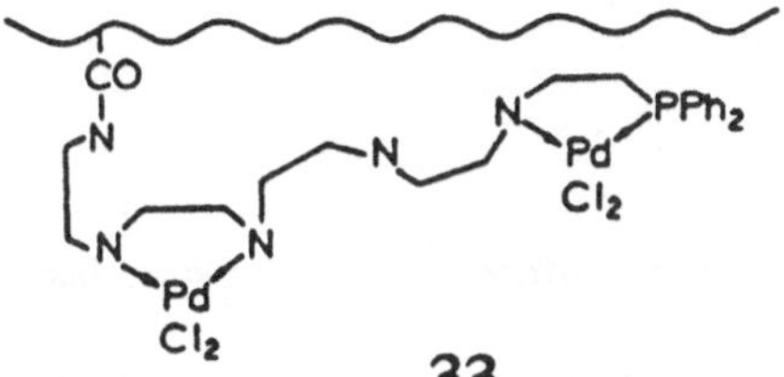

Table 13. Side chain electroactive polymers prepared via active ester synthesis (adapted from Ref. [16])

Reactive polymer	Polymer composition			Functionality (mmol/g)	Characteristic IR (cm^{-1})
	N%	Cl%	P%		
28a	–	–	–	1.82	1590, 1630–60, 1720
28b	–	–	–	3.57	1520–60, 1605, 1660
28c	7.34	11.05	–	2.74	1665, 1730
28d	6.81	–	–	1.05	Not recorded
28e[a]	5.36	1.64	6.87	–	1435, 1450, 1490, 1650
28f[a]	9.35	–	4.72	–	Not recorded

[a] After displacement of halogen with triphenylphosphine

Table 14. Polymeric transition metal complexes (chelates, 29–33) obtained from the reaction of polymeric ligand 28 with palladium (II) and copper (II) species (adapted from Ref. [16])

Polymer complex	Polymer ligand	Metal substrate	Metal in the polymeric (%)	
			Calc.	Found
29	28c	Cu(acac)$_2$	2.81	3.21
30	28d	CuSO$_4$, 5H$_2$O	5.52	5.23
31	28e	PdCl$_2$(PhCN)$_2$	9.72	12.2
32			15.14	
33	28f	PdCl$_2$(PhCN)$_2$	23.55	20.3

Electrode surface   Polymer backbone

A = Electroactive residue (cf. Figs. 28-29)

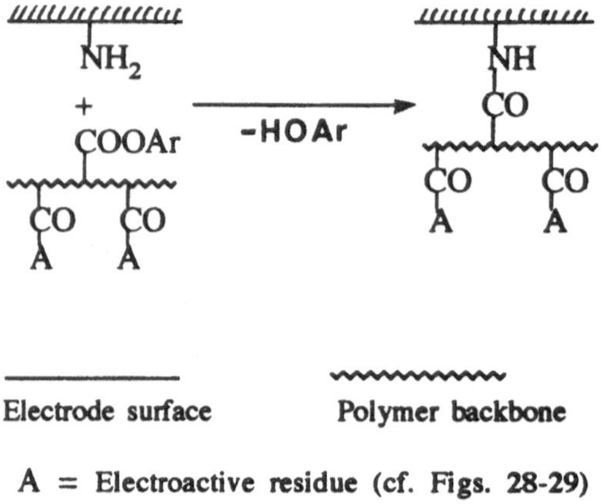

Fig. 22. Attachment of reactive polymers to electrode surface (preparation of modified electrode) via active ester synthesis

copoly(AOSu-acrylamide) in aqueous media has also been studied in some detail [48]. In this work, polymer crosslinking has been effected simultaneously with the attachment of an enzyme, and hence the formation of bioactive gels (immobilized enzymes) [48].

Side chain chiral polymers (34, in Fig. 24) have been produced via active ester synthesis, by using soluble and crosslinked samples of copoly(AOTcp-styrene) (unpublished work). The resulting chiral polymers are of considerable interest in separation technology, and as auxiliary media for asymmetric synthesis. Fluorescent labeled polymers, polyelectrolytes and hydrophobically modified ionomers (e.g. 35 and 36), are also readily available via active ester synthesis.

Side chain liquid crystalline and nonlinear optical polymers (e.g. 37 and 38), which are conventionally produced by multi-step processes, are also available very easily via active ester synthesis. A unique feature of the active ester method for this purpose is that a single activated polymer intermediate can be used for the synthesis of any number of macromolecular structures, all by a simple single-step reaction pathway. Synthesis of such polymers by copolymerization of the

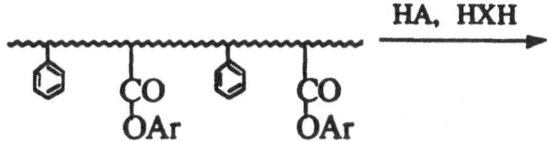

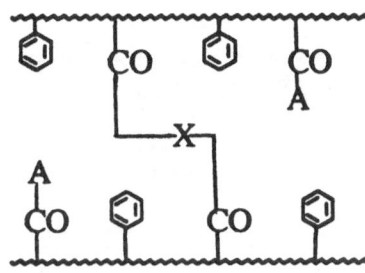

Examples of X (see also Table 19):

$HN(CH_2)_nNH$, $HN(CH_2)_2N(CH_2)_2NH$,

$N(CH_2)_6N$, $N(CH_2)_3O$, $O(CH_2)_nO$,
CH_3 CH_3

$HNCH_2CH_2$–⟨⟩–NH, HN–⟨⟩–⟨⟩–NH

Fig. 23. Crosslinking of copoly-(AOTcp-styrene) by difunctional nucleophiles

respective comonomers is not strictly feasible, because the resulting polymers may have widely different structures in terms of sequence distribution, molecular size, and/or branching.

8 Related Areas of Active Ester Synthesis

The discussion of active ester synthesis in this review focuses on polymers and copolymers of phenyl acrylates and N-acryloyloxy derivatives as "synthetic intermediates". Another interesting application of activated (meth)acrylates is the formation of photocurable oligomers [81]. Polymers and copolymers of activated (meth)acrylates have also been studied as models of macromolecular drug carriers [37, 42]. In addition, the chemistry of activated esters is equally

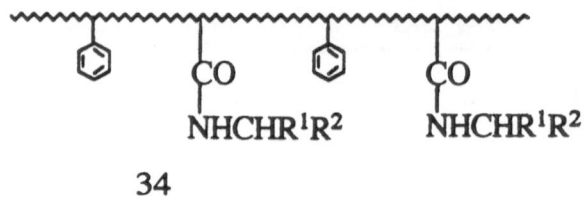

34

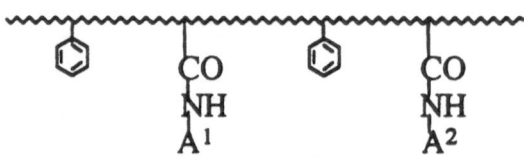

35: $A^1 = A^2 = (CH_2)_2\bar{S}\overset{+}{O}_3\overset{+}{Na}$

36: $A^1 = (CH_2)_2\bar{S}\overset{+}{O}_3\overset{+}{Na}$, $A^2 = C_8H_{17}$

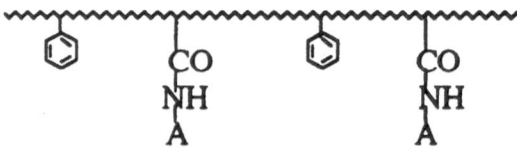

37: $A = (CH_2)_6OOC\!-\!\!\langle\bigcirc\rangle\!-\!C\!=\!C\!-\!\!\langle\bigcirc\rangle\!-\!CN$

38: $A = (CH_2)_6O\!-\!\!\langle\bigcirc\rangle\!-\!N\!=\!N\!-\!\!\langle\bigcirc\rangle\!-\!OCHC_2H_5$
CH_3

Fig. 24. Side chain specialty polymers available via active ester synthesis include structures **34** (chiral), **35** (polyelectrolyte), **36** (hydrophobically modified ionomer), and **37–38** (liquid crystalline and nonlinear optical, NLO)

applicable to polycondensation monomers and polymers. For example, preparation of polyamides via polycondensation of activated esters of diacids with diamines has been described [79]. Similarly, polycondensation of N-protected serine (3-hydroxy-2-aminopropanoic acid) via its HOBt active ester has also been recently reported [80].

9 Concluding Remarks and Future Prospects

The chemistry of activated acrylates (namely active ester synthesis) provides a uniquely versatile route for the preparation of specialty polymers. Two important features of active ester synthesis are its general applicability and practical simplicity. It involves a single-step displacement of pre-positioned activating (or leaving) groups on the polymer backbone by the desired functional and/or structural residues.

Examples of polymers described in this review include amphiphilic gels for solid phase synthesis, graft copolymers for targeting of colloidal drug carriers, and side chain chiral and potentially electroactive polymers. However, the methodology of active ester synthesis is equally applicable to other areas of basic and applied polymer research, including bioactive gels, novel polymers carrying side chain liquid crystalline and nonlinear optical residues, multi-chain and high frequency (*crowded*) graft copolymers. Similarly, novel macromolecular structures can be created from stereospecific (iso- and syndiotactic) homo- and copolymers of activated acrylates. Indeed, the general applicability and practical versatility of active ester synthesis present a fascinating new dimension of creativity throughout the realm of macromolecular chemistry.

Acknowledgements: I wish to thank my colleagues and students Benedetto Corain, Marco Zecca, Maurice H. George, Boreddy S. R. Reddy, Manesh B. Desai, Colin Butler and Fereidoon Fallah for their cooperation and diligent efforts towards the progress of this work. Financial support from the Wolfson Foundation (London, UK), and Research Funds from the University of Tabriz (Tabriz, Iran) and Kashan College of Sciences (Kashan, Iran) are also gratefully acknowledged.

10 References

1. Bergbreiter DE, Martin CR (eds) (1989) Functional polymers. Plenum, New York
2. IUPAC Microsymposium on Reactive Polymers, Prague, Czechoslovakia (1987) published in Pure Appl Chem 60: (1988); and Reactive Polymers 6: (1987)
3. Gebelein CG (ed) (1990) Biomimetic polymers. Plenum, New York
4. Warshawsky A (1987) In: Streat M, Naden D (eds) Ion Exchange and Sorption Processes in Hydrometallurgy. Wiley, New York, p 127
5. Prasad P, Ulrich D (eds) (1988) Nonlinear optical polymers. Plenum, New York
6. Arshady R (1989) Polym Eng Sci 29: 1746; (1990) Polym Eng Sci 30: 905 and 915
7. Arshady R (1991) J Chromatogr 586: 181 and 199
8. Shalaby WS, McCormic L, Butler B (eds) (1991) Water-soluble polymers: Synthesis, solution properties and applications, ACS Symp Ser 467
9. Harland RS, Prud'homme RK (1992) Polyelectrolyte gels: Properties, preparation, and applications, ACS Symp Ser 480
10. DeRossi D, Kajiwara K, Osada Y, Yamauchi A (eds) (1991) Polymer gels. Plenum, New York
11. Arshady R (1992) Polymer Preprints 33(1) : 954
12. Arshady R (1990) Colloid Polym Sci 268: 948
13. Arshady R, Atherton E, Clive DLJ, Sheppard RC (1981) J Chem Soc Perkin Trans I 1981: 529
14. Arshady R, Mosbach K (1981) Makromol Chem 182: 687
15. Arshady R, Basato M, Corain B, Lora S, Roncato M, Zecca M (1989) J Mol Catal 53: 111
16. Arshady R, Reddy BSR, George MH (1986) Polymer 27: 769
17. Arshady R, Illum L, Davis SS (1991) Polym Adv Technol 1: 193
18. Arshady R (1990) J Bioact Compat Polym 5: 315
19. Arshady R, Basato M, Corain B, Lora S, Zecca M (1990) Adv Mater 2: 412
20. Arshady R (1989–90) unpublished work

21. Arshady R (1992) JMS Rev Macromol Chem Phys C32(1): 101
22. Benham JL, Kinstle JF (1988) Chemical reactions on Polymers, ACS Symp Ser 364. ACS, Washington DC
23. Arshady R (1981) Makromol Chem Rapid Commun 2: 573
24. Arshady R (1983) Makromol Chem Rapid Commun 4: 237
25. Arshady R (1984) Makromol Chem 185: 2387
26. Arshady R, Ugi I (1982) Angew Chem Int Edn Engl 21: 374
27. Arshady R, Ledwith A (1983) Reactive Polymers 1: 159
28. Epton R, Marr G, Small PW (1981) Polymer 22: 842
29. Warshawsky A, Kahana N (1979) J Amer Chem Soc 101: 4249
30. Arshady R (1988) Makromol Chem 189: 1303
31. Darling GD, Frechet JMJ (1986) J Org Chem 51: 2270
32. Mitchell AR, Kent SB, Engelhard M, Merrifield R (1978) J Org Chem 43: 2845
33. Sparrow JT (1976) J Org Chem 41: 1350
34. Leznoff CC, Wong JY (1976) Cand J Chem 54: 3824
35. Bodanszky M (1984) The practice of peptide synthesis. Springer, Berlin Heidelberg New York
36. Pless J, Biossonas RA (1963) Helv Chem Acta 46: 1609
37. Batz HG, Franzman GF, Ringsdorf H (1973) Makromol Chem 172: 27
38. Pittman CU Jr, Stahl GA (1981) J Appl Polym Sci 26: 2403
39. Chengxun L, Chongqing W (1980) J Polym Sci Polym Chem Edn 18: 2411
40. Narasimhaswamy T, Reddy BSR (1991) J Appl Polym Sci 43: 1615
41. a) Su CP, Morawetz H (1977) J Polym Sci Polym Chem Ed 15: 185; b) Rejmanova P, Lobsky J, Kopecek J (1977) Makromol Chem 178: 2159
42. Ferruti P (1986) In: Gregoriadis G, Senior G, Poste G (eds) Targeting of drugs with synthetic systems. Plenum, New York
43. Reddy BSR, Arshady R, George MH (1983) Macromolecules 16: 1813
44. Reddy BSR, Arshady R, Georgge MH (1985) Eur Polym J 41: 511
45. Desai MDB, Reddy BSR, Arshady R, George MH (1986) Polymer 27: 96
46. Arshady R, Fallah F (1992) J Polym Sci Polym Chem 30: 1705
47. Arshady R (1990) Makromol Chem Rapid Commun 11: 193
48. Polak A, Blumenfeld H, Wax M, Baughn RL, Whitesides GM (1980) J Am Chem Soc 102: 6324
49. Yoshida M, Asano M, Yokota T (1990) Polymer 31: 371
50. Fitch RM, Scholsky KM (1988) In: Rembaum A, Tokes ZA (eds) Microspheres: Medical and biological applications. CRC Press, Boca Raton FL, p 101
51. Heidman W, Koester H (1980) Makromol Chem 181: 2495
52. Bevington JC, Melville HW, Taylor RP (1954) J Polym Sci Polym Chem Edn 12: 449
53. Limanovich AA, Papisov IM, Kabanov VA (1981) Eur Polym J 17: 981
54. Reddy BSR, Arshady R, George MH (1984) Makromol Chem 185: 1383
55. Levy GC, Nelson GL (1972) Carbon-13 Nuclear Magnetic Resonance for organic chemists. Wiley Interscience, New York, p 81
56. Inue Y, Nishioka A, Chujo R (1972) Makromol Chem 156: 207; Shaefer J (1971) Macromolecules 4: 107
57. Arshady R (1992) Colloid Polym Sci 270: 717
58. Gordon M, Miller JG, Day AR (1948) J Am Chem Soc 70: 1946
59. Gordon M, Miller JG, Day AR (1949) J Am Chem Soc 71: 1245
60. Wieland T, Schaefer W, Bokelman E (1951) Liebigs Ann Chem 99: 573
61. Bodanszky M (1984) Principles of peptide synthesis. Springer, Berlin Heidelberg New York, p 28
62. Koenig W, Geiger R (1973) Chem Ber 106: 3626
63. Koenig W, Geiger R (1970) Chem Ber 103: 788, 2024
64. Arshady R, Ledwith A, Kenner GW (1981) Makromol Chem 182: 11
65. Merrifield RB (1988) Makromol Chem Makromol Symp 19: 31
66. Merrifield B (1991) Profiles, pathways and dreams: The concept and development of solid phase peptide synthesis. American Chemical Society, Washington DC
67. Balmstrom B (ed) (1992) Noblel lectures in chemistry 1981–1990. World Scientific, London
68. Carpino LA, Mansour EME, Chung CH, Williams JR, MacDonald R, Knapszyk J, Carman M (1983) J Org Chem 48: 661
69. Carpino LA (1987) Acc Chem Res 20: 401
70. a) Kent SBH (1985) In: Derber V, Hruby V, Kople K (eds) Proc 9th Am Peptide Symp. Pierce Chem Co, Rockford Il, p 407; b) Atherton E, Sheppard RC (1985) idem p 415
71. Narita M, Tomotake Y, Isokawa S, Matsuzawa T, Miyauchi T (1984) Macromolecules 17: 1903

72. Fields GB, Fields CG (1991) J Am Chem Soc 113: 4202
73. Gross L, Ringsdorf H, Schyp H (1981) Angew Chem. Int Edn Engl 20: 305
74. McCullough RD, Lowe RD (1992) Polymer Preprints 33(1): 195
75. Patil AO, Heeger AJ, Wudl F (1989) Chem Rev 88: 183
76. Reynolds JR (1988) Chemtech 18: 440
77. Dusek K (ed) (1984) Adv Polym Sci 57
78. McCormic CL, Anderson KW, Hutchinson BH (1982–83) JMS Rev. Macromol Chem Phys C22: 57
79. Ueda M, Horada S, Aoyama S, Imai Y (1981) J Polym Sci Polym Chem Edn 19: 1061
80. Gelbin ME, Kohn J (1991) Polymer Preprints 32(1) : 241
81. Nishikubo T, Takehara E, Saita S, Matsamura T (1987) J Polym Sci Polym Chem Edn 25: 3049

Editor: Prof. A. Ledwith
Received August 26, 1992

Reactive Surfactants in Emulsion Polymerization

A. Guyot[1] and K. Tauer[2]
[1] CNRS – Laboratoire de Chimie et Procédés de Polymérisation, BP 24-69390, VERNAISON, France
[2] Max Planck Institute für Kolloid- und Grenzflächenforschung, Kantsraße 55, D-12169 Berlin, FRG

Covalent binding of the surfactants to the polymers obtained from emulsion polymerization is expected to improve some properties of the resulting latexes and also of the films formed from these latexes. This is possible if the surfactants include functional groups capable of interacting with the radical polymerization process. These functional groups may be a polymerizable function so that the surfactant is a comonomer, named in this review SURFMER. Alternatively it may be an initiator (INISURF) or a transfer agent (TRANSURF). The paper is a literature review of the state of the art and illustrates the increasing interest in this field. The main trends and some prospects are discussed in the conclusions.

Advances in Polymer Science, Vol. 111
© Springer-Verlag Berlin Heidelberg 1994

List of Abbreviations and Symbols

AIBN	Azobis isobutyronitrile
B	Butyl acrylate
CMC	Critical micellar concentration
DLVO	DerJaguin, Landau, Verwey, Overbeek
e	Charge parameter for a monomer
HLB	Hydrophilic-lipophilic balance
H-SAAS	Hydrogenated sodium alkyl allyl sulfosuccinate
ICI	Imperial Chemical Industries
KPS	Potassium persulfate
Na AAS	Sodium 9 acrylamidostearate
NIAD	Non-ionic aqueous dispersion
Q	Reactivity parameter of a polymer radical
S	Styrene
SAAS	Sodium alkyl allyl sulfosuccinate
SAU	Sodium acrylamido undecanoate
SDS	Sodium dodecyl sulfate
SPM	Sulfopropyl methacrylate
SSDSE	Styrene sodium dodecyl sulfate ether
VA	Vinyl acetate

1 Introduction

Surfactants are used in emulsion polymerization for two main purposes: the control of particle size, and the stabilization of the latexes at high solid contents. In most cases these surfactants are physically adsorbed onto the surface of the particles, in dynamic equilibrium with the water phase. Except for a small part which might be due to transfer reactions during the polymerization, the surfactants are not covalently bound to the polymer particles. Then under certain special stresses, they can be desorbed and their stabilizing properties are lost; it may be the case under shear stresses, eventually during the polymerization if the stirring conditions are not adequate. It is also the case in cycles of freezing and thawing. Flocculation occurs and peptization of the flocs is not always possible. When films are formed from the latexes, as for use in paints, textile treatments, paper coatings or adhesives, the coalescence may be partially hindered, and then delayed, and when it takes place, the surfactants remain in the concentrated water phase and, the major part of them remains in hydrophilic domains buried inside the films, even if some part is expelled toward the surface of the film. These hydrophilic domains are then responsible for water rebound when the films are exposed to high humidity conditions; in addition, the water permeability is enhanced, so that some protective properties, against corrosion for instance, sought after in the films are not satisfactorily fulfilled.

In order to overcome these drawbacks, various solutions can be considered. One would be to destroy the surfactants after polymerization and film formation. Some efforts have been made to make surfactants photosensitive [1] with a scissile covalent bond in between the hydrophilic and the hydrophobic parts of the surfactant. The second one, which is considered in this review, is to permanently anchor the surfactants onto the particle surfaces during the polymerization. In such a way, the adventitious desorption of the surfactants under stresses is no longer possible. Even if such anchored surfactants cause the coalescence process to be more difficult, their migration is not possible and the process of hydrophilic domain formation is hindered. Then it may be expected that the water solubilization inside the film is limited so that the water permeability decreases. A final advantage lies in the absence of surfactants in the serum after polymerization; then, in such applications when the polymer material is separated from the water by flocculation, the absence of surfactant in the rejected water should make its purification and recycling easier. The use of such surfactants is also claimed to reduce the volume of foams [2]; such things become increasingly important, when much more attention is paid to environmental problems.

Because the emulsion polymerization obeys a radical mechanism, the way to covalently anchor the surfactants to the polymer particles is to make them reactive in the radical process. Then one may consider three kinds of reaction: the initiation reaction, the propagation reaction and transfer reaction. The termination reaction is carried out between growing polymer radicals and there is no advantage in thinking surfactants actives act terminators of polymerization, such products are actually inhibitors of the polymerization. To enable the surfactants to interfere with one of the three reactions to be considered, it is necessary to attach to them a functional group capable of taking part in one of these reactions. If the surfactant include a radical generator function, it should be able to initiate the polymerization and such products have been named INISURFS [3]. Similarly we might give the name SURFMERS to the surfactants acting as comonomer and able to participate in a propagation reaction (polymerizable surfactants); those able to act as a transfer agent will be named TRANSURFS. Some work has been already done in this direction and the main purpose of this review is to point out the state of the art in this domain and try to raise the main problems to be solved, concerning their interference with the emulsion polymerization mechanism as well as optimizing their properties concerning their potential applications. The review will concern itself first with each of the three kinds of products and a general discussion of the problems is given after.

2 Surfmers

The oldest paper we know dealing with the polymerization of a molecule nowadays named as a surfmer is the work of Bistline et al. [4]. They obtained surface active polymers with a mean degree of polymerization of about 10 by polymerization of allylic esters of sodium salts of α-sulfo-stearinic acid and α-sulfo-palmitinic acid.

A huge number of polymerizable surfactants have been prepared with the purpose of stabilizing, upon polymerization, their micelles or eventually vesicles. Various kinds of polymerizable groups, located at all possible places in the molecules, have been used. Excellent reviews and papers have been published in recent years [5–7]. The general result is that the polymerization leads to particles much larger than the size of the micelles, i.e. around 200 nm instead of a few nm [8,9]. However, good success has been achieved in the case of vesicles [10]. These results are most likely to be associated with the dynamics of these objects: the lifetime of a micelle is very short even compared with the growth time of a polymer molecule; in addition the capture of a growing radical initially born in the water phase is a transsient event in the case of micelles because of an equilibrium between the entry and the exit rate of the radicals, and it is a more permanent event in the case of a polymer particle; then as soon as the number of polymer particle is large enough to offer a large surface area, their capture efficiency becomes close to 1. The situation is then comparable to that of emulsion polymerization of conventional monomers. On the other hand, the vesicles are much more stable objects with long lifetimes so that, once initiated, the polymerization inside the vesicles can proceed during that lifetime, and the object may be permanently stabilized, even if it contains many polymeric and oligomeric molecules from the polymerizable surfactants. Many fewer studies have been devoted to the use of polymerizable surfactants in latex synthesis. Although there have been some patents related to the topics [11, 12], papers in the open literature only appeared recently with the exception of one series [13–15]. The discussion will be divided into two parts related to the nature of the surfactants used: anionic or non ionic.

Only a few papers have been published concerning the application of cationic surfmers [16, 17] in emulsion polymerization.

Apparently nothing has been published so far in the scientific literature concerning zwitterionic surfactants.

The early work of Greene et al. [13] used as surfmer sodium 9 [and 10] acrylamido stearate (Na AAS). They prepared a base styrene-butadiene latex with a small amount of Na AAS in batch at 70 °C and obtained 134 nm diameter particles. This base latex was later covered with various amounts of Na AAS, equilibrated and heated further 1 hour at 70 °C in the presence of KPS (potassium persulfate) for in situ polymerization of the surfmer. A part of the surfmer was removed after ion exchange resin treatment, either as non-polymerized monomer or water-soluble polymer. The latex surface coverage by immobilized surfmer polymer was varied from 20 to 80%; the yield of immobilisation was almost 100% for low coverage but decreased to about 70% when high coverage was used. Up to 60% coverage, no difference could be found between monomer and polymer units for the occupancy of the latex surface; above 60%, interference between the polymer chains occurred so that some parts of the surface still accessible to small monomer molecules were forbidden for further polymer segments. The latex covered with polymerized surfmer at high coverage was shown to display definitely superior mechanical stability compared with non-polymerized surfmer at the same coverage [14]. The same is true for electrolyte stability [15]. These latexes were used to check some quantitative features of the DLVO theory.

More recently Chen and Chang [19] have used surfmer I with a vinylic end group in the hydrophobic part. The cmc at the polymerization temperature was 3.8×10^{-2} mol/l. Polystyrene latexes were prepared in the presence of KPS resulting in monodisperse particles in the range of 100 to 180 nm in diameter, increasing with solid contents, but with formation of coagulum at high solid contents. The number of particles increases as $[surfmer]^1$ and $[KPS]^{0.5}$ and is slightly sensitive to the ionic strength, passing through a maximum. The nucleation is considered to take place in a rather short period of time, because the particles are rather monodisperse, and do follow a homogeneous mechanism, because the concentration of Surfmer was under the cmc, except possibly at high ionic strength where, due to the sensitivity of the cmc to the ionic strength, micelles are present and then both micellar and homogeneous nucleation can take place. Surfmer I can be compared with sulfopropylmethacrylate (SPM), an ionogenic monomer used by Krieger et al. [20] to prepare monodisperse polystyrene latexes. In that case, the absence of surfactant properties

$$H_2C = CH - (CH_2)_8 - \underset{\underset{O}{||}}{C} - O - CH_2 - CH_2SO_3Na \qquad \textbf{I}$$

make the nucleation strongly dependent on their ionic strength, the nucleation taking place in a homogeneous medium with aggregation, the stabilization being purely electrostatic. The number of particles is proportional to $[SPM]^2$ and $[KPS]^1$. So it appears that introduction of surfactant properties to the monomers used for stabilizing the particles causes a drastic change in the nucleation mechanism.

A styrenic surfactant (**II**) has been prepared and used by Fitch and Tsaur [21], namely the styrene sodium dodecylsulfonate ether (SSDSE) with a cmc of 2×10^{-3} mol/l at room temperature. Styrene polymerization has been carried out both at 20 °C, using a photo initiator (biacetyl) and at 65 °C using KPS. In the first case, the stabilization is not good enough and sodium dodecylsulfate (SDS) must be added in concentrations larger than that of SSDSE in order to obtain monodisperse particles in the range 140–260 nm with a surface charge density in between 1.9 and 6.5 mC/m². When KPS is used as initiator, a good control of particle size is obtained between 150 and 400 nm (log Dn being proportional to log [SSDSE]) with excellent size uniformity (Dw/Dn < 1.05), the surface charge density being in between 4.0 and 14.0 mC/m². The main interest of that surfmer **II** seems to be for surface functionalization of seeded particles. A seed latex on 180–200 nm was used and polymerization of SSDSE (with styrene) was carried out in the presence of an azocarboxy compound as initiator. Monodisperse particles with no water-soluble polymer formation were obtained when [SSDSE] < cmc; the surface yield of SSDSE was between 10 and 60% so that the surface charge density was well controlled in between 10 and 100 mC/m² with a narrow charge distribution, as judged by electrophoresis experiments. When [SSDSE] was larger than cmc, polyelectrolyte was formed which can be separated from the latex upon centrifugation. It was concluded that this technique is quite useful for controlling the charge density of the latex particles.

$$\underset{O - (CH_2)_{12}SO_3Na}{\overset{CH = CH_2}{\bigcirc}} \qquad \textbf{II}$$

A carboxylato surfmer **III** has been studied by Guillaume, Guillot and Pichot [22], for the purpose of preparing latexes carrying only carboxylic surface groups. Sodium acrylamido undecanoate (SAU) has been used, with a cmc at 25 °C of 5×10^{-3} mol/l. It has been utilized in copolymerization with styrene (S) and butylacrylate (B), initiated at 70 °C by an azocarboxy compound. The reactivity ratios with S and B were measured (S) or estimated (B) from the Q, e scheme; the partition coefficient of SAU and of the comonomer between water and organic phases were also measured, so that a simulation of the copolymerization process was obtained which shows an S shape for the conversion of SAU, indicating that most of that surfmer is polymerized only at

$$H_2C = CH - \underset{\underset{O}{\|}}{C} - NH - (CH_2)_{10}COONa \qquad \textbf{III}$$

the end of the polymerization process; it results in poor stability, both at ionic strength $> 10^{-2}$ N, most of the SAU being used as adsorbed emulsifier during the process. Polyelectrolyte is formed in the later stages, causing floculation, so that the particle number Np goes through a maximum with increasing conversion. Before the maximum, Np varies as $[SAU]^1$ as in the case of Surfmer I. The surface yield in COO^- is limited to 20–30%, a part of the COO group being buried inside the particles (10 to 50%) mostly when solid contents are high (30%). The remaining part is in the water phase as polyelectrolyte or residual monomer. Seeded polystyrene latexes prepared with KPS and SPM according to Krieger [20] were used to study its coverage by SAU. The surface yield of carboxylic groups was limited to the range 25–35%.

One of the drawbacks of the surfmers with allylic, acrylic, and vinylic polymerizable groups is their tendency to produce water-soluble polyelectrolytes when they are used in too-large amounts, i.e. above their cmc. This problem may be overcome by using maleic derivatives which can copolymerize but are unable to homopolymerize at normal temperatures, their ceiling temperature being very low. Surfmer **IV** can be prepared easily by first reacting maleic anhydride with a fatty alcohol, the resulting hemiester being further reacted with propanesultone [23]. A series of such surfmers has been prepared by one of us [24] with an alkyl chain length from 12 to 18 carbon atoms and cmc from 1 to 0.1 mmol/l. These surfmers can be very efficiently bound to the latex particle surface when used in emulsion polymerization of styrene. The surface tension of the latex serum can be kept very high (above 70 mN/m) after the polymerization even if amounts of more than 100 times the cmc are used. Due to the fact that the maleic surfmers are only able to copolymerize, the surfactant is bonded covalently to the polymer chains. It has been found [25] that the incorporation of these surfmers during the third stage of the polymerization after the monomer droplets disappear is strongly increased. More than 90% of the surfmer employed is anchored at the surface at the end of the polymerization due to copolymerization. This behavior can be explained by a change in the locus of polymerization from the particle volume near to the particle water interface with increasing conversion. The results known so far concerning the application of maleic surfmers indicate that this class of surfmers with maleic copolymerizable groups is more advantageous for obtaining a high covalently bonded surface coverage than surfmers with allylic, acrylic or vinylic polymerizable groups.

In a thesis, M.B. Urquiola at Lehigh [26, 27] compares the behaviour in vinylacetate (VA) emulsion polymerization of a commercial surfmer sodium

$$-O_3S - (CH_2)_3 - O - \underset{\underset{O}{\|}}{C} \diagup^{H}_{C=C} \diagdown_{C}^{H} - O - (CH_2)_n - CH_3 \qquad \textbf{IV}$$

alkyl allyl sulfosuccinate (SAAS) and its hydrogenated equivalent non polymerizable (H-SAAS). The homopolymerization of SAAS is slow and gives
mostly oligomers. Its copolymerization with VA in a homogeneous medium
allows the reactivity ratios $R_{VA} = 0.36$ and check $R_{VA} = 0.48$ to be determined,
so that the presence of SAAS decreases the polymerization rate of vinylacetate,
due to the difference in reactivity of each monomer. In addition, SAAS is a
rather strong transfer agent with a transfer constant of 0.011. In emulsion
polymerization, oligocopolymers of SAAS and VA are found in the water phase
in amounts increasing with the concentration of the surfmer and with the
concentration of initiator. The incorporation of SAAS at the surface of polymer
particles increases with the amount of the surfmer but decreases with increasing
initiator concentration (for constant [SAAS]). The polymerization rate decreases when [SAAS] increases while the particle size decreases. The H-SAAS
has the expected effect of a normal surfactant, decreasing the size and increasing
the rate due to the higher number of particles with increasing concentration of
surfactant. In the presence of H-SAAS the initiator concentration has no effect
on the size of the particles. In contrast, in the case of SAAS, the size increases
with the concentration of the initiator. These results may be explained by
termination reactions in the water phase, producing water-soluble oligomers of
low molecular weight which do not participate in the nucleation of particles.
Competitive growth experiments were carried out with a mixture of two seed
polystyrene latexes using either SAAS or H-SAAS. The normal behavior of
growth with decreasing difference in size of the two families of particle was
observed with H-SAAS. In contrast, in the case of SAAS, the growth of the small
particles was smaller and that of the big particles was correspondingly higher,
the effect being enhanced by increasing the SAAS concentration. In addition, the
polymerization rate was decreased in the presence of SAAS. The explanation lies
in the fact that more SAAS was adsorbed on the surface of the small particle,
causing a more important retardation of the polymerization process, due to the
low reactivity of the SAAS in polymerization and its good ability to copolymerize with vinylacetate; its high transfer constant gives an additional explanation
to the retardation effect, causing an enhanced exit rate for the radical and more
efficient termination in the water phase. All these results demonstrate the active
part played by the surfmer in the polymerization process. However, it may be
expected that some of these unusual results are due to the choice of the allyl
function.

A recent paper by a Russian team [18] describes the use of a few new
surfmers, one being cationic, namely N-decylaceto-2-methyl-5 vinylpyridinium
bromide (V), and the others being anioic, namely decyl (or dodecyl), sodium
ethyl sulfonate, methacrylamides (VI), decyl (or dodecyl)-phenyl (Na or K
sulfonate) acrylate (VII), and decyl ester of sodium (or K or NH_4^+) sulphocin-
namic acid (VIII). These surfmers were used for emulsion polymerization of
styrene, butylacrylate or chloroprene, in the presence of KPS or AIBN without
any other surfactants. It should be noted that the consumption of these
surfactants take place early in the polymerization process which is faster than in

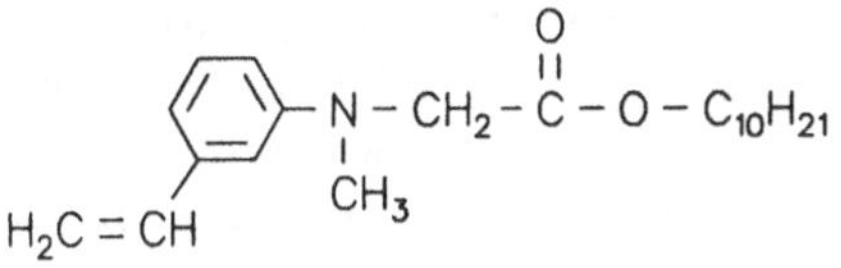

$$H_2C=CH-\langle C_6H_4 \rangle-\underset{\underset{CH_3}{|}}{N}-CH_2-\overset{\overset{O}{||}}{C}-O-C_{10}H_{21} \qquad \textbf{V}$$

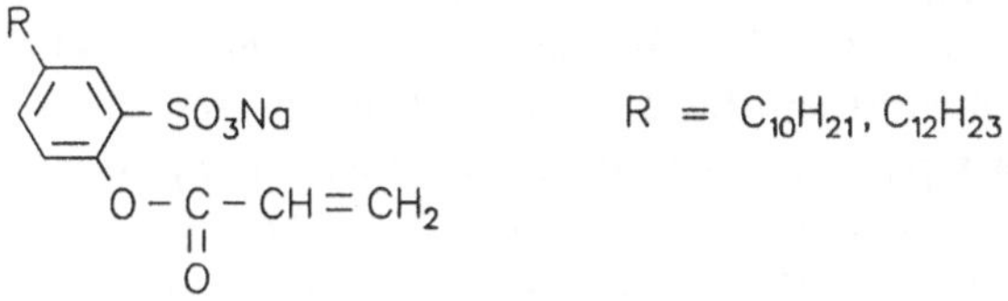

$$R-N\overset{\displaystyle C_2H_4SO_3Na}{\underset{\displaystyle \underset{O\ \ CH_3}{\overset{||\ \ |}{C-C=CH_2}}}{}} \qquad R = C_{10}H_{21},\ C_{12}H_{23} \qquad \textbf{VI}$$

$$\underset{O-C-CH=CH_2}{\overset{R}{\underset{||\atop O}{\langle C_6H_3 \rangle-SO_3Na}}} \qquad R = C_{10}H_{21},\ C_{12}H_{23} \qquad \textbf{VII}$$

$$NaO_3S-\langle C_6H_4 \rangle-CH=CH_2-COOR \qquad R = C_{10}H_{21} \qquad \textbf{VIII}$$

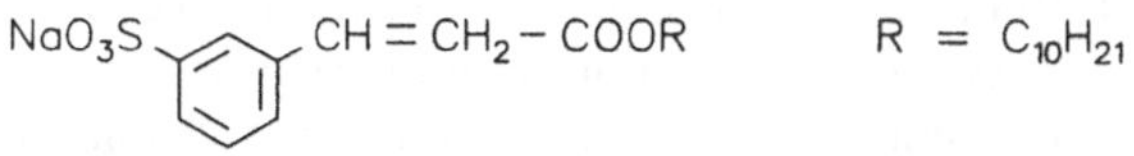

the case where SDS is used as surfactant. No emulsifier is left in the water phase, and the latexes are highly stable with regard to electrolyte, temperature and redispersion.

Finally, phospholipidic compounds carrying polymerizable groups such as **IX** and **X** have been shown very recently [28] to allow the preparation of stable monodisperse polystyrene latexes, with sizes in the range of 200–300 nm in diameter, carrying a phospholipid layer on their surface with yields higher than 75%.

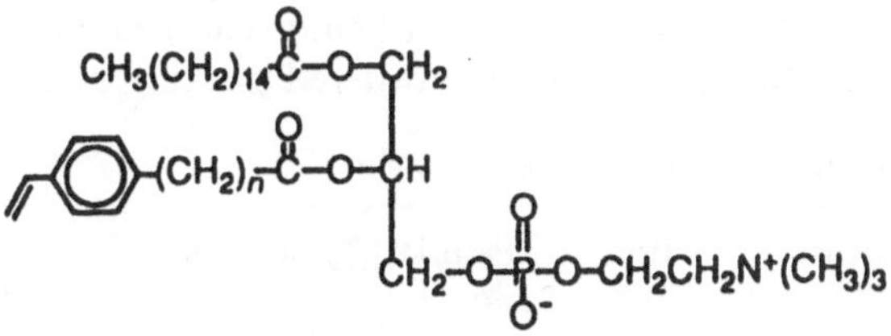

$$\begin{array}{l}
CH_3(CH_2)_{14}\overset{\overset{O}{||}}{C}-O-CH_2 \\
\langle C_6H_4 \rangle-(CH_2)_n\overset{\overset{O}{||}}{C}-O-CH \\
\qquad\qquad\qquad CH_2-O-\overset{\overset{O}{||}}{\underset{\underset{O^-}{|}}{P}}-O-CH_2CH_2N^+(CH_3)_3
\end{array} \qquad \textbf{IX}$$

$$\begin{array}{l}
\langle C_6H_4 \rangle-(CH_2)_4\overset{\overset{O}{||}}{C}-O-CH_2 \\
\qquad H-O-CH \\
\qquad\qquad\qquad CH_2-O-\overset{\overset{O}{||}}{\underset{\underset{O^-}{|}}{P}}-O-CH_2CH_2N^+(CH_3)_3
\end{array} \qquad \textbf{X}$$

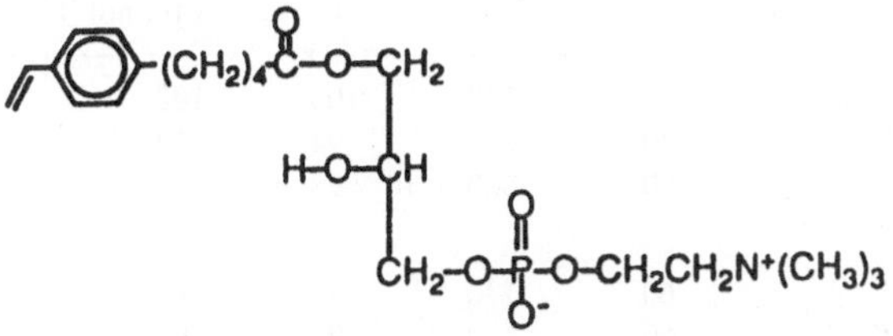

In the field of nonionic polymerizable surfactants, a pioneering work was published by the Ottewill group. They prepared two sets of polystyrene latexes [29]. The first set was initiated by KPS at 80 °C and resulted in charged latexes. The set comprised a conventional anionic surfactant, a conventional nonionic surfactant and a macromonomer of polyoxyethylene (45 units) carrying a methacrylic-polymerizable end group, the second end group being a methoxy group. The latter was used in the presence of a small amount of anionic surfactant. The second set was prepared at 60 °C using a nonionic redox initiating system (H_2O_2 and ascorbic acid). The surfactants were either the nonionic surfactant, or the same macromonomer or a combination of both. The data concerning the particle size and the glass transition temperature after dialysis against pure water or water plus ethanol mixture are reported in Table 1. In most cases rather monodisperse particles were obtained which showed a tendency to agglomerate, chiefly when the electrostatic stabilization was absent. Anionic surfactants are easily desorbed upon dialysis against water. The nonionic conventional surfactant can be desorbed when a water-ethanol mixture is used for dialysis, but a part of it remains buried inside the particles, causing a net decrease of the T_g. The macromonomer which has been incorporated cannot be desorbed by dialysis, but there is not much buried inside the particles. Burying nonionic surfactants in the latexes causes the observed drop in the T_g; in addition the T_g interval is broader. When both nonionic surfactant and macromonomer are used, the nonionic surfactant is progressively displaced from the surface and becomes increasingly buried inside the particles. Temperature is an important parameter for macromonomer incorporation; possibly because of the proximity of the cloud point, less macromonomer is incorporated at the higher temperature. Finally the polymerization of the macromonomer also produces water-soluble polymers which causes flocculation or at least aggregation of the latexes.

When the macromonomer is used as surfactant (even if it does not display actual surfactant properties) the latex obtained shows an excellent resistance to flocculation due to electrolyte addition (barium chloride) up to the highest concentration examined (0.75 mol/dm³), while conventionally charged latex flocculated at 2.1×10^{-2} mol/dm³. In addition, highly ordered packing of the

Table 1. Two sets of polystyrene latexes prepared with various surfactants

Initiator (and amount) mol/dm³ × 10⁻³	Surfactant	Amount mol/dm³ × 10³	Pzn Temp. °C	Particle size nm	T_g after dialysis	
					water °C	water + Ethanol °C
K2S2O8 (5.38)	anionic	21.2	80	190	104	103
	NP20[a]	2.88	80	212	94	99
	M.PEO[b] + anionic	4.68 21.2	80	600 + 10	98	98
H2O2 (3.56) + ascorbic acid (1.12)	NP20	6.5	60	120	70	84
	M.PEO	4.1	60	160	92	90
	M.PEO45 + NP20	4.1 6.5	60	110	77	79

[a] Nonylphenol ethoxylated (20 EO units)
[b] Macromonomer of polyethylene oxide (45 EO units)

latex can be observed even when there is high ionic strength [29]. Resistance against freeze-thaw conditions was investigated by measuring the turbidity of the suspension of latexes previously frozen at $-18\,°C$ for three days and then redispersed at room temperature [30]; two latexes, namely the first of the first set (charged latex with a surface charge density of $_4O_mC/m^2$) and the second of the second set (macromonomer only and non-charged latex) were compared. A 20-fold difference in turbidity was observed showing that the steric stabilizer, firmly anchored to the particles, efficiently protects the latex against flocculation by freezing.

In the United Kingdom, the ICI Company has developed two industrial processes using surfmers [31]. The first one, named Aquersymer is actually a dispersion polymerization process more than an emulsion polymerization process, because the monomers are soluble in the diluent mixture of water and ethanol; the polymer is not soluble and is precipitated as particles, stabilized through copolymerization of the macromonomer. The initiator is oil-soluble AIBN. A semi continuous process is used with a seed formation step followed by two successive feed steps. The particle size is well-controlled by the length of the macromonomers and also by its concentration; it increases smoothly during the feed steps. A variety of polymerizable surfactants have been used [32] including block copolymers of butylene oxide and ethylene oxide with a small number (1 to 3) of allylic pendent groups. However these compounds are not suitable for use in styrene polymerization, owing to their very low reactivity with that monomer. The nonionic latexes prepared by the Aquersymer process were compared with more conventional anionic latexes and anionic latexes stabilized with cellulosic colloids; they display superior properties of stability to $CaCl_2$ or $Fe(NO_3)_3$ addition as well as to shear and freeze-thaw stresses.

The second ICI process is the NIAD process (nonionic aqueous dispersion [31, 33]. It is a more conventional emulsion polymerization in water using a water-soluble azo initiator with steric stabilization from α-methacryloyl, ω-methoxy block copolymer of butylene oxide and ethylene oxide of low water solubility. The process is suitable for latex synthesis from monomer with low water solubility, but gives high amounts of coagulum if the monomer solubility in water is higher than 0.5%. A rather high amount of surfmer (more than 3% based on the monomer) is needed to achieve freeze-thaw stability. The particle size is very dependent on the polymerization temperature and increases greatly when the cloud point of the surfactant is approached.

Diblock polyoxyethylene-polyoxypropylene styrenic macromonomers, with the polymerizable group at the end of the hydrophobic part have been prepared and used in styrene emulsion polymerization [34]. Latexes of high stability towards added electrolyte have been obtained. However the HLB was not well-optimized so that a high amount of coagulum was formed (Surfmer **XI**).

A very recent study by Pichot and Charreyre [35] dealt with cellulosic surfactants with a methacrylic group at the end of an hydrocarbon chain (6 to

$$H_3C - O - (CH_2CH_2O)_m - (CH_2\underset{CH_3}{\overset{|}{C}HO})_n - CH_2 - \text{(ring)} - CH = CH_2 \qquad \textbf{XI}$$

10 C atoms) linked to a cellobiose hydrophylic moiety, (Surfmers **XII** and **XIII**). It involves the surface grafting of the surfactant to a polystyrene seed initially swollen with styrene or methyl methacrylate. The best grafting yield was observed if the ratio between surfactant and the swelling monomer was low, if the particle size of the seed was small and when the swelling monomer was highly water soluble; this last point may also depend on the HLB balance of the surfactant. The grafting involves two kinds of mechanism a) copolymerization of the adsorbed surfactant which gives a monolayer of grafted surfactant; b) capture of oligomers produced in the water phase: the composition of these oligomers is dependent upon the water solubility of the comonomer; in the case of styrene, their styrene contents is quite low, while in the case of methyl methacrylate the composition is close to 1/1 molar. A maximum grafting yield of 66% has been obtained, the remaining part of the surfmer being in the water phase as water-soluble oligomers. The grafting process causes a decrease of electrophoretic mobility through the screening of the charges carried by the latex on the cellobiose units; because the length of the hydrophilic cellobiose moiety is short, good steric stabilization is not achieved and the electrostatic stability is decreased due to the screening effect, so that some stabilization problems are met. However, the major interest of the grafting is to allow fixation of bioactive compounds onto the surface of the latex.

3 Inisurfs

The number of scientific papers as well as the number of patents regarding inisurfs is much smaller than those published regarding surfmers. The references [16, 24, 36, 37] contain the various inisurfs known to the authors. Inisurf molecules are composed of at least three different parts: the radical generating group, a hydrophobic part, and a hydrophilic part. Inisurfs can be subdivided into azo and peroxy compounds with regard to the chemical nature of the radical generating group. Both types of inisurf are known with nearly an equal number of papers dealing with peroxy inisurfs [36, 38–42] to those with azo inisurfs [24, 37, 43–50]. It is noteworthy that most of the papers dealing with peroxy inisurfs are published in Russian scientific papers. For more detailed information concerning publications in very special Russian journals see the papers referred to in [16].

Both types of inisurf may be monomeric or polymeric regarding the number of radical generating groups per molecule. Another way to subdivide inisurfs is based on the symmetry of the groups attached to the radical generating group. Symmetrical inisurfs have the same structural groups on both sides of the azo or peroxy group, e.g., two surface active radicals with the same structure are produced after decomposition [24, 43–45]. Non-symmetrical inisurfs have only one surface-active group attached to the radical generating group. Consequently, after decomposition they form one surface-active and one non-surface-active radical for instance a tertiary butyl or hydroxyl radical [36, 38–40, 51, 52].

A third method to subdivide surface-active initiators is based on the chemical nature of the hydrophobic and hydrophilic groups. The hydrophilic groups may be anionic or cationic ones or an oxyethylene chain of an appropriate chain length. Hydrocarbon chains (alkyl chain, alkyl phenol chains) or a propylene oxide chain are used as hydrophobic molecule parts. Oxyethylene chains are also used as hydrophobic groups if they are in the neighbourhood of ionic groups.

One advantage of inisurfs is the possibility of reducing the ingredients of an emulsion polymerization to the components: monomer, water, and initiator. The accessory content of the final latex can be decreased by this way considerably.

Problems exist with the chemical and structural purity of the inisurfs especially from the colloidal point of view. One must always bear in mind that impurities are present in most systems investigated. Nevertheless, the results known so far clearly show the pecularities of inisurfs compared to conventional initiators for emulsion polymerizations like water-soluble peroxides or AIBN.

Inisurfs behave like surfactants, e.g., they form micelles and are adsorbed at surfaces. So they are characterized by a critical micelle concentration in solution and an area per molecule in the adsorbed state. This surface activity is the most important physical property of inisurfs influencing strongly their polymerization behavior. So, for instance, the decomposition behavior of inisurfs strongly depends on whether their concentration is above or below the critical micelle concentration [52]. The ability of these initiators to form micelles or to adsorb leads to a much higher primary radical recombination due to an enhanced cage effect. Several authors estimated radical efficiency values, f, in the range of 10^{-2} to 10^{-4} which are much lower than those for conventional initiators [39, 45, 52, 54]. Such low values were found for azo inisurfs [45, 52, 54], for peroxy inisurfs [39], for symmetrical inisurfs [45, 51–54], and for non-symmetrical inisurfs [39, 51]. Up till now, it seems impossible to overcome these low radical efficiencies. However, it is possible to realize high overall polymerization rates with inisurfs in emulsion polymerization. This may be due to the following facts pointed out by Ivancev and Pavljucenko [39, 40] as results of investigations of styrene emulsion polymerization in the presence of non-symmetrical peroxy inisurfs (Inisurf 1):

$$C_{16-18}\ H_{33-37}\ (OCH_2CH_2)_{20}-O-\overset{\overset{O}{\|}}{C}-\underset{\underset{R^1-\overset{\|}{\underset{O}{C}}}{\overset{|}{CH_2}}}{CH}-\underset{\underset{R^2-\underset{OOH}{\overset{|}{C}}-OC_2H_5}{\overset{|}{CH_2}}}{CH}-\overset{\overset{O}{\|}}{C}-O^{\ominus}\ Na^{\oplus}$$

$$R^1, R^2 : -H; -CH_3$$
$$R^1 \neq R^2$$

1. The mean number of radicals per particle is greater than 0.5. The authors estimated n values in the range of 1.5.

2. Due to the adsorbed state of the initiator molecules the termination rate is lowered.

Furthermore, the authors pointed out that they obtained in the emulsion polymerization of styrene (monomer to water ratio 1 : 2) with an inisurf concentration of 5.4×10^{-6} mol/l water in the presence of an alkylated poly (oxyethylene) emulsifier (alkyl chain length C16–18 and 20 oxyethylene units; 4% by weight related to water) the same overall rate of polymerization as with water-soluble initiators in the concentration range 10^{-4} to 10^{-3} mol/l water. The polymer produced in the presence of inisurf has a molecular weight of some of 10^7 g/mol mainly due to the lowered termination rate constant.

Ivancev and Pavljuchenko published a series of papers [36–40] regarding the emulsion polymerization with surface-bonded radical generation. Their pioneering work, however, is related only to peroxy inisurfs.

Other examples of peroxy inisurfs can also be found in Russian scientific papers. As for instance in Ref. [41] Voronov et al. describe a polymeric surfactant with peroxy side chains for application as inisurfs in emulsion polymerization. They obtained the polymeric inisurf (Inisurf **2**) by copolymerization of a peroxide containing monomer (dimethyl-vinylethinyl-methyl-*tert*-butyl-peroxide) with acrylic or methacrylic acid or 2-methyl-5-vinyl pyridine with benzoyl peroxide as initiator in the presence of dodecylmercaptan as chain transfer agent. The resulting copolymers are water soluble at appropriate pH-values, surface active, and exhibit a critical micelle concentration.

$$\{CH_2-\underset{R^2}{\overset{R^1}{\underset{|}{\overset{|}{C}}}}\}_n\{CH_2-\underset{\underset{\underset{\underset{\underset{\underset{CH_3}{\overset{|}{C}}}{\overset{|}{O}}}{\overset{|}{O}}}{CH_3-\overset{|}{\underset{|}{C}}-CH_3}}{\overset{|}{\underset{\underset{\underset{\underset{CH_3}{\overset{|}{C}}-CH_3}{\overset{|}{O}}}{\overset{|}{O}}}{C}}}}{CH}\}_m$$

$$R^1 : -CH_3 ; -H$$
$$R^2 : -COOH$$

Galibei et al. [42] found that water soluble diacyl peroxides (Inisurf **3**) having surfactant properties can act as effective initiators for the emulsion polymerization of styrene in the presence of Tween 20. The authors found that the polymerization rate depends on the chain length of the hydrophobic and hydrophilic groups and was highest for $n = 7$ and $m = 9$. These initiators (inisurf **3**) have surface active properties. However, the radicals formed after decomposition are no longer surface active due to the fact that the peroxy group is the linkage between the hydrophilic and hydrophobic parts of the inisurf. That is probably the reason that the emulsion polymerization must be carried out in the presence of Tween 20 as additional emulsifier to obtain stable latexes.

$$C_nH_{2n+1}\overset{O}{\overset{\|}{C}}-O-O-\overset{O}{\overset{\|}{C}}-R-\overset{O}{\overset{\|}{C}}-O(CH_2CH_2O)_m\overset{O}{\overset{\|}{C}}-R-\overset{O}{\overset{\|}{C}}-O-O-\overset{O}{\overset{\|}{C}}-C_nH_{2n+1}$$

$R: -CH_2CH_2-, -O-C_6H_4$ $n: 6,7$
                                     $m: 1,2,3,9,13$

Surface active initiators with an azo group as radical generating functionality are another important class of inisurfs [16, 24, 37, 43–50].

Some examples of monomeric surface-active azo initiators as well as their synthesis, application in emulsion polymerization, and resulting latex properties are described in Refs. [43, 44]. These inisurfs are symmetrical and exhibit a methylene chain as hydrophobic part of the molecule and an ionic hydrophilic group (Inisurf **4**, Inisurf **5**). The main results of these investigations may be summarized as follows:

1. The inisurfs can be used in emulsion polymerization without addition of surfactants to obtain, for example, butyl acrylate-styrene-copolymer dispersions with a solid content of about 35%.

2. Particle size, polymerization rate, and solid content of the final latexes remain unchanged compared with polymerizations with the unreacted individual components of the inisurfs (water soluble non-surface-active azo-bis (diisobutyric acid amidine) and a C15-alkylmonosulfonate as emulsifier in the case of Inisurf **4** and a potassium salt of ω-aminoundecanoic acid as emulsifier in the case of Inisurf **5**, respectively).

$$Na^{\oplus}\ {}^{\ominus}O_3S(CH_2)_n SO_2-HN \overset{O}{\underset{}{\diagdown}}\overset{}{C}-\overset{CH_3}{\underset{CH_3}{\overset{|}{C}}}-N=N-\overset{CH_3}{\underset{CH_3}{\overset{|}{C}}}-C\overset{O}{\diagup}\overset{}{}\diagdown NH-SO_2(CH_2)_n SO_3^{\ominus} Na^{\oplus}$$

$$Na^{\oplus}\ {}^{\ominus}O_3S(CH_2)_n SO_2-HN \overset{NH}{\underset{}{\diagdown}}\overset{}{C}-\overset{CH_3}{\underset{CH_3}{\overset{|}{C}}}-N=N-\overset{CH_3}{\underset{CH_3}{\overset{|}{C}}}-C\overset{NH}{\diagup}\overset{}{}\diagdown NH-SO_2(CH_2)_n SO_3^{\ominus} Na^{\oplus}$$

(Idealized structure possibilities)

CH₃ structure...

$$CH_3-\langle O \rangle-NH-CO-N \diagdown \quad \underset{\underset{C_2H_5O}{\diagup}}{\overset{CH_3}{\underset{|}{C}}}-\overset{CH_3}{\underset{|}{C}}-N=N-\overset{CH_3}{\underset{|}{C}}-\underset{\underset{OC_2H_5}{\diagdown}}{\overset{CH_3}{\underset{|}{C}}} \quad \diagup N-CO-NH-\langle O \rangle-CH_3$$

Left side lower substituent:
NH
|
CO
|
NH
|
(CH₂)₁₀
|
COO⁻ K⁺

Right side lower substituent:
NH
|
CO
|
NH
|
(CH₂)₁₀
|
COO⁻ K⁺

3. The final latexes of the polymerizations with the inisurfs exhibit a lower electrolyte content as well as a lower foam stability both indicating a lower waste content in the latex serum than with polymerization with the individual components of the inisurf synthesis.

Another type of inisurf (Inisurf **6**) was investigated by the Eindhoven group [45]. The inisurf was synthetized by esterification of 4,4'-azobis (4-cyano-pentanoic acid) with poly(ethylene glycol)nony-phenol. The ethylene oxide chain had a length of 30 monomer units. The authors investigated the seeded emulsion polymerization of styrene in order to determine the radical entry rate coefficient and the initiator efficiency with varying degrees of surface coverage of the seed particles with the initiator molecules. The main result of these investigations was a very low radical efficiency in the order of 10^{-4} independent of the degree of particle surface coverage. In order to increase the radical efficiency the authors synthesized a non-symmetrical inisurf of the same type but with a *tert*-butyl group at one end of the azo group. However, in this case they also found an efficiency which was nearly equal or only just higher than with the symmetrical inisurf [51].

$$C_9H_{19}-\langle O \rangle-(OCH_2CH_2)_n\,O-\overset{O}{\overset{\|}{C}}-CH_2-CH_2-\overset{CH_3}{\underset{CN}{\underset{|}{\overset{|}{C}}}}-N=N-\overset{CH_3}{\underset{CN}{\underset{|}{\overset{|}{C}}}}-CH_2-CH_2-\overset{O}{\overset{\|}{C}}-O-(CH_2CH_2O)_n-\langle O \rangle-C_9H_{19}$$

n : 30

Polymers containing azo groups which were able to start emulsion or dispersion polymerization as well as to stabilize the polymer particles were described in a series of papers by Heitz et al. [46–49]. These polymeric inisurfs were synthesized by a two step synthesis starting with the acid-catalysed polycondensation of AIBN with α,ω-diols as for instance 1,6-hexanediol [49]. The polycondensation results in a polyazoinitiator (Inisurf **7**) with a molecular weight of some 10^3 g/mol corresponding to approximative 10 azo groups per molecule on average. The resulting polyazoinitiator is neither surface active nor water soluble as is the case with other α,ω-alkyldiols and low molecular weight poly (ethylene glycol). However, the polyazoinitiator is water soluble in the case where a poly(ethylene glycol) is used as diol with a molecular weight higher than 300 g/mol. To achieve surface activity these polymeric azoinitiators are used in a radical polymerization, as for instance, of acrylamide in *tert*-butanol limiting the decomposition of the initiator to approximately 37%. The resulting block

$$\left[\ \underset{\underset{CH_3}{|}}{\overset{\overset{O}{\|}}{C}}-\underset{\underset{CH_3}{|}}{\overset{\overset{CH_3}{|}}{C}}-N\!=\!N-\underset{\underset{CH_3}{|}}{\overset{\overset{CH_3}{|}}{C}}-\underset{\underset{CH_3}{|}}{\overset{\overset{O}{\|}}{C}}-O-R\ \right]_n$$

R : $-CH_2CH_2O$, $+CH_2$ $\frac{}{6}$

n : ~10

copolymer consists of hydrophilic and hydrophobic parts and still contains azo groups. The critical micelle concentration of this so called prepolymer depends on the polyazoinitiator acrylamide ratio – the higher the ratio the lower is the critical micelle concentration [48]. Prepolymers synthesized in this way can be used as steric stabilizers initiating the emulsion polymerization. Examples are also described where the hydrophilic blocks in the prepolymer consist of vinyl acetate or methacrylic acid [47]. The latexes are very stable whereby, the stability of the latexes is increased in the case of prepolymers with a higher content of azo groups as polymeric inisurfs. The authors obtained stable latexes with a solid content of up to 35% also by polymerizing styrene, methyl methacrylate, other acrylates, and acrylonitrile.

Prepolymers of this type can also act as inisurfs for dispersion polymerizations in organic solvents by using prepolymers soluble in the particualr solvent. This can be determined by appropriate choice of the monomer in the polymerizations forming the prepolymer. So for instance the vinyl acetate prepolymer can be used to obtain stable polyacrylamide dispersions with a solid content of upto 50% by polymerization in methanol [47]. Another application of this type of polymeric azoinitiator mentioned very briefly is the preparation of graft and block copolymers [55, 56]. The chemical composition of the blocks as well as the polymerization technique employed can be matched over a wide range to obtain polymers with desired properties.

The reaction of AIBN with diols can lead to polymeric azoester or to monomeric azoester depending on the molar ratio AIBN to diol. A 20-fold excess of the diol leads to monomeric azoester (Inisurf 9) [24, 49]. In the case of poly-(ethylene oxide) as diol the resulting azodiesters are slightly water soluble showing an increasing water solubility with increasing ethylene oxide chain length. Azodiesters with ethylene oxide chain length lower than 25 on each side of the azo group cannot be used without additional surfactants to carry out emulsion polymerizations [24]. We have found that a sulfation of the end

$$\left[\ \underset{\underset{CH_3}{|}}{\overset{\overset{O}{\|}}{C}}-\underset{\underset{CH_3}{|}}{\overset{\overset{CH_3}{|}}{C}}-N\!=\!N-\underset{\underset{CH_3}{|}}{\overset{\overset{CH_3}{|}}{C}}-\underset{}{\overset{\overset{O}{\|}}{C}}-O-R-\underset{}{\overset{\overset{O}{\|}}{C}}-\underset{\underset{CH_3}{|}}{\overset{\overset{CH_3}{|}}{C}}\ \right]_x\left[\ CH_2\!-\!\underset{\underset{R^1}{|}}{CH}\ \right]_n$$

R : $+CH_2CH_2O$ $\frac{}{4}$; $+CH_2$ $\frac{}{6}$

R^1 : $C\overset{\diagup O}{\diagdown NH_2}$; $-COOH$, $-OC\!-\!CH_3$ (with O above the last group)

$$HO-R-O-\underset{\underset{CH_3}{|}}{\overset{\overset{O}{\|}}{C}}-\underset{\underset{CH_3}{|}}{\overset{\overset{CH_3}{|}}{C}}-N\!=\!N-\underset{\underset{CH_3}{|}}{\overset{\overset{CH_3}{|}}{C}}-\underset{\underset{CH_3}{|}}{\overset{\overset{O}{\|}}{C}}-O-R-OH$$

R : $+CH_2CH_2O$ $\frac{}{m}$; $+CH_2$ $\frac{}{n}$

m : 5,15

n : 6,8,10,12,16

$$NH_4^{\oplus} \ ^{\ominus}O_3SO-R-O-\overset{\overset{\displaystyle O}{\|}}{C}-\overset{\overset{\displaystyle CH_3}{|}}{\underset{\underset{\displaystyle CH_3}{|}}{C}}-N=N-\overset{\overset{\displaystyle CH_3}{|}}{\underset{\underset{\displaystyle CH_3}{|}}{C}}-\overset{\overset{\displaystyle O}{\|}}{C}-O-R-OSO_3^{\ominus} \ NH_4^{\oplus}$$

R : $+CH_2CH_2O+_m$; $+CH_2+_n$

m : 5, 15

n : 6, 8, 10, 12, 16

standing hydroxyl groups of the diester by conventional methods [57–59], leads to true inisurfs (Inisurf **10**) [54]. These inisurfs were characterized regarding their molecular weight [24] as well as regarding their surface activity [52]. The critical micelle concentration of these surface active initiators depends on the diol used. Inisurfs prepared with poly(ethylene oxide) exhibit a higher critical micelle concentration than products prepared with α,ω-alkyldiols. Some of the features of these inisurfs have been published [24, 37, 50, 52].

The azodiestersulfates can be applied to carry out emulsion polymerizations with various monomers leading to stable latexes with a solid content of up to 40% without additional surfactants. Another advantage of these inisufs is that they allow one to prepare polymers with much higher molecular weights than those obtained with conventional initiators in emulsion polymerization [24] as well as in solution polymerization [37, 60]. In order to investigate the action of the inisurf (Inisurf **10**) regarding the reduction of the hydrophilic waste contents, we have carried out emulsifier-free emulsion copolymerization of butyl acrylate and methyl methacrylate with potassium persulphate as well as with the surface-active initiator derived from poly(ethylene glycol) with a molecular weight of 200 g/mol. The polymerizations were carried out at initiator concentrations realizing the same radical flux in both cases. The water absorption of copolymer films cast from the latexes was measured gravimetrically. The films of the inisurf latex showed much lower water absorption than the persulphate latex films indicating a lower hydrophilic waste in the inisurf latex.

The preparation of latexes as for instance with a special modified particle surface [61] and the preparation of polymers for special applications [37] are other interesting features of inisurfs. We have shown this, for instance, in the case of preparing special side-chain copolymers for non-linear optical investigations [37]. The emulsion copolymerization of methyl methacrylate with an azo-dye comonomer (Inisurf **11**) in the presence of the sulphated poly(ethylene glycol) 200 inisurf (Inisurf **10**) results in much higher conversion as well as in much higher molecular weight than solution polymerization with the same initiator in dimethyl formamide. So, for instance, the solution polymerization gives copolymers with a molecular weight in the range of 10^4 dalton whereas the emulsions polymerization copolymer has a molecular weight in the range of 10^6 dalton. The molecular weight of the copolymer prepared by emulsion polymerization is high enough for it to be possible to cast free-standing films for the optical

Dye Comonomer
$$CH_2=\overset{\overset{\displaystyle }{|}}{\underset{\underset{\displaystyle CH_3}{|}}{C}}-\overset{\overset{\displaystyle O}{\|}}{C}-O-CH_2-CH_2 \diagdown$$

$$N-\!\!\!\!\bigcirc\!\!\!\!-N=N-\!\!\!\!\bigcirc\!\!\!\!-NO_2$$

OCH_3

CH_3

investigations. Another important difference relates to the conversion in both polymerization techniques employed. The inhibition action of the dye comonomer seems to be much higher in solution than in emulsion polymerization.

In conclusion we can say that the inisurfs known today have different chemical structures and consequently different properties. Experimental data are available showing that emulsion polymerization is posssible using inisurfs without any additional emulsifiers, thus reducing the electrolyte content in the latex serum as well as foam formation. From a more technical point of view problems existing today concern the low initiator efficiencies as well as the fact that the solid content of the latexes is restricted to approximately 40% without coagulum formation.

4 Transurfs

Some transfer activity is displayed by many common surfactants. For instance, when emulsion polymerization of styrene is initiated photochemically, using biacetyle in the presence of SDS as emulsifier, the resulting latex is slightly charged ($3 \, mC/m^2$) [30] with strong acid groups. Some other surfactants show more transfer activity, for instance Transurfs **I** and **II**

Only one study has been carried out with a surfactant carrying a typical transfer agent thiol groups. Fitch and Fifield [62] reported the synthesis of Transurf **III** with a cmc of 6.1×10^{-2} mol/l at 24 °C. This surfactant was used both in polystyrene latex synthesis at concentrations lower than the cmc and in seeded polymerization experiments. Monodisperse particles of about 400 nm in diameter were obtained in latex synthesis initiated with azocyanovaleric acid. In contrast to latex prepared in the presence of SDS, these latexes are not sensitive to coagulation upon ion exchange. A very surprising result is that the particle size seems to be practically insensitive to the amount of the Transurf. Seeded experiments have been carried out with a large size seed latex (700 nm) prepared without emulsifier and carrying a small charge density ($3.4 \, mC/m^2$). The seed was first reacted with the transurf and then swollen with styrene and a solution of AIBN in dichloroethane; polymerization at 55 °C resulted in a particle size of 740 nm with a higher charge density ($53 \, mC/m^2$). Then the yield of surface location of the Transurf can be estimated to be 60%.

A program is now in progress in the author's laboratory (AG) concerning Transurfs **IV** and **V**.

I $C_{11} H_{23} \cdot CH = CH \cdot CH_2 SO_3 \, Na$

II $C_{11} H_{23} \, \underset{\underset{OH}{|}}{C}H \, (CH_2)_2 \, SO_3 \, Na$

III $HS \cdot C_{10} H_{22} \cdot SO_3 Na.$

IV $HS \cdot C_n H_{2n} \cdot (EO)_m \cdot OH$

V $CH_3 (EO)_n \cdot COO CH_2 SH$

5 Conclusions

From the various studies already published a few trends can be observed. The stabilizing efficiency of the reactive surfactants needs a rather high hydrophilic character: for the anionic surfactants, it is clear that carboxylic groups can be easily buried inside the growing polymer, and then sulfonic groups or sulfates are much better. For the non-ionic surfactants, based on polyethylene oxide sequences an adequately long sequence (i.e. more than 20 units) is necessary. On the other hand, a hydrophilic character that is too high may lead to the production of water-soluble polymers, which may cause flocculation of the latex. Then an adequately hydrophobic character is also desirable. Therefore it can be concluded that the reactive surfactants must have a true amphiphilic character. However in the field of surfmers. macromonomers of polyoxyethylene have justified interest in them.

A lot of mechanistic problems remain to be solved. It is not so clear, at the moment, why most of the inisurfs studied up to now have such a low efficiency, whatever their structure. On the other hand it is quite remarkable that, even with that low efficiency, they are able to allow the preparation of stable latexes under acceptable experimental conditions. Another problem is the control of the nucleation: both with inisurfs and transurfs, there are indications that the particle number and size may not be very sensitive to the amount of reactive surfactants, and more dependent on the amount of monomer used. Such behavior, up to now, has not been explained. Very few studies have been devoted to the reactivity of these surfactants, i.e., reactivity ratios for the surfmers, transfer constants for the transurfs and initiator efficiency for the inisurfs. Both the reactivities and the partition coefficient between the water and the organic phase have to be determined. In addition the reactivities may be dependent on the other components of the recipe, for instance due to the effect of the ionic strength on the cmc.

The colloidal properties of the reactive surfactants should also be known. In addition to the determination of the cmc in the presence of monomers and other additives, the dynamics of the reactive surfactants and their adsorption on the polymer surface are important factors. In the case of the non-ionic surfactants the knowledge of the cloud point and its distance from the polymerization temperature is also an essential parameter.

Finally one has to be careful about all the side effects, such as the production of water soluble polymers, the sensitivity of the surfactants to transfer reactions in addition to their main character (the case of allylic surfmers for instance), or the effect of the nonionic surfactants on the T_g of the polymers, which may make coalescence of the particles easier.

In spite of all these problems, a huge number of possibilities are open through the variation of the reactive surfactant structure, nature and sequence length of each part of the amphiphilic molecule, position of the reactive group, eventual polymeric character. For instance the concept of polymeric inisurfs offers possibilities for a wide variation in chemical structures, properties and applications. However, it should be pointed out that, regarding the monomeric inisurfs, the variety in their structures is relatively low especially in comparison with the wide variety of surfactant or emulsifier structures known today. Further progress in this field seems to be possible in enhanced research activities regarding the synthesis of inisurfs with new chemical structures as well as in more investigations regarding their properties and the polymerization mechanism. Most of these considerations can also be applied to surfmers and transurfs.

It would be unwise to conclude this review without mentioning the potential of these reactive surfactants in dispersion polymerization. In this process, the reaction medium is initially homogeneous, but the polymer which is produced is not soluble; suitable surfactants are introduced to stabilize the precipitated polymer into particles of controlled size; such surfactants may be block copolymers with one sequence compatible with the polymer and the second one with the surrounding medium, but it is generally used to introduce a polymer soluble in the reaction medium (often a mixture of water and alcohol) on which the growing polymer can be grafted, so that the actual surfactant is produced throughout the process. Although such a process is very efficient for producing model colloids of rather large size (1 to 20 µm) and monodisperses, it is difficult to really control the process, because the exact structure and amount of the surfactant produced is not known. The use of suitable reactive surfactants might be a more flexible way than the use of block copolymers to get a more controlled stabilization. Pioneering work in that direction has been recently done in Japan. Kobayashi et al. [63] have been able to produce monodisperse particles of polymethyl methacrylate in aqueous methanol solution in the range of 1 to 3 µm using polyoxazoline styrenic macromonomers as stabilizer. As compared with the use of polyoxazoline homopolymer (acting only by grafting through a transfer process), it was shown that much lower amount of macromonomer 0.006 g instead of 0.18 g of lower molecular weight (4000 dalton instead of 50 000) was able to give a comparable result. The same authors [64] have prepared block type amphiphilic macromonomers of 2 methyl and 2-n-butyl oxazolines and used them in the emulsion polymerization of vinyl acetate to get sub-micron size latexes. They have observed that the size of the particles are lower if the polymerizable group is located at the end of the hydrophobic block (n-butyl) instead of the end of the hydrophilic block (methyl); on the other hand, with increasing length of the hydrophilic block, the size of the latex particle decreases.

6 References

1. Strub-Leconte MP, Riess G (1991) Bull Soc Chim Belg 100: 137
2. German Patent DE 3 239 527 1984 to Rohm GmbH
3. Kusters JMH, Kampman CJ, Verwerden TMM, German AL (1990) 33rd IUPAC Symp on Macromolecules MONTREAL 1990, Sect. 2-5-6
4. Bistline RG, Stirton AJ, Weil JK, Port WS (1956) J Amer Oil Chem Soc 33: 44
5. Regen SL, Czech B, Singh A (1980) J Am Chem Soc 102: 6638
6. Fendler JH, Tundo P (1984) Acc Chem Res 17: 3
7. Bader H, Dorn K, Hashimoto K, Hupfer B, Petropoulos JH, Ringsdorf H, Sumitomo H (1989) In: Gordon M (ed) Polymeric membranes. Springer, Berlin Heidelberg New York, p 1
8. Nagai K, Ohishi Y, Irraba H, Kudo S (1985) J Polym Sci Polym Chem Ed 23: 1221
9. Hamid SM, Sherrington DC (1987) Polymer 28: 325, 332
10. Regen SL (1987) In: Fontanille M, Guyot A (eds) Recent advances in mechanistic and synthetic aspects of polymerization. Reidel, Dordrecht NATO ACI C215, p 317
11. Jaan Kokai (1974) 74 40 338 to Nippon Oil
12. Kokai JP, Tokkyo Koho (1987) (87) 10 102; (1989) (89) 174511 and (89) 74512 (to Sanyo Chemical Ind.)
13. Greene BW, Sheetz DP, Filer TD (1970) J Colloid Interf Sci 32: 90
14. Greene BW, Sheetz DP (1970) J Colloid Interf Sci 32: 96
15. Greene BW, Saunders FL (1970) J Colloid Interf Sci 33: 393
16. Tauer K, Goebel KH, Kosmella S, Stähler K, Neelsen J (1988) Plaste Kautschuk 35: 373
17. Egorov VV, Zubov VP (1987) Usp Chim (russ) 56: 2076
18. Malyukova YB, Navmava SV, Gritskova IA, Bondarev AN, Zubov VP (1991) Vysokomolek Soed A33: 1469; (1991) Polym Sci 33: 1361
19. Chen SA, Chang HS (1985) J Polym Sci, Polym Chem Ed 23: 2615
20. Juang MS, Krieger IM (1976) J Polym Sci, Polym Chem Ed 14: 2089
21. Tsaur SL, Fitch RM (1987) J. Colloid Interf Sci 115: 450
22. Guillaume JL, Pichot C, Guillot J (1990) J Polym Sci, Polym Chem Ed 28: 137
23. GB Patent 1 427 789 (1976) to Kanegafuchi Kagatu Kogyo Kabushiki Kaisha
24. Tauer K, Goebel KH, Kosmella S, Stahler K, Neelsen J (1990) Makromol Chem Macr Symp 31: 107
25. Gobel K-H, Stähler K, unpublished results
26. Urquiola MB, Dimonie VL, Sudol ED, El Aasser MS (1992) J Polym Chem Ed 30, 2619, Report no 33–37 1990–92
27. Urquiola MB, Dimonie VL, Sudol ED, El Aasser MS (1992) J Polym Sci Polymer Chem Ed 30: 2631
28. Watanabe S, Ozaki H, Mitsuhashi K, Nakahama S, Yamaguchi K (1992) Makromol Chem (1992) 193: 2781
29. Ottewill RH, Satgurunathan R (1987) Colloid Polym Sci 265: 845; (1988) ibid 266: 547
30. Ottewill RH, Satgurunathan R, Waite FA, Westby MJ (1987) Brit Polym J 19: 435
31. Palluel AL, Westby MJ, Bramley CWA, Davies SP, Backhouse AJ (1990) Makromol Chem Makromol Symp (1990) 35/36: 509
32. Leary B, Lyons CJ (1989) Austr. J. Chem. 42: 2055
33. Bromley CWA (1986) Coloids and Surfaces 17: 1
34. Schechtman LA (1990) 33rd IUPAC Symp. on Macromolecules, Montreal, Section 2.5.6.
35. Charreyre MT, Boulanger P, Delair Th, Maudrand B, Pichot C (1993) Colloid Polym Sci (in press)
36. Ivancev SS, Pavljucenko VN (1979) Plaste Kautschuk 26: 314
37. Tauer K, Wedel A, Morozova EM (1992) Makromol Chem 193: 1387

38. Pavljucenko VN, Ivancev SS, Roschkova DA, Dikaja NN, Domniceva NA, Budtov VT (1978) Colloid J (russ) 40: 64
39. Ivancev SS, Pavljucenko VN (1981) Acta Polymerica 32: 407
40. Ivancev SS, Pavljucenko VN, Byrdina NA (1987) J Pol Sci Part A Polym Chem 25: 47–62
41. Vornov SA, Kiselev EM, Tokarev VS, Pucin VA (1980) Colloid J (russ) 42: 452
42. Galibei VI, Dudnik LV, Tolpigina TA, Petrova AB, Sokolova VI (1977) Vysokomol. Soedin (russ) Ser. A 19: 131
43. German patent, DE 3118372 A1 (9.5. 1981) to Bayer AG, Germany, Schmid A., Roos E.
44. German patent, DE 3118373 A1 (9.5. 1981) to Bayer AG, Germany, Schmid A., Roos E.
45. Kusters IMH, Napper DH, Gilbert RG, German AL (1992) Macromol 25: 7043
46. Dicke HR, Heitz W (Proc. IUPAC Macromol Symp 28th Intern. Union Pure Appl Chem, Oxford, UK, p 134
47. Dicke HR, Heitz W, (1982) Colloid Polymer Sci 260: 3
48. Dicke HR, Heitz W (1981) Makromol Chem Rapid Commun 2: 83
49. Walz R, Bömer B, Heitz W (1977) Makromol Chem 178: 2527
50. Aslamazova TR, Eliseeva VI, Tauer K, Wiener K, Pimenova JV, Kasulke U (1992) Vysokomol Soedin (russ.) 32: 1387
51. Kusters et al. (private communication)
52. Tauer K, Kosmella S (1993) Polymer International 30: 253
53. Kasargod P Fitch RM (1980) In: Fitch RM (ed) Polymer colloids II. Plenum Press, (1980) p 457
54. Tauer K, Kosmella S (to be published)
55. Nyuken O, Weidner R (1986) Adv Polymer Sci 73/74: 145–1499
56. Wodka T (1987) Polimery (Pol) 32: 140
57. Czichoki G, Vollhardt D, Seibt H (1981) Tenside Detergents 18: 6
58. Gawalek G (1962) "Washn- und Netzmittel", Akademieverlag, Berlin
59. Bueren H, Grossmann H (1971) In: Foerst W, Grünwald H (eds) Grenzflächenaktive substanzen, Chem T aschenbücher 14. Verlag Chemie
60. Zierke M., Tauer K, Kosmella S (to be published)
61. Paulke BR, Tauer K, Kosmella S (to be published)
62. Fifield CC (1985) Thesis, University of Connecticut (Univ Microfilms Internat 8607856)
63. Kobayashi S, Uyama H, Choi JH, Matsumoto Y (1993) Polym Intern 30: 265
64. Uyama H, Honda Y, Kobayashi S (1993) J Polymer Sci A Polymer Chem 31: 123

Editor: Prof. Ringsdorf
Received March 1993

Recent Advances in Styrene Polymerization

D. B. Priddy
The Dow Chemical Company, Midland, MI 48667, USA

Over 200 references describing spontaneous, and chemically initiated styrene polymerization chemistry are reviewed with special emphasis on advances taking place in the past decade. The review is limited to chemistry useful for making amorphous high molecular weight polystyrene in solution polymerization processes. Chemical initiators have been categorized into three basic groups as follws: 1) anionic; 2) mono-radical; and 3) diradical. Analytical techniques used for determination of free radical polymerization kinetics and mechanisms are also discussed.

Advances in Polymer Science, Vol. 111
© Springer-Verlag Berlin Heidelberg 1994

List of Symbols

FR	free radical
S	styrene
PS	polystyrene
A^-	active anion concentration
EB	ethylbenzene concentration
EB^-	ethylbenzene anion concentration
H^+C^{-1}	contaminant concentration
$[H^+C^{-1}]$	contaminant concentration in feed
I	initiator concentration
I_0	initiator concentration in feed
I^-	initiator anion
C_t	chain transfer constant
k_{cts}	rate constant for chain transfer to ethylbenzene
k_i	rate constant of n-butyl lithium initiation
k_p	rate constant of polystyrene propagation
k_t	rate constant of termination
k_{tic}	rate constant of contaminant reacting with initiator
k_{tpc}	rate constant of contaminant terminating anion
k_{tt}	rate constant of thermal termination of anions
Li^+	lithium cation concentration
M_n	number average molecular weight
PS	terminated polystyrene concentration
PS^-	growing polystyrene anion concentration
R_{cts}	rate of chain transfer to ethylbenzene
R_i	rate of n-butyl lithium initiation
R_p	rate of polystyrene propagation
R_t	rate of anion termination
S	styrene monomer concentration
S_0	styrene monomer concentration in feed
Q	residence time in CSTR

1 Introduction

Styrene is one of the oldest and most studied monomers. It spontaneously generates free radials upon heating above 100 °C and polymerizes yielding amorphous polystyrene (PS). Styrene can also be polymerized by other mechanisms (anionic, cationic, or Zeigler-Natta) with the aid of chemical initiators. Commercially, over twenty billion pounds of PS are produced annually worldwide. All of this polystyrene is produced via free radical (FR) chemistry, and mostly via continuous solution polymerization processes. The commercial preference for the continuous solution process is due mainly to economic factors. Non-solution polymerization processes (suspension and emulsion) have lower reactor efficiency (product/reactor volume) due to reactor volume occupied by the water which adds to the manufacturing cost.

Polystyrene was first manufactured commercially (1938) by The Dow Chemical Company. Styrene was non-continuously bulk polymerized, without the aid of a chemical initiator, to high conversion by heating it in metals cans. The cans were opened and the solid PS ground into small pieces. Over the next 35 years, most of the research focused on understanding the mechanism of self-initiated (spontaneous) polymerization of styrene and developing continuous solution polymerization processes. In recent years, solution polymerization research emphasis has focused upon understanding the chemistry of chemical initiators. Today, most PS is produced via continuous solution polymerization with the aid of peroxide initiation.

During the 1980s Dow researchers developed continuous anionic polymerization technology they hoped would compete economically with the free radical process [1]. Their research approach was to utilize existing continuous solution free radical polystyrene reactor configurations to conduct the polymerization experiments. If successful, existing plants could be converted from FR to anionic polymerization chemistry for minimal cost. Also, they hoped the polymer property advantages of anionically produced PS (better thermal stability and very low residual styrene monomer and oligomers) would offset the slightly higher manufacturing cost due to more stringent monomer purity requirements of the anionic chemistry.

In general, three reactor types are used for continuous solution styrene polymerization (Fig. 1) [2]. These three reactor types are characterized as being either plug-flow or backmixed. Continuous plug-flow reactors (CPFR) typically have excellent radial mixing but virtually no backmixing, unless they are recirculated. These reactors are usually described as stratified agitated tower reactors. Continuous stirred tank reactors (CSTR), on the other hand, have high degrees of backmixing. They are usually single staged and operated isothermally and at constant monomer conversion. CPFR type reactors, on the other hand, are multi-staged having a temperature profile, typically 100–170 °C. Two general configurations of CSTR reactors are utilized commercially; i.e., recirculated coil and ebullient.

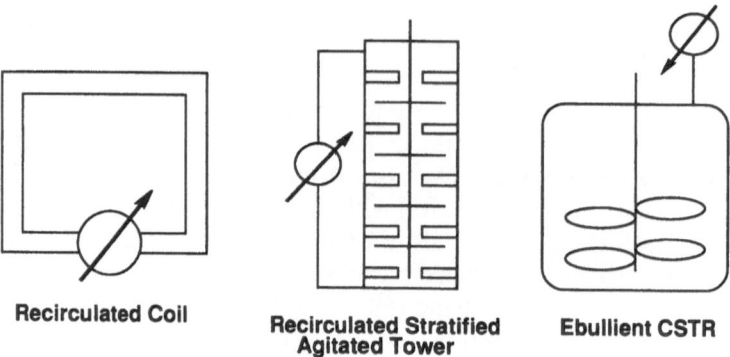

Recirculated Coil   **Recirculated Stratified**   **Ebullient CSTR**
                                Agitated Tower

Fig. 1. Two CSTR and a CPFR configuration utilized for continuous mass polymerization of styrene

This review will focus on chemistry useful for making amorphous high molecular weight PS in continuous solution polymerization processes. Non-mass processes (suspension and emulsion) are excluded along with chemistries that result in low molecular weight (cationic) and stereo-regular (Ziegler-Natta) polystyrene. Research on the use of chemicals to initiate polymerization, as well as the spontaneous initiation mechanism of styrene will be summarized. Chemical initiators have been categorized into three basic groups as follows: 1) anionic; 2) mono-radical; and 3) diradical. Analytical techniques used for determination of FR polymerization kinetics and chemistry are also discussed.

2 Analytical Methods for Determination of Kinetics and Mechanisms

A variety of analytical techniques [e.g., radiochemical labeling [3–7], UV-VIS [8, 9], ^{1}H NMR [8, 9], and ^{13}C NMR [10–15] have been used to determine the initiator residues incorporated into vinyl polymers during their preparation. In general, these techniques are not sufficiently sensitive to the environment of the initiator derived functionality to allow assignment of its mode of incorporation into the polymer. However, by using ^{13}C labeled initiators, NMR has been successfully utilized to analyze the nature of initiator derived residues and functionalities in polystyrene.

The use of FR trapping reagents has allowed the elucidation of initiation mechanisms in both spontaneous and chemically initiated styrene polymerizations. These reagents trap the carbon centered FR formed by reaction of an initiating radical with monomer before any significant propagation has occurred. One of the most commonly employed FR traps used in investigations of styrene polymerization are the C-nitroso spin traps. Many FR trapping studies have been performed in styrene using 2-methyl-2-nitrosopropane with a variety of initiating systems [16–25]. However, the trapped radicals could only be analyzed by ESR spectroscopy, which led to very ambiguous results. The use of nitroxide FR traps [26–31], on the other hand, leads to stable products which can be characterized by a variety of techniques. The FR trapping reagents 2,2,6,6-tetramethylpiperidine-1-oxyl (TMPO) (1) and, 1,1,3,3-tetramethylisoindolin-2-oxyl (2) [28, 29, 32] have been found to be useful. 2 is especially useful because it yields adducts which are usually crystalline, stable, and readily separable by HPLC. One example of the use of reagent 2 includes measuring the ratio of 3 and 4 formed during the solution polymerization of styrene using alkoxy radicals. In the case of cumyloxy radicals, the ratio of 3 to the fragmentation product 4 was 1 : 15 [32] while the ratio for *tert*-butoxy radicals was 1 : 76 [30].

where R=Ph or Me

Kinetic analysis of FR styrene polymerization is difficult due to competing initiation, propagation, and termination reactions all occuring simultaneously. For several decades researchers have dealt with this problem utilizing pulsed illumination (rotating sector) [33] or spacially intermittant polymerization (SIP) [34] techniques to achieve "pseudostationary polymerization." The illumination quickly generates a discontinuous supply of FR. The rate at which the radicals decay and the length of the polymer chains produced gives the propagation (k_p) and termination rate constants (k_t). Other techniques more recently developed include: laser-flash initiation [35], pulsed low-temperature initiator [36], and ESR [37].

Recently, an international group of polymer chemists met under the auspices of the IUPAC Working Party on "Modeling of Free Radical Polymerization Kinetics and Processing" [38]. They expressed concern over the wide divergence of literature values [39] of the kinetic parameters for FR vinyl polymerizations. The absence of consistent values for rate coefficients poses major problems for modeling of polymerization processes. They discussed each of the techniques utilized for measuring kinetic parameters and tried to reach a concenses regarding the values of the kinetic parameters for styrene. They felt that once concenses was reached for styrene (the most studied vinyl monomer), it could serve as a standard or "benchmark" system for developing new kinetic techniques. They concluded that the correct propagation rate constant for styrene at 25 °C lies between 74 and 110 L mole^{-1} sec^{-1} since this is the range of values

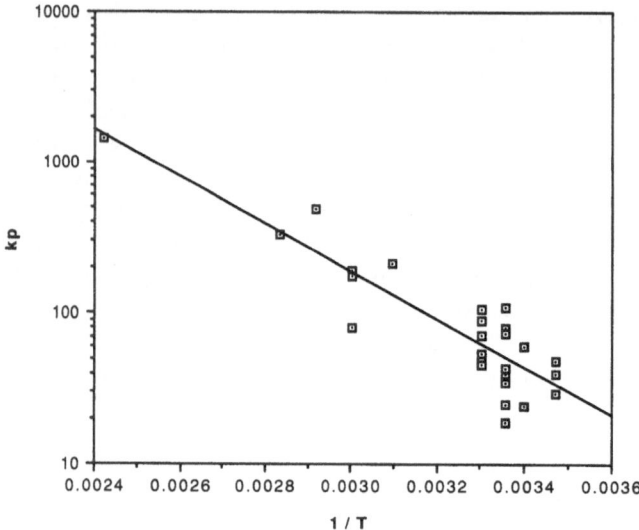

Fig. 2. Arrhenius plot of literature values of solution styrene polymerization rate constants (solid line is a least squares fit)

obtained using three independant and diverse methods. An Arrhenius plot of solution styrene polymerization rate constant literature data [2, 39] is shown in Fig. 2.

However, the group preferred ESR as the best technique for obtaining kinetic data because it can be applied up to very high conversions. Unfortunately, ESR kinetic values for styrene significantly disagree with data from the

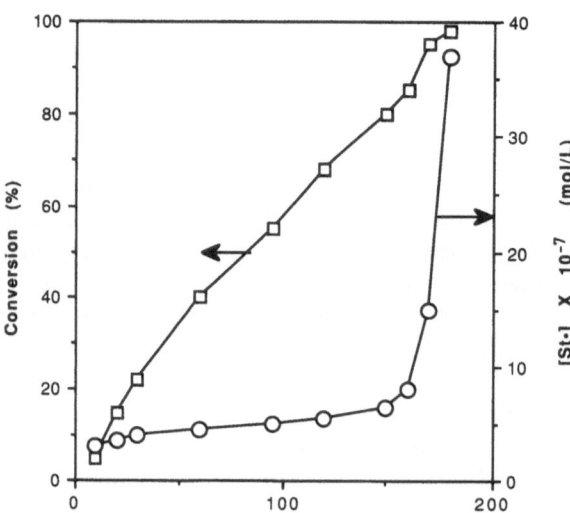

Fig. 3. Concentration of polystyryl radicals versus conversion and time

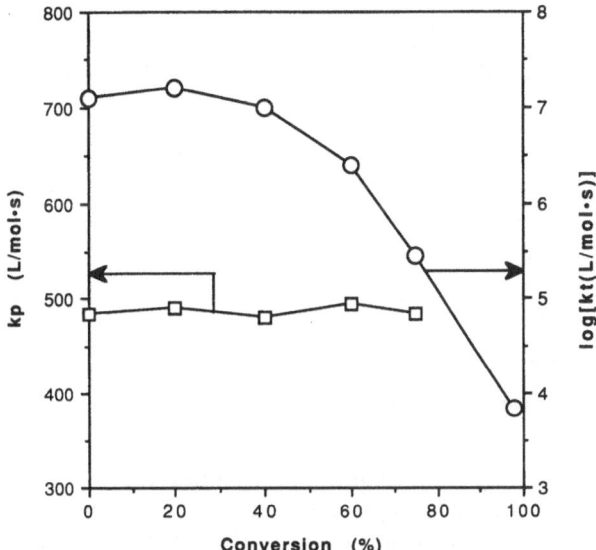

Fig. 4. Change in k_p and k_t with styrene conversion at 70 °C

other techniques. An example of the value of ESR for conducting kinetic studies over a wide conversion range has recently been published by Yamada et al. [37]. Values of k_p and k_t for styrene polymerization (Figs. 3 and 4) were determined from 0 to 100% conversion initiated using dimethyl 2,2'-azobis(isobutyrate) at 70 °C. These figures show the concentration of polystyryl radicals versus conversion and time, and the change in k_p and k_t with styrene conversion.

This ESR data allowed the following conclusions regarding styrene polymerization to be drawn: 1) k_p remains relatively constant up to high conversion during styrene polymerization; 2) FR concentration greatly increases at high conversions; 3) k_t decreases at high conversion due to high viscosity (Trommsdorff effect) [40] resulting in acceleration of the polymerization rate.

3 Spontaneous Polymerization

The self-initiated (spontaneous) polymerization of styrene has challenged researchers and has received considerable attention over the past fifty years. Two mechanisms explaining spontaneous styrene polymerization have been proposed and supported by considerable circumstantial evidence. The oldest mechanism, first postulated by Flory [41] (Scheme 1), involves a bond forming reaction between two molecules of styrene (S) to form a 1,4-diradical ($\cdot D \cdot$). However, experiments to test the mechanism showed that there was no signifi-

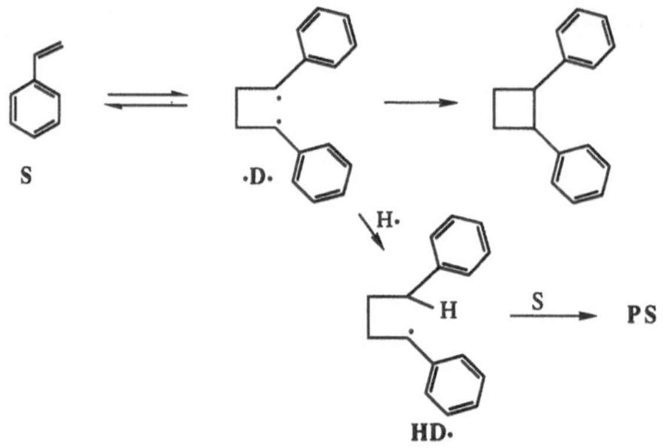

Scheme 1. Flory mechanism for spontaneous initiation of styrene polymerization

cant difference in the molecular weight of PS initiated by monoradicals compared with the spontaneously initiated polymerizations taken to the same
monomer conversion [42]. It became clear that the initiating species was not a
diradical. The mechanism thus had to be modified whereby the diradical
abstracts a hydrogen atom forming a mono-radical initiator (HD·).

Pryor [43, 44] showed that the spontaneous polymerization kinetics of
styrene were bimolecular and again postulated the ·D· intermediate. Further
evidences favoring this mechanism include: 1) the identification of *cis* and *trans*
1,2-diphenylcyclobutanes as the major dimers [45], and 2) the large differences
between spontaneous and chemically initiated (azobisisobutyronitrile) styrene
polymerizations in the presence of the FR scavenger 1,1′-diphenyl-2-
picrylhydrazyl (DPPH). The rate of consumption of DPPH is 25 times faster
than that expected from rate of polymerization measurements. This difference
was explained by the spontaneous formation of ·D·, many of which become self-
terminated before initiating polymer radicals [46].

The second mechanism (Scheme 2), was proposed by Mayo [47]. The
mechanism involves the Diels-Alder reaction of two styrene molecules to form a

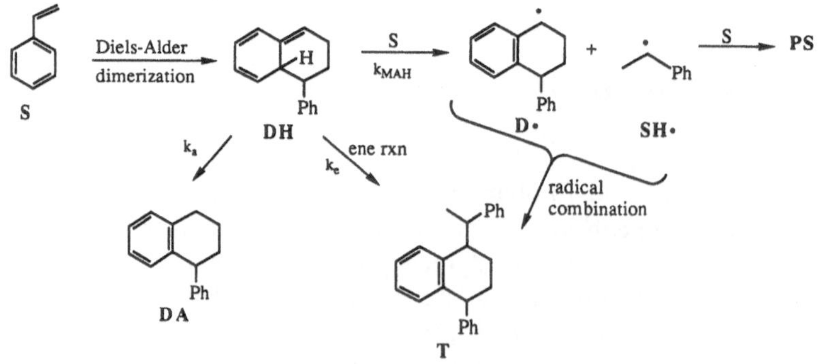

Scheme 2. Mayo mechanism for spontaneous initiation of styrene polymerization

reactive dimer (DH) (*alpha* regio-selective) followed by a molecular assisted homolysis (MAH) between DH and another styrene molecule. The Mayo mechanism has been generally preferred even though critical reviews [43, 44] have pointed out that the mechanism is only partially consistent with the available data. Also, the postulated intermediate DH has never been isolated. Evidences supporting the mechanism include: kinetic investigations [48, 49], isotope effects [43], and isolation/structure determination of oligomers [43, 50]. Even though the reactive dimer intermediate DH has never been isolated, the aromatized derivative DA has been detected in polystyrene [50]. Also, D· has been indicated as an end-group in polystyrene using ¹H NMR and UV spectroscopy [51].

The Flory and Mayo proposals could be combined to some extent by the common diradical ·D·, which collapses to either DH or 1,2-diphenylcyclobutane (Scheme 3). Non-concerted Diels-Alder reactions are permissible for two non-polar reactants [52].

Trimer T possibly could be formed by two routes: 1) MAH followed by radical combination or 2) a concerted Alder-ene reaction between S and DH. Frontier molecular orbital (FMO) calculations between the HOMO of DH (ene) and the LUMO of styrene (enophile) predict that an Alder ene reaction would yield the β-phenethyl derivative (Scheme 4) as the major product [53]. However, only the α-phenethyl derivative has been reported indicating that radical combination, rather than the Alder ene reaction, is the predominant path.

Several models of DH have been synthesized and tested as MAH initiators. They include: 5-methylene-1,3-cyclohexadiene (MCHD) [54], the adduct of 4-

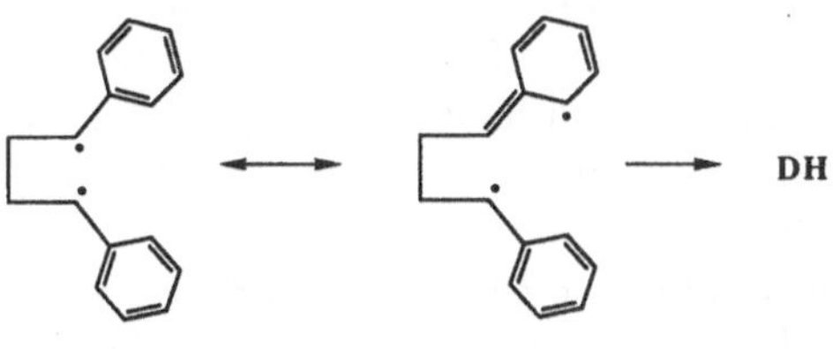

Scheme 3. Formation of Mayo Dimer via the Flory Diradical

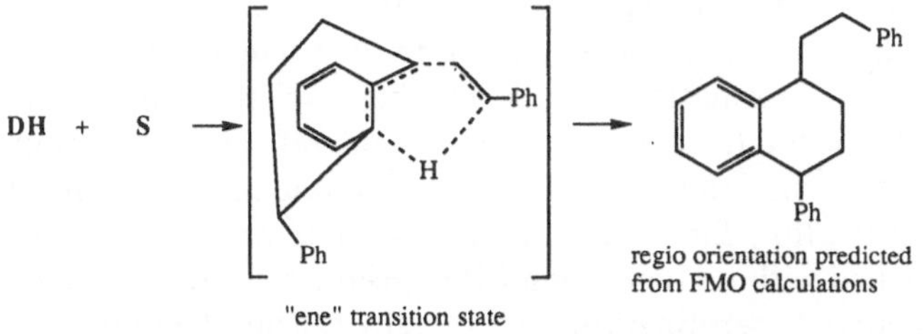

"ene" transition state

regio orientation predicted
from FMO calculations

Scheme 4. Predicted regeochemistry (FMO theory) of trimer formed via "ene" mechanism

phenyl-1,2,4-triazoline-3,5-dione with 2-vinylthiophene (PTVT) [55], and
1,2,3,10-tetrahydrophenanthrene-1,2-dicarboxylic anhydride (THPA) [56].

Both PTVT and THPA were found to be MAH initiators for styrene;
however, MCHD gave almost entirely the "ene" adduct with styrene. Also,
MCHD was found to rapidly aromatize to toluene in the presence of certain
acids [57].

The relative amounts of DA reported in the various studies differ greatly.
Levels ranging from 2% [50] to 70% [58] of the total dimeric fraction have
been reported. The rapid isomerization of DH to DA by differing levels of
adventitious acid has been used to explain these large differences [59].

Recently the spontaneous polymerization of styrene was studied in the
presence of various acid catalysts [60] to see if the postulated reactive intermediate DH could be intentionally aromatized to form inactive DA at a faster
rate than the MAH (i.e. $k_a > k_{MAH}$). The results showed that the rate of
polymerization of styrene is significantly retarded by acids (e.g. camphorsulfonic
acid) accompanied by increases in the formation of DA. This finding gave
further confirmation for the intermediacy of DH since acids would have little
effect on the cyclobutane dimer intermediate in the Flory mechanism.

A potentially important commercial benefit of adding an acid catalyst to the
spontaneous FR polymerization of styrene is that a significant shift results in the
rate/molecular weight curve for PS. This shift (Fig. 5) is most pronounced at
high molecular weights allowing formation of high molecular weight PS at a
much faster polymerization rate. An explanation for this phenomenon is that
the rate of formation of initiating radicals is reduced in the presence of acid. To
return the rate of initiating FR formation back to a higher level, the temperature
must be increased. The increased temperature increases the rate of propagation.
The main mechanism of termination is chain coupling, the rate of which is most
affected by radical concentration. Since the polymerization temperature can be
raised in the presence of acid without increasing FR concentration, the propagation rate is increased relative to termination rate, thereby raising the molecular
weight.

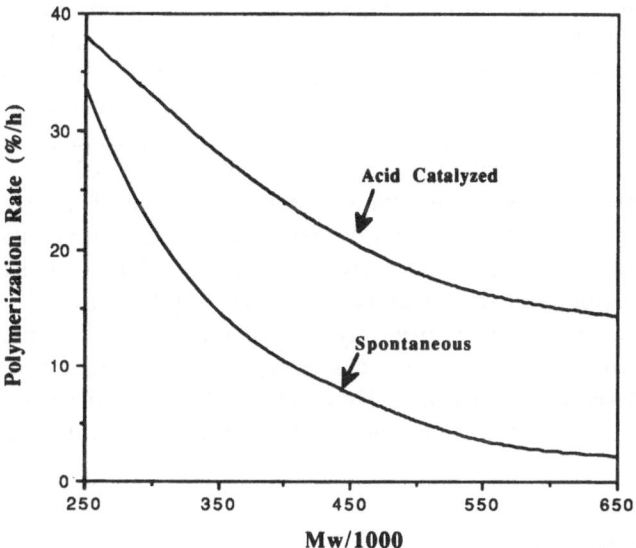

Fig. 5. Comparison of the rate/Mw relationships for spontaneous and camphorsulfonic acid (500 ppm) catalyzed styrene polymerization in CSTR type reactors

Clearly then, a major portion of the primary initiating radicals are derived by the Mayo mechanism. However, the formation of styrene dimers 1,2-diphenylbutane and 1,3-diphenyl-1-butene must be accounted for. Otsu and Sato [61] believe that the formation of primary initiating radicals from styrene

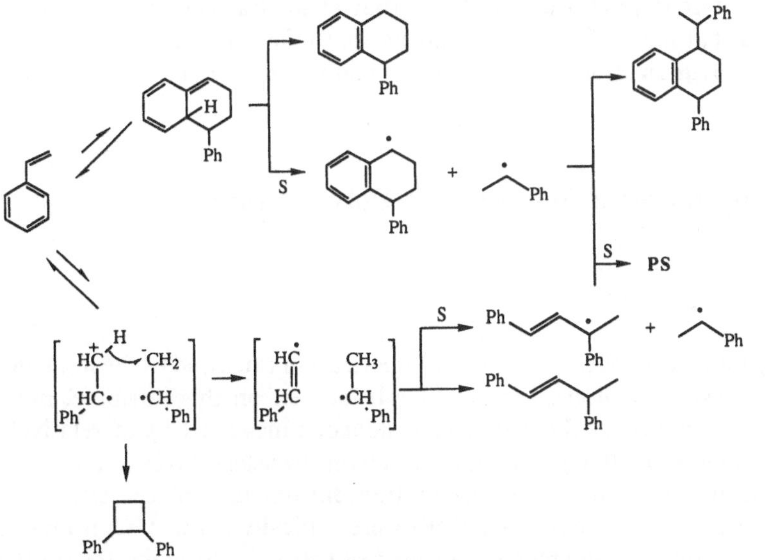

Scheme 5. Multiple initiation mechanism for styrene proposed by Otsu and Sato

cannot be explained by just one mechanism. They combine the Mayo approach with the formation of an electron-transfer complex which collapses to form radicals and the cyclobutane and butene dimers (Scheme 5). Although formation of electron-transfer complexes between donor-acceptor vinyl monomer pairs (e.g. methylvinylsulfide-acrylonitrile) is common, it is felt that the formation of such a complex between two molecules of styrene is highly unlikely [62].

Olaj et al. [63] expand the Mayo mechanism focusing on the chemistry of the dimer intermediate. They performed UV spectroscopic measurements (315–365 nm) on polymerizing styrene and presented evidence to support the formation of two stereoisomers of the "Mayo dimer" (DHa and DHb). They suggest that both isomers are consumed during styrene polymerization. Possible consumption pathways are: copolymerization, chain transfer, and formation of initiating radicals by MAH with monomer. They believe that only the axial phenyl isomer DHa is capable of generating initiating radicals by reaction with monomer.

DHa   DHb

Kinetic analyses indicate that DHb is consumed during polymerization about twenty times slower than DHa. The main consumption pathway for DHb appears to be copolymerization. From kinetic analysis of photoinitiated polymerization of styrene [64], they conclude that chain transfer to monomer is negligible and that most of the chain transfer that takes place during spontaneous styrene polymerization is due to DHa (chain transfer constant ~ 100).

4 Chemically Initiated Bulk Styrene Polymerization

4.1 Anionic Initiation

Anionic polymerization offers very fast polymerization rates due to the long lifetime of polystyryl carbanions. Early research focused on this attribute, most studies being conducted at short reactor residence times (< 1 h), at relatively low temperatures (10–50 °C), and in low chain transfer solvents (typically benzene) to insure that premature termination did not take place. Also, relatively low degrees of polymerization (DP) were typically studied. Continuous commercial FR solution polymerization processes (Fig. 1) to make PS on the other hand, operate at relatively high temperatures (> 100 °C), at long residence times (> 1.5 h), utilize a chain transfer solvent (ethylbenzene), and produce polymer in the range of 1000 to 1500 DP.

Since the polymerization rate for anionic polymerization is extremely fast due to the high concentration of active chain ends, the necessity for fast mixing kinetics is paramount. It is well known that mixing has profound effects in continuous flow reactors. In recent years many investigators have tried to analyze and develop a fundamental understanding of these effects by developing numerous mixing models. These models utilize simplified concepts of micro-mixing and segregation and describe many, but not all conditions of mixing. Most notable of these are the two-environment models of Keairns and Manning [65], and Chen and Fan [66], and the dispersion models of Spielman and Levenspiel [67]. Since polymerization reactions in continuous processes are carried out under conditions of high viscosity, both theoretical and experimental investigations have been carried out to determine the effects of different types of mixing on the molecular weight distribution of various polymerization systems [68].

Manning, Wolf, and Keairns [69] describe a CSTR as being divided into two zones: a small region surrounding the impeller of the agitator where violent mixing takes place ("micro-mixing") and the remaining reactor volume where relatively mild mixing occurs ("macro-mixing"). The mixing in the micro-mixing region is assumed to be perfect while the viscous syrup emanating from the impeller undergoes no micro-mixing but is only recirculated throughout the vessel; that is, the syrup behaves as if it were a mixture of segregated elements which maintain their identity until they again pass through the impeller region. The micro-mixing region is typically quite small (< 1% of the total reactor volume) with the flow rate between the two regions controlled by the impeller pumping capacity.

Under ideal condition, when an isothermal CSTR is operating at steady state (SS) in a totally micro-mixing condition, the PS produced will be homogeneous having a perfect 2.0 polydispersity (the Schultz-Flory distribution). Deviations from ideality have been theoretically studied by operating CSTR processes under non-steady state conditions with forced periodic operation [70]. The effects of an independent sinusoidal forcing of the monomer and the initiator feed concentrations results in theoretical control of the polydispersity.

In a totally segmented CPFR of the stratified agitated tower design, the feed stream is considered to enter the reactor as macro-molecular capsules which maintain their integrity throughout their life-time inside the reactor. The capsules each act as though they were batch reactors and theoretically should produce a monodisperse polystyrene having a Poisson distribution.

In the case of the anionic polymerization of styrene in CSTR designs, two highly reactive fluids of greatly differing viscosities must be quickly mixed, whereas in stratified agitated tower reactor designs, radial mixing is required mainly to achieve uniform temperature control. In either case, if the kinetics of mixing are much slower than the rate of polymerization, a product having a macroscopic non-homogeneity will be produced. The difficulty of mixing poly-styrenes of greatly differing DP to achieve a homogeneous mixture is well known. A non-homogeneous product manifests itself with the appearance of lumps and inconsistencies in extruded strands and blown films and by the lack of a smooth surface in injection molded parts.

The initiation and propagation mechanism of the anionic polymerization of styrene using NBL has been the subject of considerable investigation and has been found to be very complex [71]. This is due to the associated complexes of the initiator and polystyryllithium as well as cross association between the two species. The simultaneous initiation and propagation that occurs in CSTR processes adds further complications. Tryberg and Anthony [72] simplified this problem by initiating the polymerization in a separate reactor. They con-added the low molecular weight polystyryllithium solution to a CSTR polymerization reactor.

As mentioned previously, most continuous anionic polymerization studies have been conducted at relatively low temperatures ($< 50\,°C$). Even then, mixing kinetics have been of considerable concern due to the fast polymeriz-ation kinetics. In the recent anionic polymerization studies of Priddy and Pirc [1, 73], the polymerization temperature range of $80-140\,°C$ was studied (typical free radical temperature range). At these temperatures, the polymerization kinetics are extremely fast. Also, the high polymerization temperature results in significant thermal termination of active polystyryl chains. Kern et al. [74] found that the termination reaction involved liberation of lithium hydride (1) and was first order. They found the apparent rate constant K_{tt} at 65, 93, and $120°C$ are 0.15, 0.78, and $1.3\,h^{-1}$ respectively.

$$\qquad\qquad\qquad\qquad\qquad\qquad\qquad\qquad\qquad (1)$$

Also, Priddy and Pirc used a chain transfer solvent (ethylbenzene) in a CSTR operating at $> 99\%$ monomer conversion, and at high polymer solids (40–50 w/w). Under these conditions, they found that chain transfer to solvent (CTS) was extremely high (2) since the high monomer conversions achieved under steady state (SS) operation resulted in a large ratio (typically, 500:1) of solvent to monomer. Gatske [75] has shown that the CTS increases expo-nentially with conversion as shown in Fig. 6.

$$\qquad\qquad\qquad\qquad\qquad\qquad\qquad\qquad\qquad (2)$$

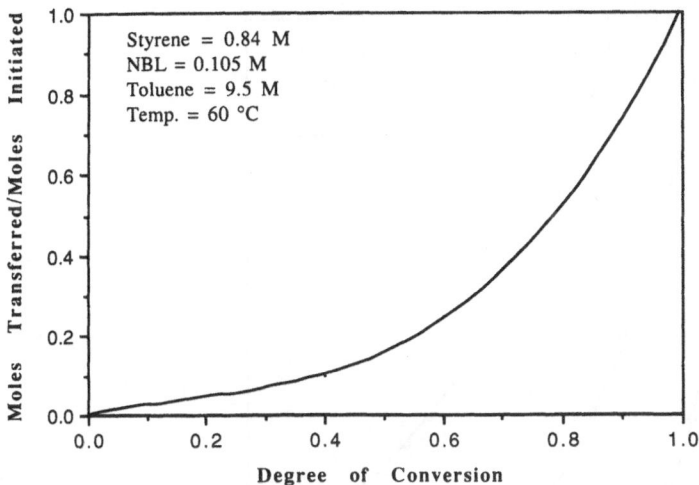

Fig. 6. Effect of monomer conversion on CTS during anionic polymerization of styrene

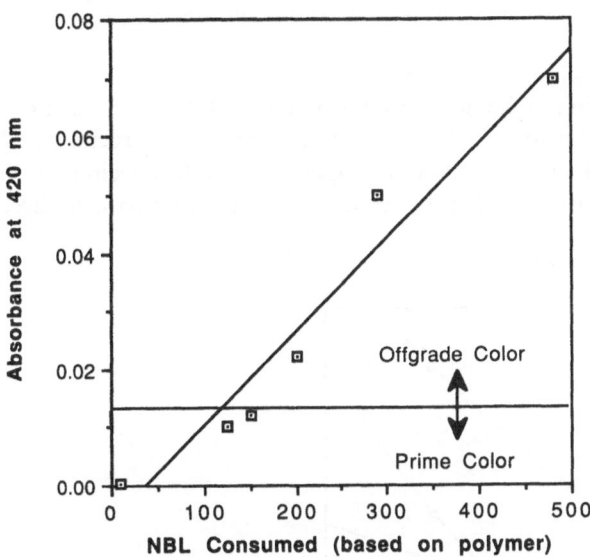

Fig. 7. PS color vs amount of NBL consumed for its production in a CSTR

The result of high CTS had both positive and negative aspects. The positive aspects are that very high yields of PS based on NBL (the most costly raw material) and PS having high clarity (low color) are produced (Fig. 7). In fact, under very stringent feed purification conditions, as high as an 8000% yield based on NBL initiator [76], can be achieved. According to Fig. 7, without high levels of CTS, it would be impossible to make a PS having sufficient clarity to meet the current color requirements to be sold as "prime" resin (with no CTS, 640 ppm of NBL is required to produce a PS of 1000 DP).

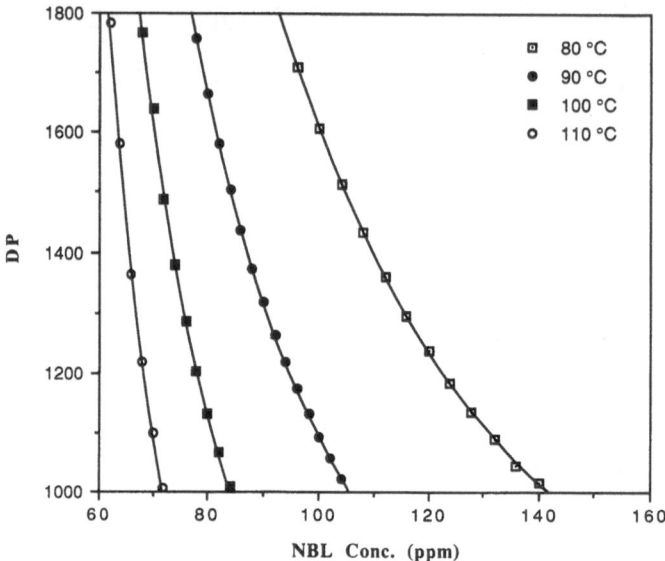

Fig. 8. Effect of NBL concentration on DP of the polymer. This graph assumes 10 ppm of NBL consumed by impurities

The importance of mixing at higher temperatures (> 100 °C) was demonstrated by Priddy et al. [77] using a CSTR polymerizer set up in a recirculated coil configuration (Fig. 1). The coil reactor was operated at a fast recirculation rate of 100 times/h to achieve at ± 1 °C temperature spread between three

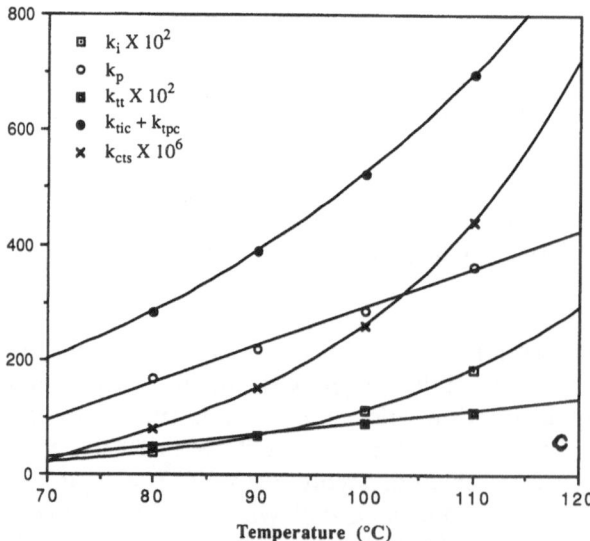

Fig. 9. Rate constants (estimated) used in model for calculation of Fig. 8

Initiation $Li^+ I^- + S \xrightarrow{k_i} IS^- Li^+$

Propagation $IS^- Li^+ + S \xrightarrow{k_p} PS^- Li^+$

Termination

Thermal: $PS^- Li^+ \xrightarrow{k_{tt}} PS + LiH$

Contaminant: $Li^+ I^- + H^+ C^- \xrightarrow{k_{tic}} Li^+ C^- + HI$

     $PS^- Li^+ + H^+ C^- \xrightarrow{k_{tpc}} PSH + Li^+ C^-$

Chain Transfer $PS^- Li^+ + EB \xrightarrow{k_{cts}} PSH + EB^- Li^+$

Total SS Anion Conc. $[A^- Li^+] = [PS^- Li^+] + [EB^- Li^+] + [IS^- Li^+]$

For a well mixed CSTR at SS with a residence time θ:

Initiator Concentration $I = \dfrac{I_o}{1 + k_i S\theta + k_{tic}[H^+C^-]\theta}$

Styrene Concentration $S = \dfrac{S_o}{1 + k_p[A^- Li^+]\theta}$

Active Anion Conc. $A^- Li^+ = \dfrac{k_i [I^- Li^+]S\theta}{1 + k_{tpc}[H^+C^-]\theta + k_{tt}[PS^- Li^+]\theta}$

Contaminant Conc. $H^+C^- = \dfrac{[H^+C^-]_o}{1 + k_{tic}[I^- Li^+]\theta + k_t[A^- Li^+]\theta}$

These four equations are solved simultaneously for I, S, $A^- Li^+$, and H^+C^-

The four rate equations are:

$R_i = k_i [Li^+ I^-]S$

$R_p = k_p [A^- Li^+]S = \dfrac{S_o - S}{\theta}$

$R_{cts} = k_{cts} [PS^- Li^+][EB]$

$R_t = k_{tt} [PS^- Li^+] + k_{tic} [Li^+ I^-] [H^+C^-] + k_{tpc} [PS^- Li^+][H^+C^-]$

The M_n of the polymer produced is given by: $M_n = \dfrac{104 (S_0 - S)}{(R_t + R_{cts}) \theta}$

Fig. 10. Equations used to calculate the graph shown in Fig. 8

probes located at the pump inlet, outlet, and about midway through the loop. This reactor represents a good example of the mixing model described by Keairns and Manning [65]; i.e., the region inside the recirculation pump is the micro-mixer and the loop is the macro-mixer. Since maximum mixing takes place inside the recirculation pump, the monomer and initiator feeds were introduced at the inlet to the pump. While the polymer syrup travels around the loop, very little mixing takes place. The polymer produced from this reactor gave smooth extruded strands and a 2.0–2.2 polydispersity (PD) [weight average molecular weight/number average molecular weight (M_w/M_n)] at polymerization temperatures below 100 °C. As the temperature was raised over 100 °C, the PD began to broaden until at 110 °C, the extruded polymer strands became non-uniform. Also, upon injection molding tensile specimens, it was noted that their surface was textured. As the temperature was increased further, molecular weight control became very difficult and the shape of the GPC curves began displaying non-Gaussian shapes.

A third source of termination of active polymer chains is termination by contaminants (3). The list of potential contaminants includes: water, oxygen, carbon dioxide, benzaldehyde, acetophenone, phenethyl alcohol, and more. Although the bulk of the contaminants can be removed by simple feed purification techniques (e.g. alumina beds), there will always be remaining traces which can only be removed using highly reactive absorbant bed techniques (e.g. lithium aluminum hydride/ion exchange resin) [76].

$$\left[\begin{array}{c} \text{polystyryl anion} \end{array} \right] Li^+ \;+\; H^+C^- \;\xrightarrow{k_{tpc}}\; \left[\begin{array}{c} \text{polystyryl–H} \end{array} \right] \;+\; Li^+C^- \qquad (3)$$

In the study of Treyberg and Anthony [72], the anionic polymerization of styrene was carried out in benzene (15–30% styrene w/w) in a small (50 ml) laboratory CSTR. They studied the 25–38 °C temperature range at residence times between 15 and 32 min. Monomer conversions and DP as high as 70% and 600, respectively, were achieved. Even at this relatively low temperature range and short residence times, they had to add a dead polymer fraction term to their calculations in order to match the theoretical and experimental molecular weights.

Reactors of the CSTR design (Fig. 1) are typically operated either hydrauically full with heat removal using heat exchangers or operated partially full with heat removal by ebullition [2]. Investigations by Priddy et al. [77] of anionic polymerization at 100 °C in a non-ebullient CSTR polymerizer resulted in the production of a non-uniform polystyrene as evidenced by large lumps in the stranded polymer after being extruded through a die. Also the PD of the polymer was > 2.3. Temperature monitoring devices at various points inside the reactor indicated that stratification existed after operation under continuous conditions for several hours. Increasing the speed of the turbine agitator did not change this condition. However, when a vacuum was applied to the reactor to create ebullition, the temperature probes quickly unified their readings to within 1 °C and after a few hours of SS operation, the extruded strands became smooth and the PD approached 2.0 when making a 200 000 M_w polystyrene. As the NBL pumping rate was decreased to increase the M_w, the PD increased slightly to a maximum of 2.2 at 300 000 M_w. These experiments demonstrated the importance of uniform heat removal, temperature control, and mixing to the production of high quality PS using anionic initiation.

A study by Priddy et al. [77] of continuous solution anionic polymerization of styrene in a reactor of the CPFR design, but with some backmixing (outlet to inlet recirculation), showed a temperature gradient due to the relatively slow recirculation rate (reactor contents were recirculated at a rate of 8 times/h). Thus, the jacket temperature of each of the three zones was adjusted to maintain isothermal polymerization. When SS was achieved at a 2 h residence time, the jacket temperatures of the three zones required to achieve isothermal conditions inside the reactor, in sequence from the inlet to the outlet, were 101, 67, and 85 °C, respectively. At these SS conditions, the strands and the GPC curves of the PS produced were smooth due to the excellent radial mixing which takes place throughout the reactor. Even though the agitator RPM was relatively slow (50 RPM), the mixing was excellent due to the shear resulting from the close distance between the agitator and the wall mounted pins. The PD of the PS produced in this reactor was 1.7–1.9. The < 2.0 PD was due to the hybrid nature of the polymerization being intermediate between CSTR and CPFR kinetics.

Another attribute of anionic polymerization chemistry is that high molecular weight PS can be produced at high solvent concentrations. This feature allows

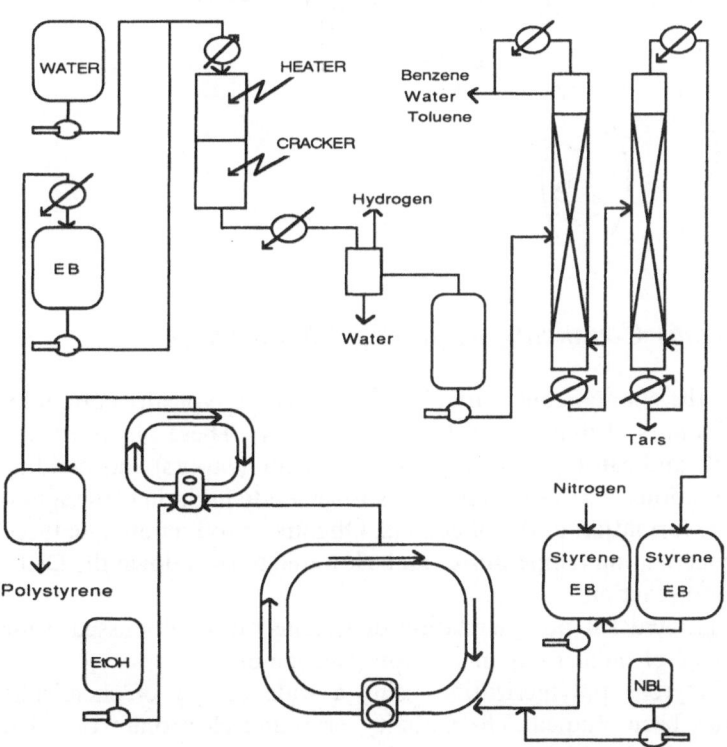

Fig. 11. Integrated monomer/polymer process demonstrated for anionic PS

ethylbenzene dehydrogenation and styrene polymerization operations to be integrated at a cost savings since the ethylbenzene/styrene mixture from the dehydrogenator does not need to be fractionated. Furthermore, the distallation unit can be designed to provide polymerization feed of high "anionic purity" eliminating the feed clean-up cost required for plants utilizing purchased styrene. In a recent study [73], anionic polymerization and ethylbenzene dehydrogenation were carried out in the laboratory utilizing a CSTR of the recirculated coil design (Fig. 11). The dehydrogenator was operated at 43–55% conversion with continuous distillation of the ethylbenzene/styrene mixture to remove by-products which would interfere with the subsequent anionic polymerization. The anionic CSTR polymerization operated at > 99% conversion of styrene. The volatiles were recovered from the polymer syrup and recycled back to the dehydrogenator. During four months of continuous operation, the integrated process showed no detrimental build-up of impurities which effected the anionic polymerization or dehydrogenation. The PS produced had excellent color, clarity, thermal stability and PD = 2.1 to 2.4. The ability to control weight average molecular weight was within a range of 20 000 using an on-line GPC in concert with a colorimeter [78].

4.2 Miscellaneous Comments Regarding FR Initiation

PS produced by the spontaneous initiation mechanism is typically contaminated by large amounts of dimers and trimers (1–2 w/w). These oligomers are somewhat volatile and cause problems during extrusion (vapors) and molding (mold sweat) operations. The use of chemicals to generate initiating FR significantly reduces the formation of the oligomers. Oligomer production is reduced because the polymerization temperature can be lowered to slow down the Diels-Alder dimerization reaction.

Chemically initiated FR polymerization of styrene will be discussed using the classical scheme of radical polymerization (Scheme 6).

Initiation of styrene polymerization using a wide variety of chemically generated FR has been studied. The stability, size, and electronics of a FR greatly affects the rate at which it adds to vinyl monomers.

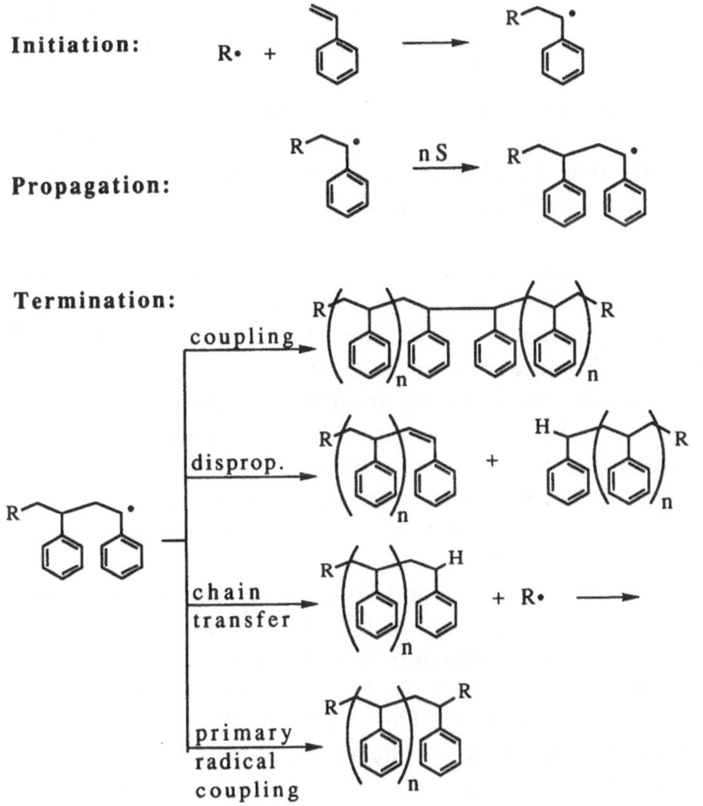

Initiation: R• + [styrene] ⟶ [R–CH₂–•CH–phenyl]

Propagation: [R–CH₂–•CH–phenyl] $\xrightarrow{nS}$ [R–CH₂–CH(phenyl)–CH₂–•CH–phenyl]

Termination:

coupling ⟶

disprop. ⟶ +

chain transfer ⟶ + R• ⟶

primary radical coupling ⟶

Scheme 6. Chemical initiation of FR styrene polymerization

Table 1. Bond dissociation energies of various bonds

	ΔH
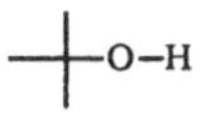 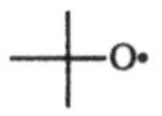 $\equiv\!C\!-\!O\!-\!H \longrightarrow \equiv\!C\!-\!O\bullet + \bullet H$	104
$H_3C\!-\!H \longrightarrow H_3C\bullet + \bullet H$	104
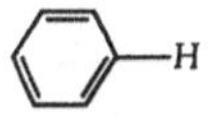 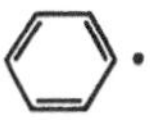 [phenyl]$-H \longrightarrow$ [phenyl]$\bullet + \bullet H$	110
$CH_3CH_2\!-\!H \longrightarrow CH_3CH_2\bullet + \bullet H$	97
[phenyl]$-CH_2\!-\!H \longrightarrow$ [phenyl]$-CH_2\bullet + \bullet H$	87

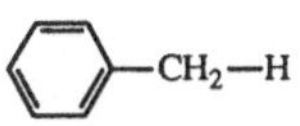

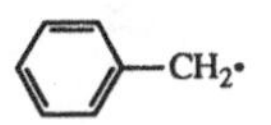

The relative reactivity of a FR can be estimated from its homolytic bond dissociation energy (DH) if it is bonded to hydrogen [79]. Examples are shown in Table 1.

From Table 1 it is estimated that the radical stability of methyl and *tert*-butoxy radicals are about equivalent. However, their size and electronegativity or polarity are quite different. All of these factors interplay with each other to affect the rate and mode of addition (e.g. head versus tail) of a radical with a monomer.

Also, the electronics of the monomer greatly affect the rate of radical addition. For example, the relative rate of addition of *tert*-butoxy radical to various monomers has been found to correlate well with the e value of the monomers [62] (Fig. 12). For example, acrylonitrile has an electron deficient vinyl group (e = 1.2) while the styrene vinyl group is electron rich (e = − 0.8). Methyl radicals (electron rich) add two times faster to acrylonitrile than to styrene while *tert*-butoxy radicals (electron poor) add 57 times more rapidly to styrene. Therefore, most of the FR used for styrene polymerization tend to be of the electron deficient type (i.e. *tert*-butoxy, benzoyloxy, and cyanoisopropyl).

Also, *tert*-butoxy radicals have been shown to have a strong propensity to abstract H-atoms. This propensity has found utility when making high impact PS. For example, the use of initiators which generate *tert*-butoxy radicals increases the level of grafting of PS onto polybutadiene rubber. The increase in grafting is generally thought to be due to H-abstraction from the rubber backbone by *tert*-butoxy radicals [80]. This raises questions regarding the extent of H-abstraction from the PS backbone during polymerization. The use of initiators that generate *tert*-butoxy radicals may cause some long chain branching. This question has been investigated by decomposing bis(*tert*-butylperoxy)oxalate in benzene solutions of polystyrene [81, 82]. Bis(*tert*-butylperoxy)oxalate decomposes upon heating to form two *tert*-butoxy radicals

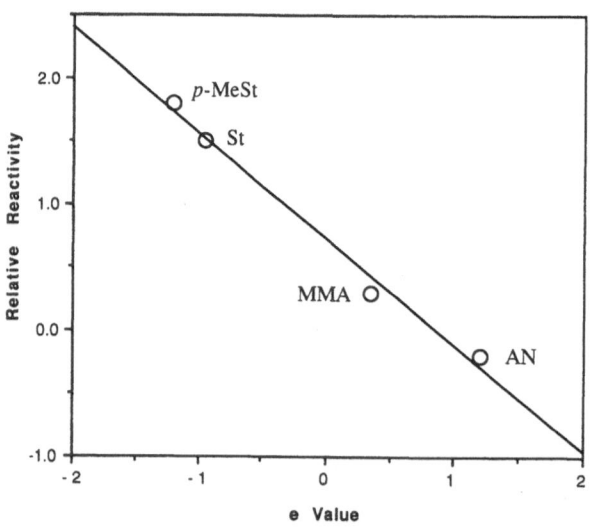

Fig. 12. Correlation of the relative reactivity of vinyl monomers having different e values toward *t*-butoxy radicals

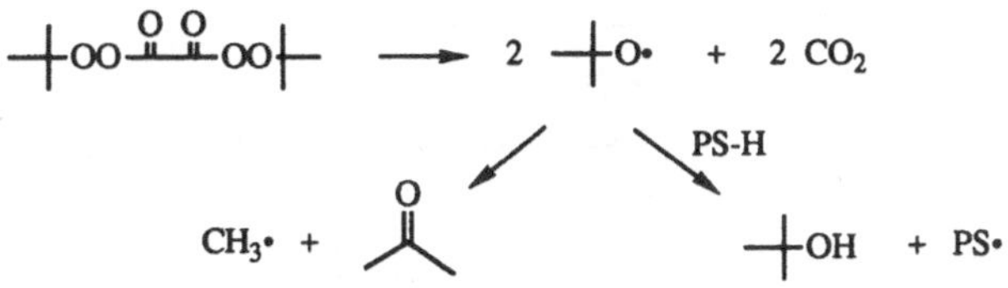

Scheme 7. Formation of *t*-butoxy radicals and their use to measure H-abstraction

and carbon dioxide. Once the *tert*-butoxy radicals are formed they either abstract a H-atom or decompose to form acetone and a methyl radical (Scheme 7).

The extent of H-abstraction was determined by measuring the ratio of *tert*-butanol (TBA)/acetone produced. Using cumene as a model for PS, a high level of H-abstraction was observed. However, PS showed a very low level of H-abstraction which decreased further as the DP of the PS was increased (Fig. 13). This was explained by the coil configuration of the PS chains restricting access of the *tert*-butoxy radicals to the labile tertiary benzylic H-atoms on the PS backbone.

Evidence clearly shows that the thermal stability of PS is dependent on the polymerization mechanism. Upon heating at 285 °C for 2.5 h under different vacuum levels, anionic PS loses less molecular weight (Fig. 14) and generates less styrene monomer (Fig. 15) than FR PS; both produced using continuous solution polymerization processes [73].

A significant amount of research has been carried out to determine the nature of the "weak links" in FR PS [83–102]. Various initiation of degradation

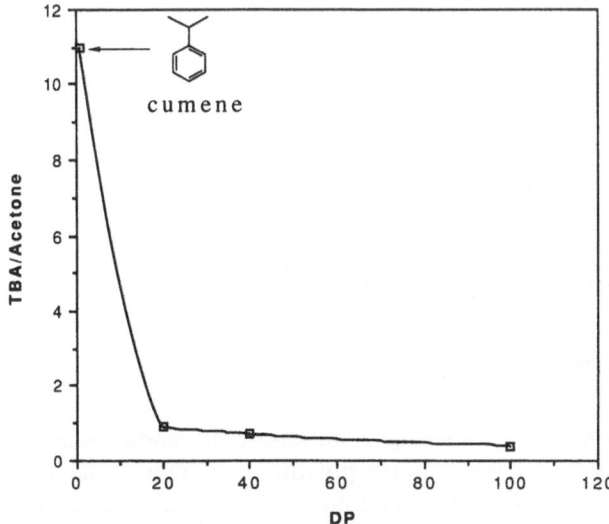

Fig. 13. H-abstraction vs DP of PS

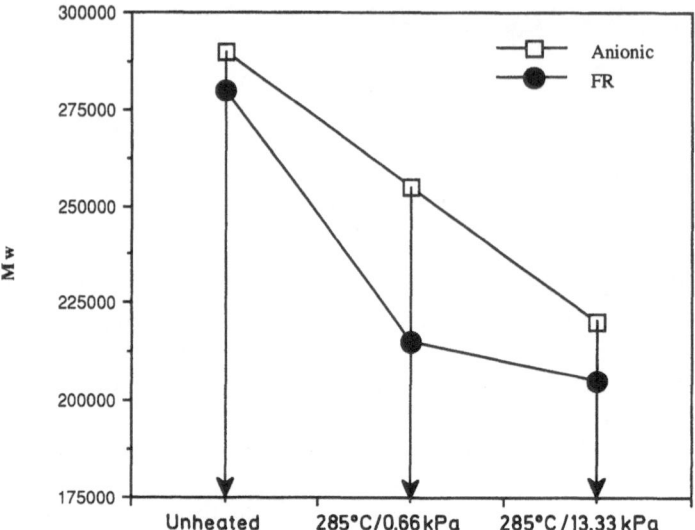

Fig. 14. Comparison of the thermal stabilities of anionic and FR PS produced in CSTR processes. Loss of Mw upon heating under different levels of vacuum

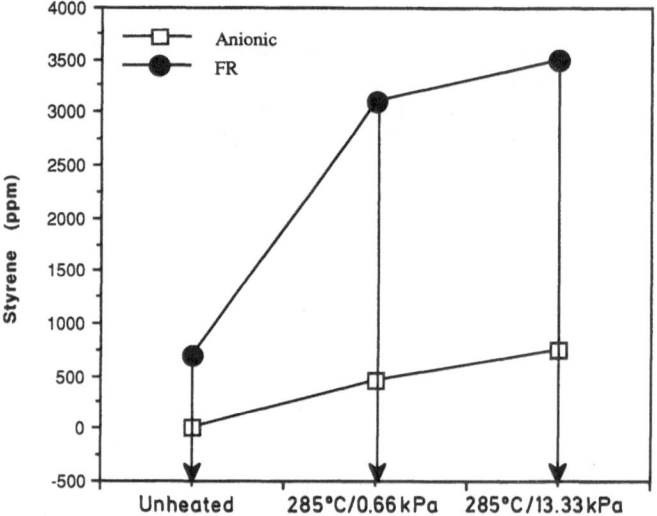

Fig. 15. Comparison of the thermal stabilities of anionic and FR PS produced in CSTR processes-formation of styrene monomer upon heating under different levels of vacuum

mechanisms have been proposed: 1) chain-end initiation; 2) random scission initiation; and 3) scission of "weak links" in the polymer backbone. It has been suggested that chain-end initiation is the predominant mechanism at 310 °C while random scission produces stable molecules. Evidence for "weak link" scissions comes mainly from studies showing loss of molecular weight vs degradation time. These plots usually show a rapid initial drop in molecular weight indicating initial rapid weak link scission. However, the picture is also complicated by the fact that the mechanism of degradation is temperature dependent. It appears that weak link scissions taking place at high temperatures initiate depolymerization while at lower temperatures scissions simply cause a decrease in molecular weight. In any case, a clear difference in thermal stability has been shown between PS produced using AIBN and benzoyl peroxide initiators. This difference is due mainly to the initiator derived fragments that remain in the polymer after isolation [93–95].

The FR initiators that have been used to make PS can be generally categorized into three types: carbon-carbon, peroxide and azo.

4.3 Mono-Radical Initiators

4.3.1 Carbon-Carbon Initiators

C–C bonds can be strained by the connection to large bulky substituents or by placing them in a small ring.

Over 40 years ago Schulz [103–105] attempted to initiate styrene polymerization with 1,1,2,2-tetraphenyl-1,2-dicyanoethane (5). The point of his work was to prove that highly hindered C–C compounds could be used as FR initiators for vinyl polymerization. It was established that 5 is consumed relatively quickly at 100 °C in styrene and that the polymer molecular weight is inversely proportional to initiator concentration.

Later, Braun and Becker [106, 107] showed that initiation of styrene using benzopinacole (Scheme 8) takes place mainly by H-transfer and that the overall rate is proportional to the 1.5th order of monomer concentration.

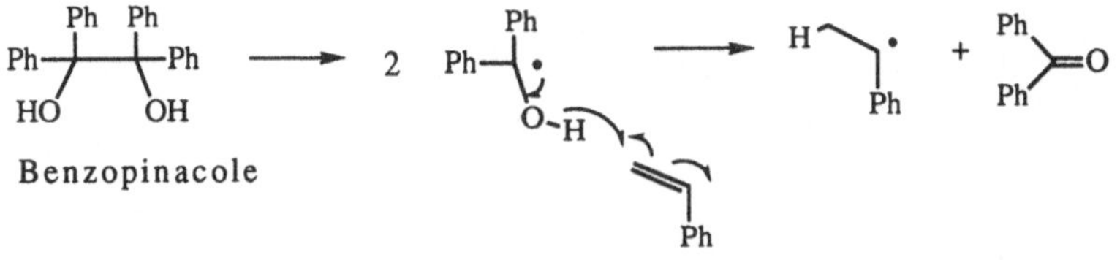

Scheme 8. Initiation of styrene polymerization using benzopinacole

Later, Akzo researchers [108] studied the polymerization kinetics of styrene initiated by a variety of tetrasubstituted 1,2-diphenylethanes in the temperature range of 60–110 °C. The polymerization kinetics were found to be similar to initiation by peroxides.

More recently, the related benzopinacole ethers 6 [109] and 7 [110, 111] have been evaluated as FR initiators.

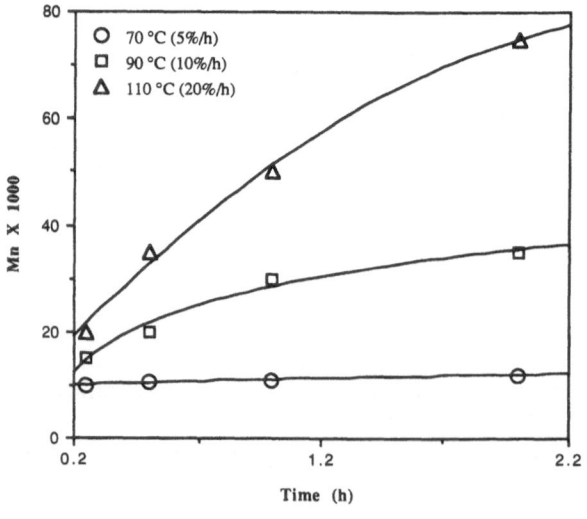

Structures: **V**, **VI**, **VII**

Bledzki and Braun [109] found that initiation of FR styrene polymerization using **6** does not take place at a relatively constant rate as with the classic peroxide initiators, but rather like a dead-end (primary radical termination) polymerization as reported for AIBN at 100 °C by Tobolsky [112, 113]. Furthermore, styrene polymerizations initiated using **6** had a unique feature. Unlike normal initiators, the polymer molecular weight was found to increase with increasing temperature and rate (Fig. 16). This effect was explained by the amount of primary radical termination decreasing as the temperature is increased.

Bledzki and Braun [109] analytically confirmed that primary radical termination was the main mode of termination during low temperature (1 h at 80 °C) polymerization of styrene initiated using **5** as initiator. The polymer was extracted with methanol to isolate the oligomers. The oligomer mixture was analyzed using Field Desorption Mass Spectrosopy (FD-MS). The spectra showed a molecular weight series corresponding to 384 + n(104) indicating that the oligomers were both initiated and terminated by diphenylcyanomethyl radicals.

However, these results are also consistent with greater extents of "living polymerization" taking place as the temperature is increased. This is supported by the findings of Crivello et al. [114]. At temperatures > 100 °C, when styrene was polymerized in the presence of **7**, conversion and M_w continue to rise with

Fig. 16. Polymerization of styrene using 1,1,2,2-tetraphenyl-1,2-diphenoxyethane (**6**); effect of temperature on polymerization rate and molecular weight

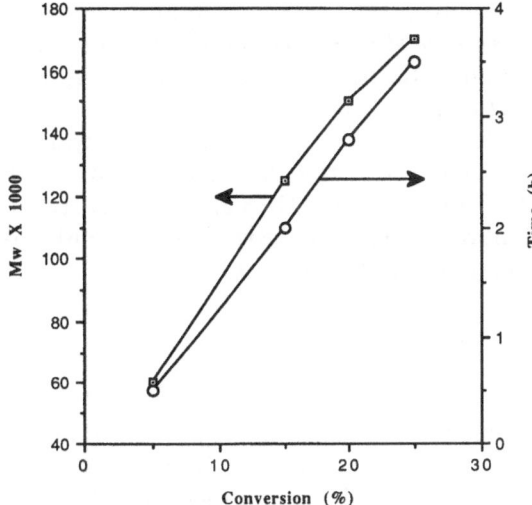

Fig. 17. Polymerization of styrene at 100 °C initiated using 0.1 M bis(trimethylsilyl) benzopinacole (7)

increasing time and conversion, respectively (Fig. 17) exhibiting "living polymerization" behavior; i.e. 1) during the initial period, oligomers containing terminal initiator fragments are formed by primary radical termination; 2) the C–C bond formed during the primary radical termination process is labile and later cleaves to reform the initiator fragment and a macro-radical; and 3) after all of the initiator radicals have been consumed, the normal propagation and termination processes occur.

This behavior is best explained by a "living" FR polymerization mechanism (Scheme 9).

Braun et al. [115] also found that initiation of the polymerization of methyl methacrylate (MMA) at 80 °C using **5** had characteristics of a living polymerization. However, in styrene, under the same conditions, the C–C bond formed during the primary radical termination process is inert resulting in a "dead-end"

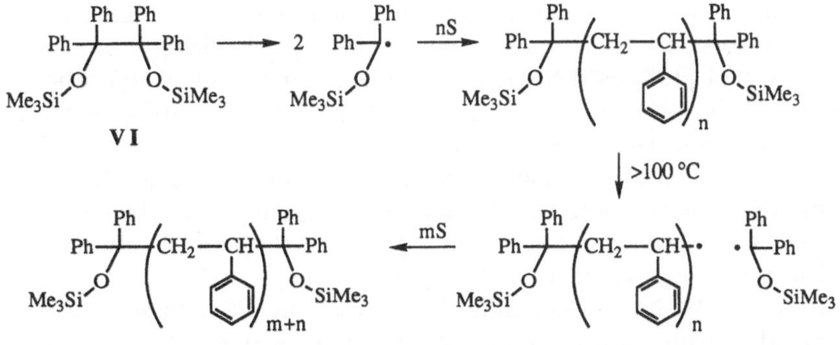

Scheme 9. Living FR polymerization of styrene using bis(trimethylsilyl)benzopinacole (**7**)

polymerization. The difference in lability of the C–C bond formed during the primary radical termination process in MMA and styrene polymerizations is likely due to the degree of substitution on the newly formed ethane linkage; i.e., hexasubstituted versus pentasubstituted for MMA and styrene, respectively. Based on this assumption, one might predict that α-methylstyrene polymerization initiated by C–C initiators would polymerize similarly to MMA. This indeed was found to be the case [116]. Oligomers of MMA and α-methylstyrene containing these labile C–C bonds have been found to be very effective initiators for styrene polymerization at temperatures as low as 50 °C.

4.3.2 Azo Initiators

It is well known that certain azo compounds initiate the FR polymerization of styrene. The most common azo initiators are the azobisalkylnitriles. However, there has been some concern since, in vivo, azobisalkylnitriles give HCN and heating polymers produced using these initiators also liberates HCN [117]. To circumvent these problems, attempts have been made to develop azo initiators which do not contain nitrile groups [118–121]. However, it is difficult to obtain non-nitrile azo initiators having decomposition kinetics satisfactory for styrene polymerization.

The most studied azo initiator is AIBN. The literature [7, 122–133] suggests a number of ways by which the cyanoisopropyl radicals (CIP) from AIBN might become incorporated into the polymer during FR polymerization. These include the following processes: 1) initiation (addition to monomer); 2) primary radical termination (coupling with a growing polymer radical); and 3) formation of methacrylonitrile (MAN) [by disproportionation of CIP [134] or by H-abstraction from AIBN followed by β-scission [122, 126, 135] which can undergo copolymerization. These processes along with possible cage reactions of AIBN are shown in Scheme 10.

There is some evidence that CIP radicals, which can be considered a hybrid of two resonance forms, can react in its unstable keteneimine form [123].

The overall initiator efficiency of AIBN is about 50–60% [136]. The instantaneous efficiency as a function of conversion is shown in Fig. 18 [137].

The main loss of CIP is self-reaction in the cage. Utilizing ^{13}C-labelled AIBN, Moad et al. [138] used NMR to determine the nature of incorporation of AIBN into PS by solution polymerization at both low (< 10%) and high (> 50%) conversion. This investigation showed that the only detectable AIBN initiator residues in "low conversion" PS were CIP groups. Evidence for MAN fragments copolymerized into the PS chain were only found in "high conversion" samples. No evidence for incorporation of keteneiminyl radicals was found. This was in agreement with Bevington [12–14].

Moad et al. [138] has found that in "low conversion" PS initiated by AIBN, > 85% of the chains are terminated by coupling of two growing polymer chains. They found no evidence for primary radical termination (process 2)

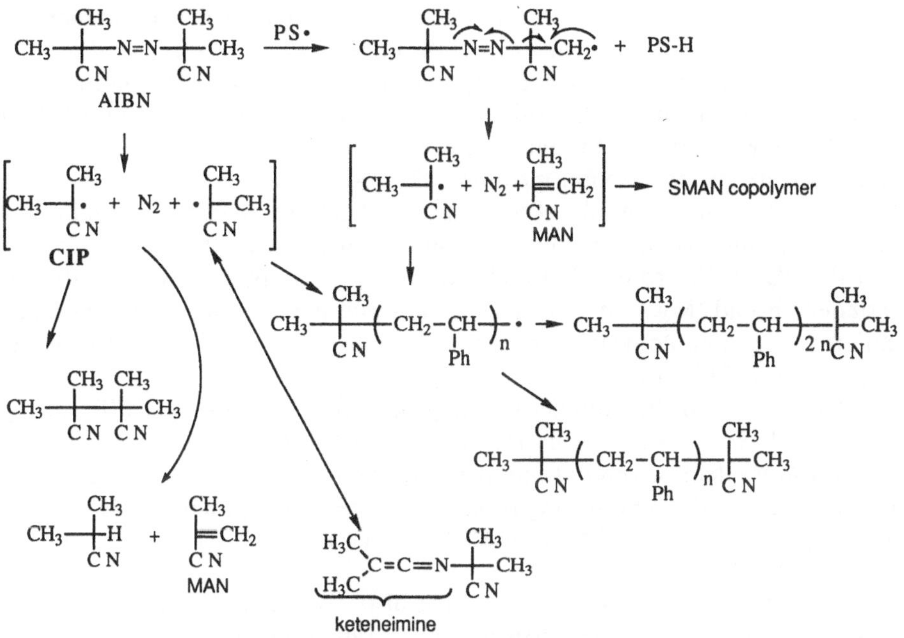

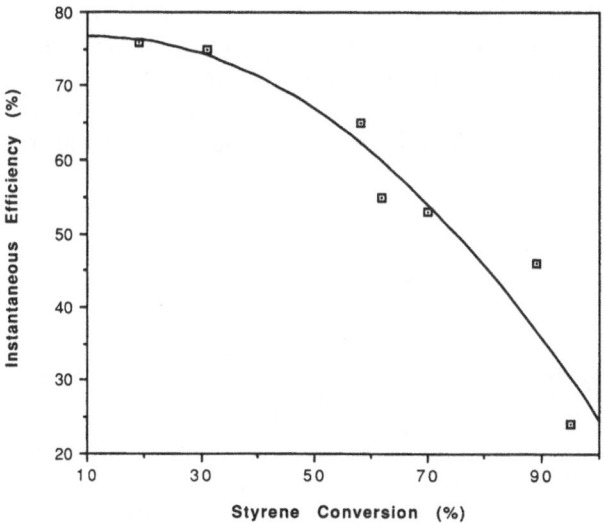

Scheme 10. Initiation and cage reactions of AIBN in solution styrene polymerization

Fig. 18. Instantaneous initiation efficiency of AIBN during solution styrene polymerization

which is in contradiction to a number of other researchers [123, 125, 129–132]. Moad et al. [136] also found little evidence for incorporation of CIP via chain transfer to initiator followed by β-scission (as suggested by some researchers) [122, 126, 135] in "low conversion" PS. They estimate the chain transfer

constant for AIBN in styrene polymerization to be < 0.01 which is considerably
lower than other literature values (0.02–0.16) [122, 127, 132, 139].

High conversion solution styrene polymerization using AIBN, on the other
hand, leads to quite different results. The onset of the Trommsdorff or gel effect
results in lower initiator efficiency because of radical cage reactions due to the
high viscosity of the reaction medium [139–142]. Also, termination by chain
coupling (the primary mode of termination at low conversion) is diminished
opening up opportunities for termination by other processes. The net result is
that the PD of PS produced at high conversion using bulk AIBN initiation is
extremely broad (Mw/Mn > 5). This same trend is not observed for peroxide
initiators (e.g. high conversion styrene polymerization using benzoyl peroxide
gives PD ~ 2) [143, 144].

Measurement of the number of end-groups in AIBN initiated PS utilizing
^{13}C NMR [138] and a radiochemical technique [126] are in agreement that
there are more than two end-groups per polymer molecule (indicates branching).
The mechanism of the branching reaction is uncertain, especially in light of the
relatively poor ability of CIP radicals to abstract H-atoms from the polymer
backbone [138].

Unsymmetrical azo compounds have also been used to initiate styrene
polymerization. For example, Otsu et al. [145, 146] studied phenylazotriphenyl-
methane as an initiator. They found that both a phenyl and a trityl radical are
generated. The phenyl radical initiates polymerization while the trityl radical
does not. Instead, the trityl radical acts as a radical trap and efficiently
terminates polymerization by primary radical coupling (Scheme 11). As a result
of steric crowding between the pendant groups on the polymer chain and the
phenyl groups of the trityl moiety, the C–C bond can redissociate at elevated
temaperature and add more monomer. This is another example of a "living" free
radical polymerization.

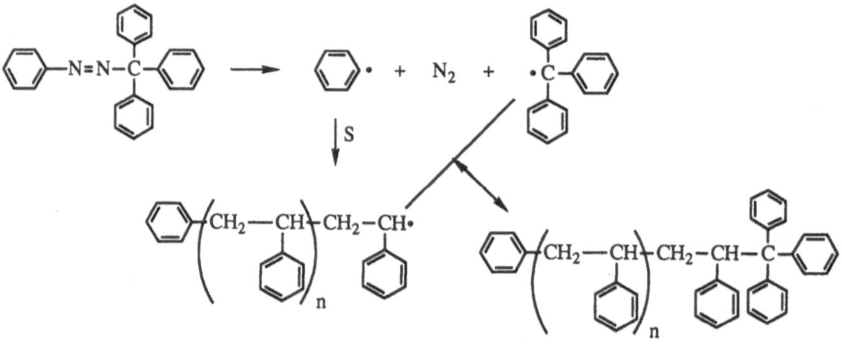

Scheme 11. Initiation of styrene polymerization using phenylazo-triphenylmethane

4.3.3 Peroxide Initiators

The addition of FR to styrene is usually depicted as a selective process involving exclusive addition to the unsubstituted terminus (tail) of the double bond [2]. While this is undoubtedly the primary mode of addition for most FR during the initiation step, it is well recognized that side reactions can occur during the initiation step.

The most common peroxide initiators for solution styrene polymerization yield either *tert*-alkoxy or acyloxy radicals or both. *tert*-Butoxy radicals are extremely selective in their reaction with styrene giving only addition to the less substituted end of the double bond (tail addition). This is in marked constrast to the behavior of benzoyloxy radicals which add to the aromatic ring of styrene, as well as showing both head and tail addition [147].

The relative ratio of tail, head, and aromatic addition was determined by decomposing benzoyl peroxide (BPO) in styrene containing the radical trapping agent 1 (Scheme 12) [148]. The tail addition product 9a accounted for 80% of the benzoyloxy derived products while the head addition 11a only accounted for 5%. Aromatic substitution products 13a accounted for the other 15% of the benzoyloxy radicals. The ratio of these products was however somewhat dependant upon the concentration of 1. This is likely due to the relative rates of addition to styrene and of 1 to benzyl radical 8a and primary alkyl radical 10a.

The trapping of cyclohexadienyl radicals 12a by 1 oxidation (hydrogen abstraction) to afford the corresponding styrylbenzoate (13a) and the hydroxy-

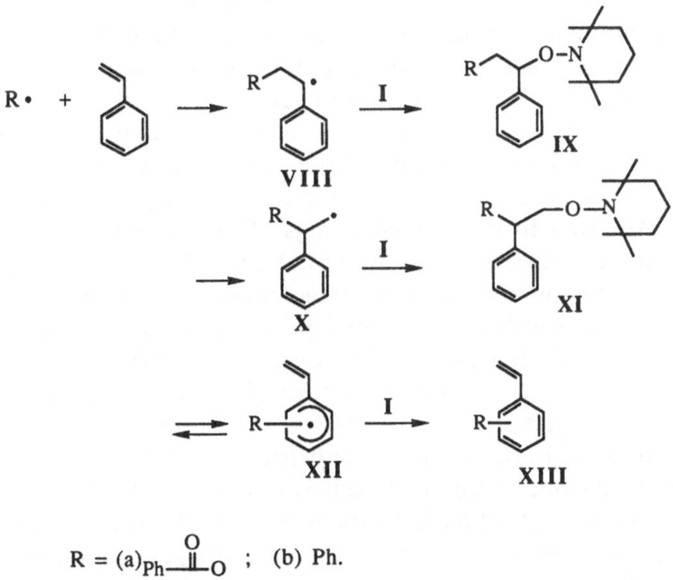

Scheme 12. Initiation of styrene polymerization using BPO in the presence of TMPO

lamine [148]. At low concentration of **1** a low yield of **13a** was obtained indicating that the formation of **12a** is reversible and that the rate of the reverse reaction is competitive with oxidation by **1**. This finding is in agreement with other studies showing the addition of benzoyloxy radicals to benzene is reversible [149, 150].

Phenyl radical derived products **9b**, **11a**, and **13b** were quite low ($\sim 5\%$). This is in contrast to radiochemical studies [151, 152] which showed about equal amounts of phenyl and benzoyloxy end-groups in BPO-initiated PS.

Head addition to styrene by benzoyloxy radicals is remarkable in that it results in the formation of a significant amount of primary alkyl radicals in preference to the resonance-stabilized benzylic radical. In constrast, the reaction of styrene with most other radicals yields almost exclusively tail addition. The key difference in selectivity between benzoyloxy and *tert*-butoxy radicals is likely due primarily to steric factors [153]. *tert*-Butoxy radicals are significantly larger than benzoyloxy radicals and therefore would be expected to add to the sterically less hindered tail position. Also, benzyloxy radicals are likely more electrophilic than *tert*-butoxy radicals (due to the electron withdrawing effect of the carbonyl) making them more reactive and therefore less selective in their addition to the electron rich styrene molecule.

Within the cage, both *tert*-alkoxy and acryloxy radicals undergo side reactions. For example, *tert*-butoxy radicals decompose to form acetone and methyl radicals while benzoyloxy radicals liberate carbon dioxide to form phenyl radicals. Fink [154] has studied the decomposition of benzoyl peroxide in benzene and analyzed the products using gas chromatography and mass spectroscopy. The three products produced were: biphenyl (72%), phenyl cyclohexadiene (28%), and phenyl benzoate (0.3%). Using deuterated benzoyl peroxide, it was determined that these products arose from reaction with the benzene rather than radical combination within the cage.

The reaction of methyl radicals with styrene gives only tail addition while phenyl radicals also show addition to the phenyl ring of styrene. It is uncertain whether the benzoyloxy and phenyl radicals add to the phenyl ring of monomer or polymer or both. Addition to the aromatic rings of the polymer become increasingly important at higher conversions [8]. Using UV spectroscopy calibrated against model compounds, the percentage of aromatic substitution, as a function of styrene conversion for benzoyl peroxide initiated solution polymerization, was quantitatively measured. The results are shown in Fig. 19.

The effects that initiator derived residues and functional groups have on polymer properties is an area that needs further study. For example, as discussed above, benzoyl peroxide leads to the formation of benzoyloxyphenyl groups in PS. It is known that the polymerization of *p*-benzoyloxystyrene [155] and its copolymerization with styrene [156] leads to the formation of photo-reactive polymers. Upon irradiation these polymers undergo a facile photo-Fries Rearrangement resulting in the conversion of the benzoyloxyphenyl groups to hydroxybenzophenone moieties [157]. Other side reactions also lead to the formation of phenolic groups and free benzoyloxy radicals [155].

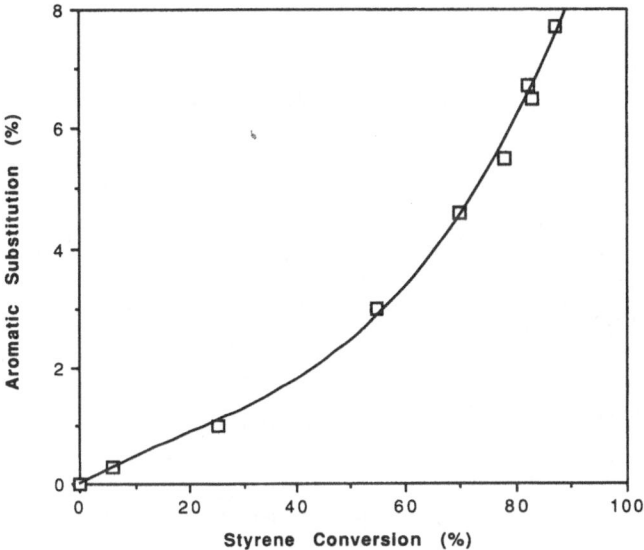

Fig. 19. The extent of aromatic substitution as a function of conversion during solution styrene polymerization using BPO initiator at 90 °C

The benzoyloxy end-groups in polystyrene produced using BPO as initiator have been shown to play an important role in the initiation of thermal degradation of free radically produced PS. Moad et al. [158] initiated styrene polymerization to high conversion (80%) with ^{13}C-labeled BPO. The polymer was degraded at 300 °C. Analysis of the polymer after heating showed that all of the resonances due to secondary benzoate groups had disappeared while the primary benzoate resonances remained unchanged. GPC analysis showed that the molecular weight and PD remained virtually unchanged. This indicates that the secondary benzoate end-groups (Scheme 13) resulting from 1) head addition to monomer; 2) transfer to initiator; or 3) primary radical termination are thermally less stable than the primary benzoate end-groups resulting from tail addition.

They suggest that the loss of secondary benzoate end-groups on thermolysis likely involves elimination of benzoic acid (BA) to form an unsaturated end-group. Unsaturated end-groups have long been known to contribute to thermal instability of polystyrene [83, 91, 159]. Thus the secondary benzoate end-groups are "weak links" in PS. The unsaturated end-groups, resulting from elimination of BA from the secondary benzoate end-groups formed by processes 1–3 (Scheme 13), have different structures. It is uncertain how these unsaturated end-group structures differ in respect to their ability to initiate further degradation. The primary benzoate end-groups appear to be at least as stable as the PS backbone. After thermal degradation to 50% weight loss, primary benzoate end-groups still remained. This finding is further supported by Cameron et al. [95] using infrared spectroscopy.

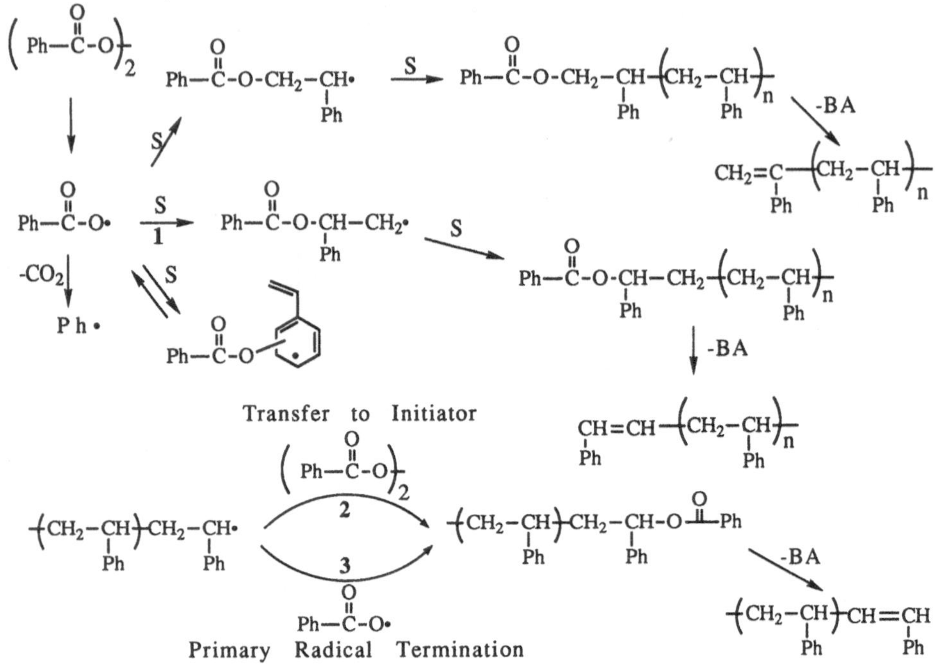

Scheme 13. End-groups in PS produced using BPO as initiator

Polymers prepared in academic studies are typically made at low conversion. However, in industry polymers are prepared under conditions favoring high conversion to maximise economics [2]. As already pointed out, high conversion increases initiator side reactions which leads to an increase in aromatic substitution, transfer to initiator and primary radical coupling. Therefore, initiator derived "weak links" are most important in commercial PS produced using FR initiators.

Another method which has been utilized in the study of initiators in FR styrene polymerization is the use of peroxides containing highly absorbing azobenzene substituents. Olaj et al. [9, 160] synthesized di(3-phenylazo) BPO and studied its fate during styrene polymerization. However, they found that the chain transfer constant of the initiator was an order of magnitude higher than BPO. Measurements showed that each chain had a phenylazo BPO unit on each end but the mechanism of incorporation was chain initiation followed by chain transfer to initiator rather than by chain coupling. The electron withdrawing nature of the phenylazo substituent increases the chain transfer activity of the initiator.

Another class of common peroxide initiators for solution styrene polymerization are the *gem*-bis(*tert*-alkyldioxy)alkanes (14). Yenal'ev et al. [161, 162] found peroxide linkages in PS produced using these initiators. They concluded that the initial decomposition of the bisperoxide occurred at one of the peroxide

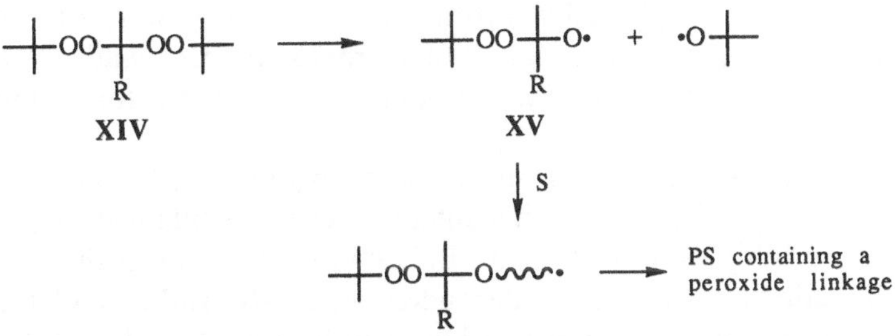

XIV   →   XV   +   •O—|

XV   ↓ S

—|OO—|O~~~•   →   PS containing a peroxide linkage

Scheme 14. Proposed non-synchronous mechanism for decomposition of peroxyketals

bonds rather than synchronous scission of both peroxide bonds. They further suggested that the PS chains that contained peroxide linkages were produced via initiation by the intermediate peroxide containing alkoxyl radical (15) as shown in Scheme 14.

Watanabe et al. [163] also found peroxide linkages in PS produced using certain *gem*-bis(*tert*-alkyldioxy)alkanes. Iodometric titration showed that about half of the polymer chains contained a peroxide linkage when using 2,2-bis(*tert*-butylperoxy)propane (**14**, R = Me) as an initiator. However, when R = Et or *i*Pr, < 5% of the chains contained peroxide linkages. They pursued the investigation further to elucidate the mechanism of decomposition of **14** and initiation of styrene polymerization using **14** where R = Me, Et, *i*Pr. Comparison of the decomposition of **14** in cumene [164, 165] and styrene [163] gave interesting results. Analysis of the liquid products left after the decomposition

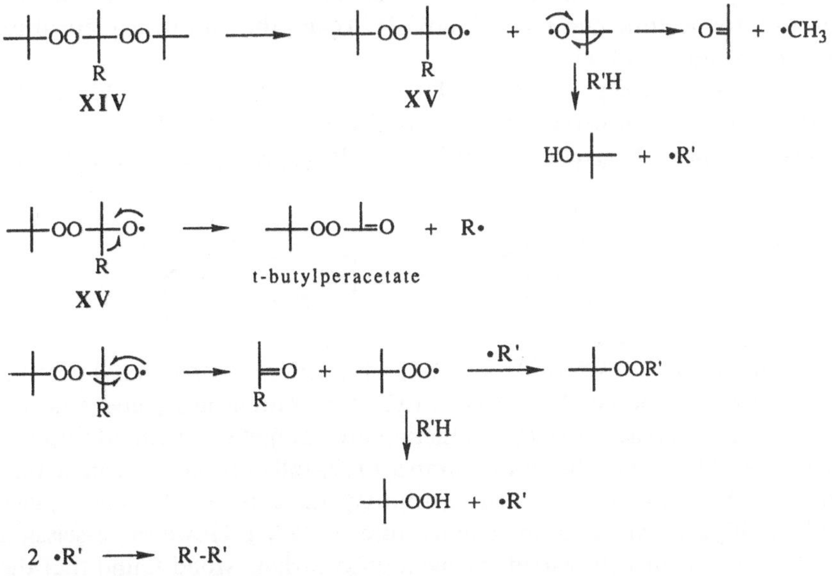

XIV   →   XV   +   •O—| →   O=| +   •CH$_3$

↓ R'H

HO—| +   •R'

XV   →   —|OO—|=O   +   R•
          t-butylperacetate

—|OO—|O•   → |=O   +   —|OO•   —•R'→   —|OOR'

↓ R'H

—|OOH   +   •R'

2 •R'   →   R'-R'

Scheme 15. Decomposition of *gem*-bis(*t*-alkyldioxy)alkanes in cumene in cumene showed acetone, alkyl methyl ketone, *tert*-butyl alcohol, *tert*-butyl hydroperoxide, *tert*-butylperacetate, *tert*-butylcumylperoxide, and 2,3-dimethyl-2,3-diphenylbutane. Scheme 15 was proposed to explain the formation of these products.

The yield of *tert*-butylperacetate formed by decomposition of **14** in cumene varied greatly with R (Me ≪ Et < *i*Pr). This difference can be explained on the basis of the relative stability of alkyl radicals. Bailey [166] has shown that the rates of scission reactions of radicals is dependent upon the stability of the expelled radicals. For example, copolymerization of cyclic ketal **16** with styrene results in simple vinyl polymerization while copolymerization of **17** with styrene results in 100% scission to form the ring-opened copolymer.

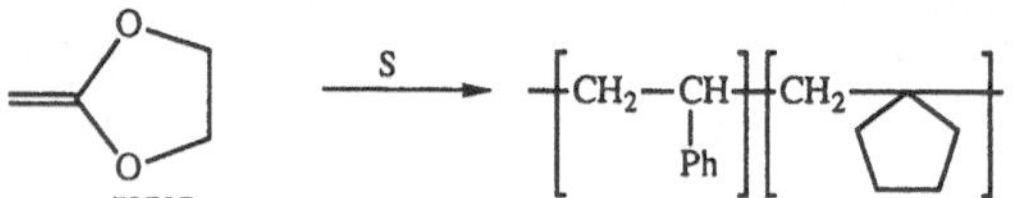

Decomposition of **14** in styrene produced very similar products. About the same amount of *tert*-butylperacetate was formed in both styrene and cumene indicating that the scission reaction of **15** is much faster than its rate of addition to the styrene double bond. This is surprising since the analogous addition of *tert*-butoxy radical to styrene is very fast relative to the analogous scission reaction to form acetone and methyl radicals. Again, this can be explained by the relative stability of alkyl radicals.

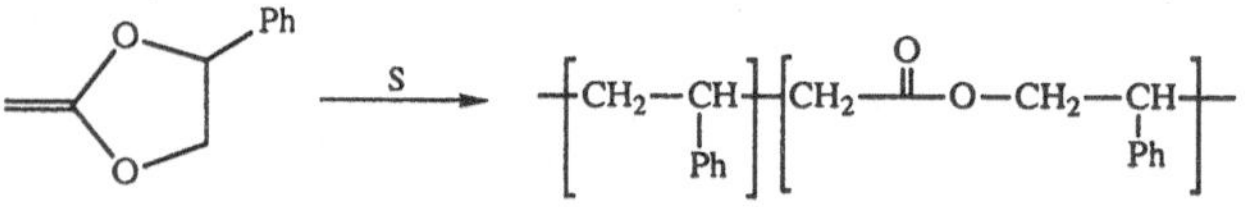

Watanebe et al. [163] noticed the formation of styrene oxide during the polymerization of styrene using *gem*-bis(*tert*-alkyldioxy)alkanes. They suggested radical expoxidation of styrene by expelled *tert*-butylperoxy radicals as shown below.

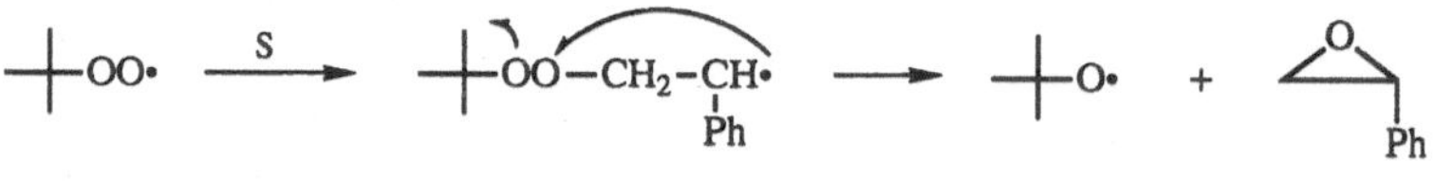

The formation of considerable amounts of acetone and *tert*-butyl alcohol during the decomposition of 2,2-bis(*tert*-butylperoxy)butane in styrene indicates that the rates of β-scission and H-abstraction are competitive with addition to the styrene double bond. Niki and Kamiya [167, 168] are in agreement that considerable H-abstraction between *tert*-butoxy radicals and PS takes place especially at high polymerization temperatures (125 °C). However, β-scission rate is in disagreement with Moad. As mentioned earlier, Moad found that the rate of vinyl addition by *tert*-butoxy radicals is 76 times faster than β-scission

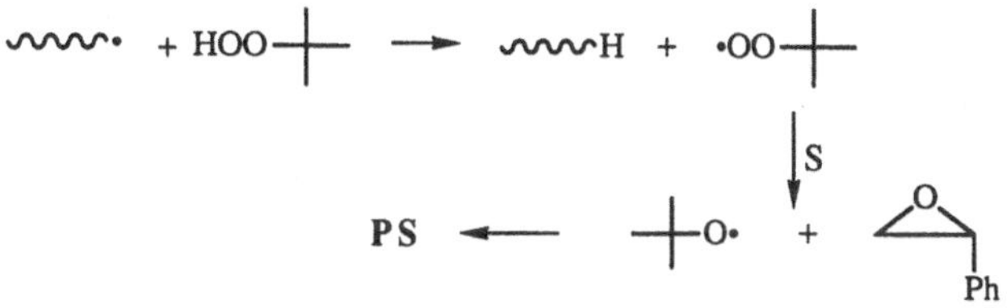

Scheme 16. Possible chemistry of chain transfer and initiation using organic hydroperoxides

(60 °C). The difference in these results may be the temperature at which the experiments were performed. It appears that the ratio of addition and β-scission of alkoxy radicals during styrene polymerization is temperature dependent with the scission reaction accelerating faster as temperature increases.

A class of peroxides which has received little attention in styrene polymerization is hydroperoxides. In inert solvents, hydroperoxides are relatively stable. However, in a FR environment, they undergo an induced decomposition. The induced decomposition results from the relatively high chain transfer constant (C_t) of the hydroperoxide hydrogen ($C_t \sim 0.05$). Abstraction of the hydrogen by a growing polystyryl radical produces a peroxy radical which adds to styrene. As mentioned peviously, the styrene adduct likely decomposes to form styrene oxide and a *tert*-alkoxy radical which subsequently initiates polymerization.

4.4 Diradical Initiators

4.4.1 Carbon–Carbon Diradicals

The earliest work [42] aimed at forming C–C diradicals for styrene polymerization was performed in an effort to test the Flory diradical mechanism (Scheme 1). However, early on, there was controversy over the ability of diradical initiation to produce higher molecular weight polymers [169, 170]. Later work [171, 172] however, showed successes in producing higher molecular weight vinyl polymers using biradical initiators. For example, when comparing the monoradical C–C initiator **18** and the diradical C–C initiator **19**, Borsig et al. [173] found (polymerization of methyl methacrylate at 60 °C) that significantly higher molecular weight was formed at the same monomer conversion using the diradical initiator. However, this data is complicated by the fact that at the same molar initiator concentration, the polymerization rate was slower for **19**. The difference in initiating efficiency of the two initiators was explained in terms of cage reactions. The monoradicals formed from initiator **18** diffuse quite efficiently from the cage and react with monomer. The diradicals, on the other hand, are in a permanent cage in that they can not diffuse away from each other. Therefore, significant disproportionation of the diradical (4) takes place to compete with initiation.

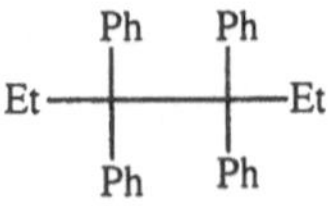

XVIII 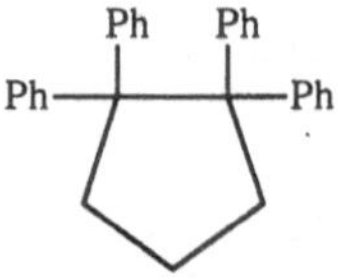 **XIX**

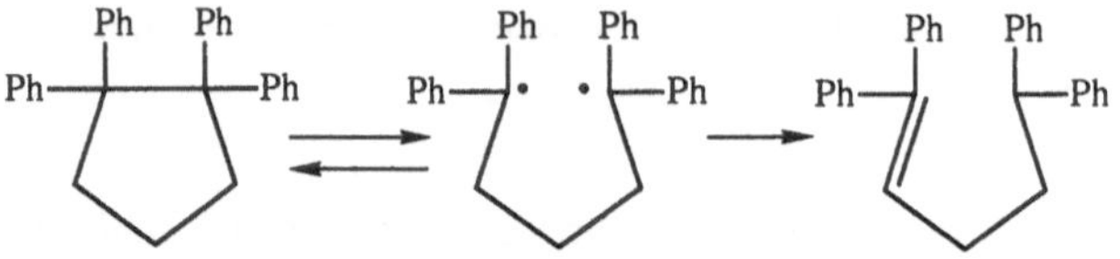

(4)

Later research on diradical C–C initiators [174, 175] focused on oligomeric silyl pinacol ethers (**20**) which were used to prepare PS-co-polysiloxane block copolymers.

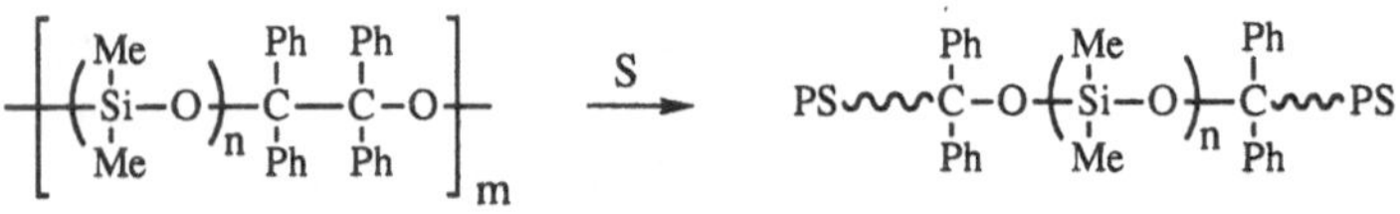

XX

More recently, Crivello et al. [176–179], investigated C–C diradicals produced during styrene polymerization using cyclic pinacol ethers **21** and **22**.

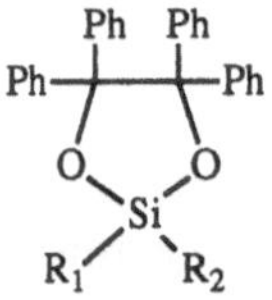

XXI 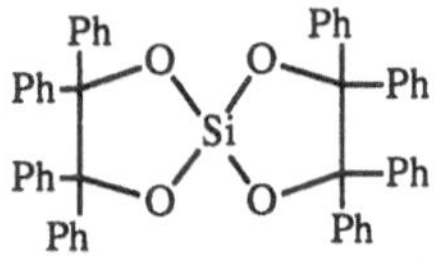 **XXII**

Comparison of the performance of **6**, **21**, and **22** which yield mono, di, and tetra carbon radicals respectively, clearly shows a relationship between functionality and molecular weight during styrene polymerization (Fig. 20). Because of the predominance of termination by chain coupling, polymerizations using the tetrafunctional initiator **22** gave crosslinked PS when taken to high conversion.

Recently, Hall et al. [180, 181] have investigated the use of a donor-acceptor substituted cyclopropane (**23**) to form carbon–carbon diradical initiators. Gel permeation chromatography showed that the PS produced had a bimodal molecular weight distribution [180]. The low molecular fraction is attributed to self-termination of the growing diradical. Once a growing diradical chain becomes long enough, the diradical ends become far enough apart that they do not terminate each other. Besides the normal dimers and trimers of styrene

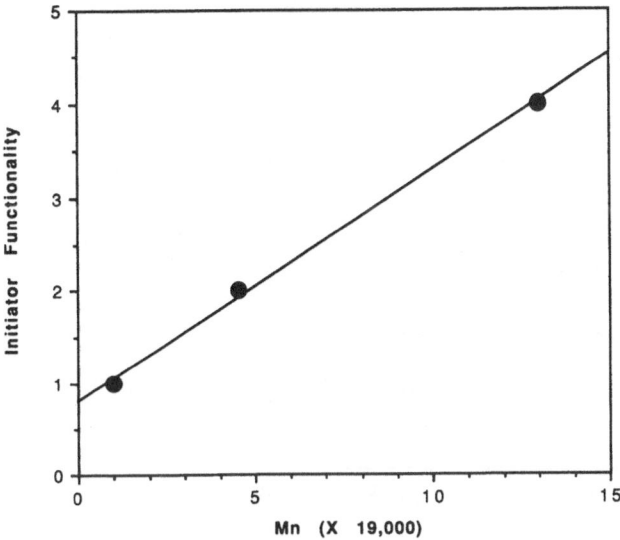

Fig. 20. Variation of Mn as a function of initiator functionality during styrene polymerization

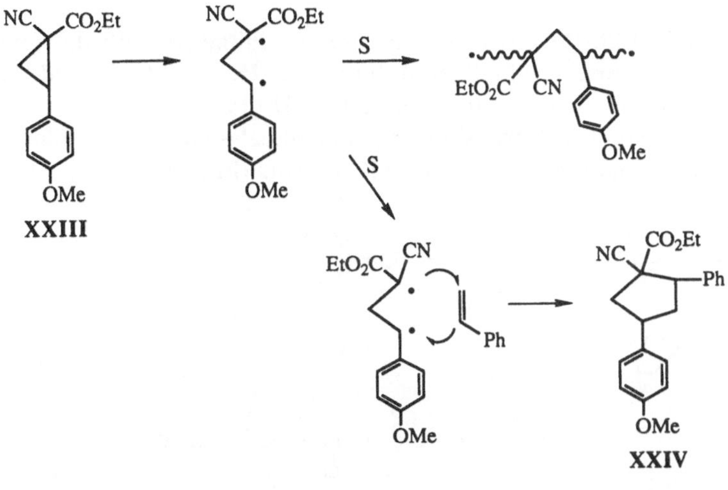

Scheme 17. Initiation of styrene polymerization with a capto-dative substituted cyclopropane

normally encountered in styrene polymerization, the 1:1 adduct of styrene and **23** was produced (**24**) (Scheme 17).

4.4.2 Peroxide Diradicals

There has been much interest in recent years regarding kinetic modeling of solution FR styrene polymerization. Several kinetic models have been developed for styrene polymerization without added initiators [182–184], using

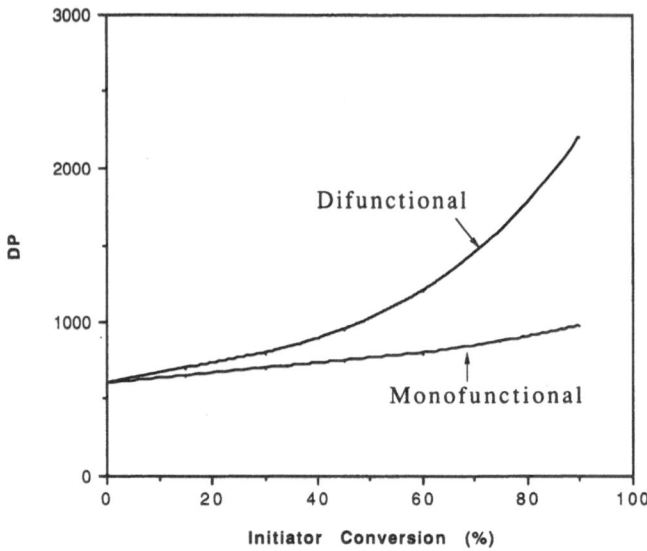

Fig. 21. Comparison of theoretical PS molecular weight for mono- and di-functional initiators

one monofunctional [185–189], two monofunctional initiators with different half-lives [190], symmetrical difunctional initiators **25**, **26** [191], and **27** [192], and the unsymmetrical difunctional initiators **31** [193, 194] and **32** [195, 196]. These modeling studies clearly show the theoretical advantage of difunctional over monofunctional initiators in CSTR reactors (Fig. 21) [197].

Since difunctional initiators contain more than one labile group, the kinetic schemes become quite complicated, especially for unsymmetrical difunctional initiators. One reason for the complicated kinetic modeling is the large number of polymer chain structures present during the polymerization. At least ten theoretical chain structures (Fig. 22) must be taken into account when modeling unsymmetrical difunctional initiators [193].

1) ·——   live polymer without any undecomposed peroxide

2) ·——A   live polymer with undecomposed peroxide A

3) ·——B   live polymer with undecomposed peroxide B

4) ·——·   double ended living polymer

5) ——   dead polymer without any undecomposed peroxide

6) ——A   inactive polymer with a single undecomposed peroxide A

7) ——B   inactive polymer with a single undecomposed peroxide B

8) A——B   inactive polymer with undecomposed peroxides A and B

9) A——A   inactive polymer with undecomposed peroxides A

10) B——B   inactive polymer with undecomposed peroxides B

Fig. 22. Ten polymer chain structures taken into account when modeling styrene polymerization initiated by A-R-B where A and B are peroxide groups having different decomposition rates

The rate advantage achievable using initiators which yield both mono and di-radicals is, at best, only incremental. To gain major rate increases requires the use of initiators which yield almost entirely di-radicals. Polymeric peroxides should theoretically yield almost entirely diradicals and therefore theoretically should allow the production of high Mw PS at extremely fast rates. However, studies on multifunctional initiators has shown that efficiency decreases with an increasing number of labile functional groups in the molecule [193, 198, 199]. This effect becomes a key detriment to the development of initiators for generation of high levels of diradicals. Even so, a significant amount of literature has appeared in recent years describing polymeric free radical initiators for increasing styrene polymerization rates [200–217]. Most of these references describe oligomeric peroxides (DP ~ 5–10) made by reacting diacid chlorides with hydrogen peroxide to form polydiacylperoxides. Most polydiacyl peroxides are extremely hazardous to handle due to their friction sensitivity and poor solubility characteristics. These problems have been minimized by using long chain (C_{18}) diacid chlorides or using bishydroperoxides (instead of hydrogen peroxide) to make polyperesters.

Another class of peroxides that should theoretically form diradicals are cyclic peroxides. However, anomalous results have been reported. For example, cyclic peroxides **30–32** decompose without initiating of polymerization [218].

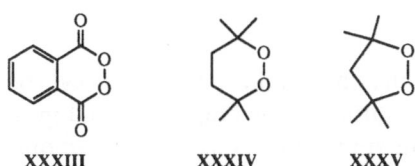

XXXIII   XXXIV   XXXV

The cyclic perketal **33** initiates styrene polymerization but with very low efficiency [219]. The molecular weight of the PS produced was no higher than PS prepared at the same rate of polymerization using di-*tert*-butylperoxide. A

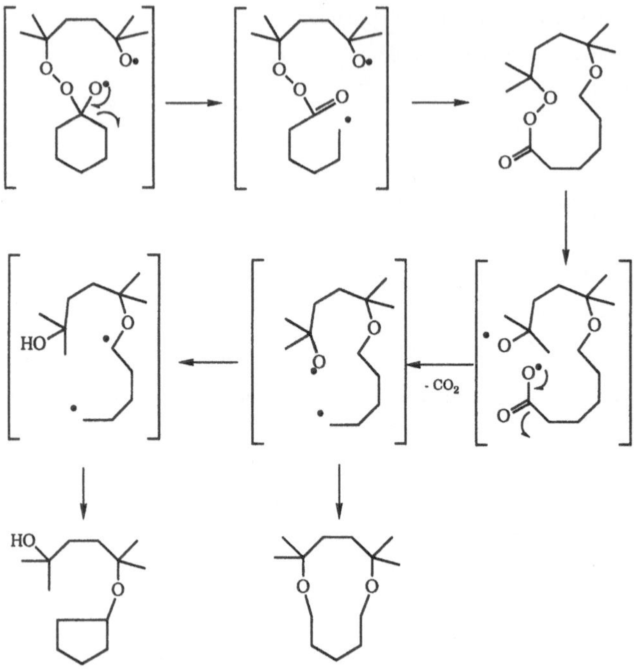

Scheme 18. Possible cage reaction of cyclic perketal **33**

study of the decomposition chemistry of **33** in ethylbenzene (EB) showed that
less than 20% of the intermediate radicals formed escape the EB solvent cage to
abstract hydrogen to form EB dimer. Analysis of the EB solution of decomposed
33 using GC-MS showed several products arising from in-cage reactions [220].
The main products were the hydroxyether **34** and the macrocyclic diether **35**. A
possible in-cage decomposition mechanism of **33** showing formation of **34** and
35 is shown in Scheme 18.

5 Telechelic Polystyrene

There has been considerable interest recently in making telechelic polymers
[221, 222] (reactive end-groups) due to their utility in making block and graft
[223] copolymers. Telechelic polymers are most often made using anionic
polymerization techniques due to the absence of side reactions. However,
telechelic PS having end-group functionality close to two can be produced in
solution FR polymerization using functionalized initiators if the conditions are
such that the main mechanisms of termination involve radical coupling. For

example, Konter et al. [224] used high levels of AIBN to make cyano end-functional PS having an end-group purity of 1.99. The nitrile end-groups were subsequently converted to carboxy and amino groups by hydrolysis and hydrogenation, respectively. The amino groups were also further converted to isocyanate groups by reaction with phosgene. These telechelomers (oligomers with reactive end-groups) were then used to make a variety of copolymers containing PS blocks. Because of the high end-group purity, block copolymers having Mw as high as 50 000 were demonstrated.

Kawabe et al. have made telechelic PS having carboxyl end groups using both succinic acid peroxide [225] and 4,4'-azobis(4-cyanovaleric acid) [226] as initiators. Other FR initiators utilized to make telechelic PS are shown below.

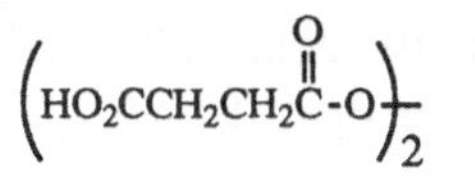

succinic acid peroxide [225]

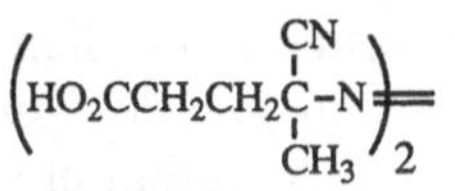

4,4'-azobis(4-cyano-
valeric acid) [226]

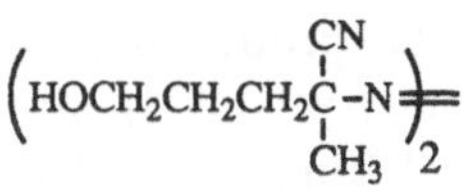

4,4'-azobis(4-cyano-
pentanol) [227]

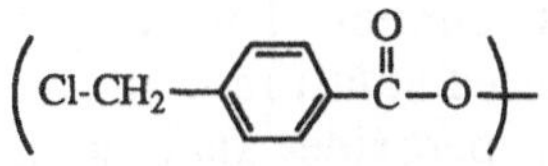

bis(4-chloromethyl-
benzoyl)peroxide [228]

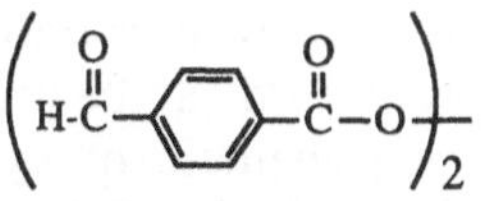

bis(4-formylbenzoyl)
peroxide [228]

Tong and Priddy [229] recently developed a new class of initiators containing benzocyclobutene (BCB) moieties. These initiators were used to prepare polystyrenes containing BCB end-groups. The BCB functional polystyrenes were found to chain extend upon heating to temperatures over 200 °C. Heating a solution containing BCB functional PS and a polymer containing unsaturation results in the formation of a graft copolymer (Scheme 19).

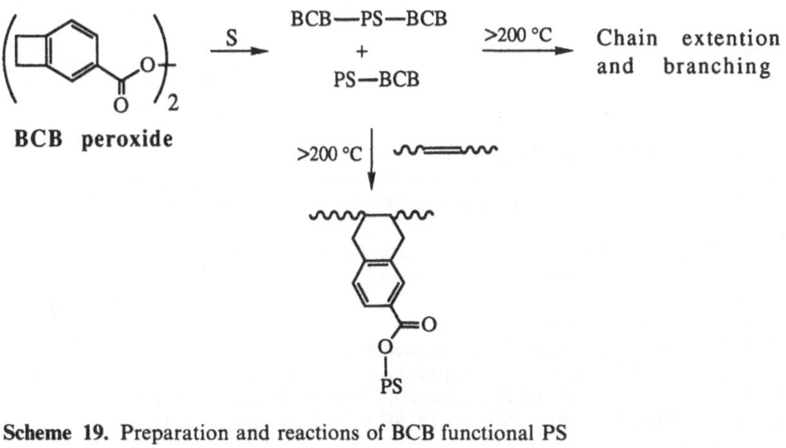

BCB peroxide

Scheme 19. Preparation and reactions of BCB functional PS

6 Conclusions and Future Directions

Styrene polymerization is the oldest, most studied, and best understood of all vinyl polymerizations. PS has been produced commercially for over fifty years and has reached an annual worldwide sales volume of over twenty billion pounds. All of this PS is produced using FR chemistry and mostly by continuous solution polymerization processes.

Anionic polymerization of styrene may soon become commercially important due to the following considerations: 1) the increasing demand for PS containing low residual styrene monomer; 2) the inherent quality of the PS produced using anionic chemistry (no weak links); 3) high molecular weight PS can be produced at much faster rates than currently achievable using FR chemistry; and 4) monomer and polymer production operations can be integrated eliminating the need for separation of styrene from ethylbenzene.

Research efforts will likely continue aimed at the development of highly efficient initiators that will initiate double ended polystyryl radicals. Since the main mechanism for termination in PS is radical chain coupling, double ended polystyryl radicals will continue to grow after radical coupling. The net result should be that high molecular weight PS can be produced at fast rates. Success will require the development of polymeric or cyclic peroxides that do not undergo cage reactions which reduce their efficiency.

Telechelic PS are of interest for making block copolymers. These materials are usually prepared in the laboratory using batch anionic polymerization since FR polymerization usually leads to less than the desired end-groups functionality of 2. However, commercial production of telechelic PS likely will require the lower cost of solution FR polymerization chemistry. The success of solution FR chemistry for making these polymers requires the development of highly efficient functionalized initiators.

7 References

1. Priddy DB, Pirc M (1989) J Appl Polym Sci 37: 393
2. For an excellent discussion of continuous reactor designs for mass styrene polymerization see Meister BJ, Malanga MT (Moore ER, Editor) (1989) Ency Polym Sci Eng 16: 48–52
3. Pryor WA, Fiske TR (1969) Macromolecules 2: 62
4. Bevington JC, Troth HG (1962) Trans Faraday Soc 58: 186
5. Berger KC, Meyerhoff G (1975) Makromol Chem 176: 1983
6. Berger KC (1975) Makromol Chem 176: 3575
7. Henrici G, Olive S (1960) J Polym Sci 48: 329
8. Rubio LHG, Ro N, Patel RD (1984) Macromolecules 17: 1998
9. Olaj OF, Breitenbach JW, Hofreiter I (1966) Makromol Chem 91: 264
10. Hatada K, Kitayama T, Yuki H (1980) Makromol Chem Rapid Commun 1: 15
11. Moad G, Solomon DH, Johns SR, Willing RI (1982) Macromolecules 15: 1182
12. Bevington JC, Ebden JR, Huckerby TN, Hutton NWE (1982) Polymer 23, 163

13. Bevington JC, Huckerby TN, Hutton NWE (1982) J Polym Sci Polym Chem Ed 20: 2655
14. Bevington JC, Huckerby TN, Hutton NWE (1982) Eur Polym J 18: 963
15. Moad G, Solomon DH, Johns SR, Willing RI (1984) Macromolecules 17: 1094
16. Chalfont GR, Perkins MJ, Horsfield A (1968) J Am Chem Soc 90: 7141
17. Kunitake T, Murakami S (1974) J Polym. Sci 12: 67
18. Sato T, Hibino K, Otsu T (1975) J Macromol Sci Chem 9: 1165
19. Sato T, Abe M, Otsu T (1977) Makromol Chem 178: 1061
20. Sato T, Otsu T (1977) Makromol Chem 178: 1941
21. Sato T, Abe M, Otsu T (1977) Makromol Chem 178: 1951
22. Sato T, Abe M, Otsu T (1979) Makromol Chem 180: 1165
23. Sato T, Metsugi M, Otsu T (1979) Makromol Chem 180: 1175
24. Sato T, Fukumura N, Otsu T (1983) Makromol Chem 184: 431
25. Lane J, Tabner BJ (1989) Eur Polym J 25: 677
26. Rizzardo E, Solomon DH (1979) Polym Bull 1, 529
27. Moad G, Rizzardo E, Solomon DH (1982) J Macromol Sci Chem 17: 51
28. Griffiths PG, Rizzardo E, Solomon DH (1982) J Macromol Sci Chem 17: 45
29. Griffiths PG, Rizzardo E, Solomon DH (1982) Tetrahedron Lett 23: 1309
30. Moad G, Rizzardo E, Solomon DH (1982) Macromolecules 15: 909
31. Howard JA, Tait JC (1978) J Org Chem 43: 2493
32. Rizzardo E, Serelis AK, Solomon DH (1982) Aust J Chem 35: 2013
33. Elias HG (1977) Macromolecules 2: Synthesis and Materials Plenum, New York
34. O'Driscoll KF, Mahabadi HK (1976) J Polym Sci Polym Chem Ed 11: 869
35. Olaj OF, Schnoll-Bitai I (1989) E Polym J 25: 635
36. Olaj OF, Schnoll-Bitai I (1991) Macromol Chem Rapid Commun 12: 373
37. Yamada B, Kageoka M, Otsu T (1991) Macromolecules 24: 5236
38. Buback M, Garcia-Rubio LH, Gilbert RG, Napper DH, Hamielec AE, O'Driscoll KF, Olaj OF, Shen J, Solomon D, Moad G, Stickler M, Tirrell M, Winnik MA (1988) J Polym Sci Part C Polym Lett 26: 293
39. Berger KC, Meyerhoff G (Braundup J, Immergut EH editors) (1989) Polymer Handbook, 3rd edn. Wiley, New York p II-81
40. Trommosdorff E, Kohle H, Lagally P (1947) Makromol Chem 1: 169
41. Flory PJ (1937) J Am Chem Soc 59: 241
42. Russel KE, Tobolsky AV (1954) J Am Chem Soc 76: 395
43. Pryor WA, Lasswell LD (1975) Advances in Free-Radical Chemistry 5: 27
44. Pryor WA (1978) ACS Symp Ser 69: 33
45. Brown WG (1969) Makromol Chem 128: 130
46. Barr NJ, Bengough WI, Beveridge G, Park GB (1978) Eur Polym J 14: 245
47. Mayo FR (1968) J Am Chem Soc 90: 1289
48. Mayo FR (1953) J Am Chem Soc 75: 6133
49. Hiatt RR, Bartlett PD (1959) J Am Chem Soc 81: 1149
50. Kaiser R, Kurze J, Sinak P, Stein DJ (1970) Angew Makromol Chem 12: 25
51. Chong YK, Rizzardo E, Solomon DH (1983) J Am Chem Soc 105: 7761
52. Mulzer J, Kuehl U, Huttner G, Evertz K (1988) Chem Ber 121: 2231
53. James Tanko (1991) private communication.
54. Graham WD, Green JG, Pryor WA (1979) J Org Chem 44: 907
55. Pryor WA, Coco JH, Daly WH, Houk KN (1974) J Am Chem Soc 96: 5591
56. Sato T, Abe M, Otsu T (1977) Makromol Chem 178: 1061
57. Kopecky KR, Lau MP (1978) J Org Chem 43: 525
58. Muller KF (1964) Makromol Chem 79: 128
59. Kopecky KR, Hall MC (1981) Can J Chem 59: 3095
60. Buzanowski WC, Graham JD, Priddy DB, Shero E (1992) Polymer 33: 3055
61. Otsu T, Sato T (1986) Mem Fac Eng Osaka City Univ 27: 129
62. Hall Jr HK (1991) private communication
63. Olaj OF, Kauffmann HF, Breitenback JW (1977) Makromol Chem 178: 2707
64. Olaj OF, Kauffmann HF (1976) Makromol Chem 177: 3065
65. Keairns DL, Manning FS (1969) AIChE J 15: 660
66. Chen MSK, Fan LT (1971) Can J Chem Eng 49: 704
67. Spielman LA, Levenspiel O (1965) Chem Eng Sci 20: 247
68. Tadmor Z, Biesenberger JA (1966) I&EC Fundamentals 5: 336
69. Manning FS, Wolf D, Keairns DL (1965) AIChE J 11: 723

70. Frontini GL, Elicabe GE, Meira GR (1987) J Appl Polym Sci 33:2165
71. Bywater S, Worsfold DJ (1969) Adv in Chem Ser 52:36
72. Treyberg MN, Anthony RG (1979) ACS Symp Ser 104:295
73. Priddy DB, Pirc M (1989) J Appl Polym Sci 37:1079
74. Kern WJ, Anderson JN, Adams HE, Bouton TC, and Bethea TW (1972) J Appl Polym Sci 16:3123
75. Gatske AL (1969) J Poly Sci Part A-1 7:2281
76. Priddy DB, Michael Pirc (1990) J Appl Polym Sci 40:41
77. Priddy DB, Pirc M, Meister BJ (1992) Polym Reaction Eng (to appear)
78. Priddy DB, Pirc M (1986) US Pat 4.572,819
79. William A. Pryor (1966) Free Radicals McGraw-Hill, New York
80. Riess C, Locatelli JL (1975) Adv Chem Ser 142:186
81. Niki E, Kamiya Y (1973) J Org Chem 38:1403
82. Niki E, Kamiya Y (1975) J Chem Soc Perkin Trans 1975:1221
83. Lehrle RS, Peakman RE, Robb JC (1982) Eur Polym J 18:517
84. Knight GJ (1967) J Polym Sci B5 5:855
85. Jones CER, Reynolds GEJ (1967) J Gas Chromat 5:25
86. Wall LA, Straus S, Florin RE, Fetters LJ (1972) J Res Natn Bur Stand 77A:157
87. McNeill IC, Makhdumi TM (1967) E Polym J 3:637
88. Grassie N, Kerr WW (1959) Trans Faraday Soc 55:1050
89. Cameron GG (1967) Makromol Chem 100:255
90. MacCullum JR (1965) Makromol Chem 83:129
91. Cameron GG, Meyer JM, McWalter IT (1978) Macromolecules 11:696
92. Singh M, Nandi US (1979) J Polym Sci Polym Let Ed 17:121
93. Cascaval CS, Straus S, Brown DW, Florin RE (1976) J Polym Sci 57:81
94. Moad G, Solomon DH, Willing RI (1988) Macromolecules 21:855
95. Cameron GG, Bryce WA, McWalter IT (1984) Eur Polym J 20:563
96. Cameron GG, Kerr GP (1970) Eur Polym J 6:423
97. Cameron GG, Grassie N (1962) Makromol Chem 53:72
98. Cameron GG, McWalter IT (1982) Eur Polym J 18:1029
99. Chiantore O, Camino G, Costa L, Grassie N (1981) Polym Deg Stab 3:209
100. Chiantore O, Guaita M, Grassie N (1985) Polym Deg Stab 12:141
101. Cameron GG, Grassie N (1961) Polymer 2:367
102. Guaita M, Chiantore O, Costa L (1985) Polym Deg Stab 12:315
103. Schulz GV (1937) Naturwissenschaften 27:387
104. Schulz GV (1941) Z Elektrochem 47:265
105. Schulz GV (1973) Kunstoffe 33:224
106. Braun D, Becker KH (1969) Angew Makromol Chem 6:186
107. Braun D, Becker KH (1971) Makromol Chem 147:91
108. de Jong HAP, de Jong CRHI, Huysmans WGB, Sinnige HJM, de Klein WJ, Mijs WJ, Jaspers H (1972) Makromol Chem 157:279
109. Bledzki A, Braun D (1986) Makromol Chem 187:2599
110. Rudolph H, Traeckner HJ (1976) US Pat 3.931,355
111. Vio L (1974) US Pat 3.792,126
112. Tobolsky AV (1958) J Am Chem Soc 80:5927
113. Tobolsky AV, Rogers CE, Brickmann RD (1960) J Am Chem Soc 82:1277
114. Crevello JV, Lee JL, Conlon, DA (1986) J Polym Sci Polym Chem Ed 24:1251
115. Braun D, Linder HJ, Tretner H (1989) Eur Polym J 25:725
116. Bledzki A, Braun D (1986) Polymer Bulletin 16:19
117. Onishi Y, Kodaira K, Ito K (1982) Polymer 23:630
118. Cohen SG, Wang CH (1955) J Am Chem Soc 77:2457
119. Cohen SG, Groszos SJ, Sparrow DB (1950) J Am Chem Soc 72:3937
120. Cohen SG, Wang CH (1953) J Am Chem Soc 75:5504
121. Kopecky KR, Gillan T (1969) Can J Chem 47:2371
122. Pryor WA, Fiske TR (1969) Macromolecules 2:62
123. Bevington JC, Troth HG (1962) Trans Faraday Soc 58:186
124. Berger KC, Meyerhoff G (1975) Makromol Chem 176:1983
125. Berger KC (1975) Makromol Chem 176:3575
126. Athey RD (1977) J Polym Sci Polym Chem Ed 15:1517
127. Braks JG, Huang RYM (1978) J Appl Polym Sci 22:3111

128. Band BB, Tobolski AV (1952) J Polym Sci 8:529
129. Pryor WA, Coco JH (1970) Macromolecules 3:500
130. Deb PC, Gaba ID (1978) Makromol Chem 179:1559
131. Manaba T, Utsumi T, Okamura S (1962) J Polym Sci 58:121
132. Mahabadi HK, O'Driscoll KF (1977) Makromol Chem 178:2629
133. May JA Jr., Smith WB (1968) J Phys Chem 72:2993
134. Serelis AK, Solomon DH (1982) Polym Bull 7:39
135. Ayrey G, Haynes AC (1965) Makromol Chem 175:1463
136. Avrey G (1963) Chem Rev 63:645
137. Moad G, Rizzardo E, Soloman DH, Johns SR, Willing RI (1984) Makromol Chem Rapid Commun 5:793
138. Moad G, Soloman DH, Johns SR, Willing RI (1984) Macromolecules 17, 1094
139. Cardenas JN, O'Driscoll KF (1976) J Polym Sci Polym Chem Ed 14, 883
140. Sol SK, Sundberg DC (1982) J Polym Sci Polym Chem Ed 20:1345
141. Nishimura NJ (1966) Macromol Chem 1:257
142. Valiquette G, Weir NA (1972) J Chem Soc Chem Commun 1972:1071
143. Moad G, Soloman DH, Johns SR, Willing RI (1982) Macromolecules 15:1182
144. Podosenova NG, Zotikov EG, Bovunenko OP, Mel'nichenko VI (1979) J Appl Chem USSR (Engl Transl) 53:1513
145. Otsu T, Yoshida M, Tazaki T (1982) Makromol Chem Rapid Commun 3:133
146. Otsu T, Yoshida M, Kuriyama A (1982) Polym Bull 7:45
147. Moad G, Rizzardo E, Solomon DH (1982) Macromolecules 15:909
148. Moad G, Rizzardo E, Solomon DH (1982) J Macromol Sci Chem A17:51
149. Kochi JK (1973) Free Radicals. Wiley, New York 1:231
150. Waters WA (1975) Int Rev Sci Org Chem Ser 2 10:25
151. Bevington JC, Brooks CS (1956) J Polym Sci 22:257
152. Berger KC, Deb PC, Meyerhoff G (1977) Macromolecules 10:1975
153. Ruchardt C (1980) Top Curr Chem 88:1
154. Fink JR (1983) J Poly Sci Polym Chem Ed 21:1445
155. Bellus D, Slama P, Hrdkovic P, Manasek Z, Durisinova L (1969) J Polym Sci Part C, 22:269
156. Susuki H, Tanaka Y, Ishii Y (1979) Sen'i Gakkaishi 35:296
157. McKellar JF, Allen NF (1978) Photochemistry of man-made polymers. Applied Science, London
158. Moad G, Solomon DH, Willing RI (1988) Macromolecules 21:855
159. Costa L., Camino G, Guyot A, Bert M, Chiotis A (1982) Polym Degrad Stab 4:425
160. Olaj OF, Breitenbach JW, Hofreiter I (1969) Rec Chem Prog 30:87
161. Yenal'ev VD, Zaitseva VV, Sadovskii YS, Sadovskaya TN, Nazarova ZF (1965) Vysokomol Soedi. (Engl Transl) 7:275
162. Zaitseva VV, Yenal'ev VD, Yurzhenko AI (1967) Vysokomol Soedin Ser A (Engl Transl) 9:1958
163. Watanabe Y, Ishigaki H, Okada H, Suyama S (1991) Bull Chem Soc Jap 64:1231
164. Komai T, Suyama S (1985) Bull Chem Soc Jap 58:3045
165. Suyama S, Watanabe Y, Sawaki Y (1990) Bull Chem Soc Jap 63:716
166. Bailey WJ (1985) Polym J 17:85
167. Niki E, Kamiya Y (1973) J Org Chem 38:1403
168. Niki E, Kamiya Y (1975) J Chem Soc Perkin Trans 2:1221
169. Haward RN (1950) Trans Faraday Soc 46, 204
170. Overberger CG, Lapkin M (1955) J Amer Chem Soc 77:4659
171. Hahn W, Fischer A (1956) Makromol Chem 21:77
172. Grauber H, Hrubesch A (1964) Makromol Chem 72:38
173. Borsig E, Lazar M, Capla M (1973) Collect Czech Chem Commun 38:1343
174. Wolfers H, Rudolph H, Rosenkrantz HJ (1978) Ger Offen 2.632,294
175. Reuter K, Dhein R (1983) Ger Offen 3.151,444
176. Crivello JV, Lee JL, Conlon DA (1986) Polym Bull 16:95
177. Crivello JV, Lee JL, Conlon DA (1986) J Polym Sci Polym Chem Ed 24:1197
178. Crivello JV, Lee JL, Conlon DA (1986) J Polym Sci Polym Chem Ed 24:1251
179. Crivello JV (1987) US Pat 4.675,426
180. Li T, Willis TJ, Padias AB, Hall HK Jr (1991) Polymer Bull 25:537
181. Li T, Willis TJ, Padias AB, Hall HK Jr (1991) Macromolecules 24:2485
182. Schmidt AD, Ray WH (1981) Chem Eng Sci 36:1401

183. Hammer JW, Akramov TA, Ray WH (1981) Chem Eng Sci 36: 1897
184. Ito K, Aoyama T (1987) Eur Polym J 23: 955
185. Marten FL, Hamielec AE (1982) J Appl Polym Sci 27: 489
186. Bogunjoko JST, Brooks BW (1983) Makromol Chem 184: 1603
187. Bamford CH (1990) Polymer 31: 1720
188. Hsu KY, Chen SA (1984) Polym Eng Sci 24: 1253
189. Blavier L, Villermaux J (1984) Chem Eng Sci 39: 101
190. Kim KJ, Choi KY (1991) Polym Eng Sci 31: 333
191. Villalobos MA, Hamielec AE, Wood PE (1991) J Appl Sci 42: 629
192. Choi KY, Liang WR, Lei GD (1988) J Appl Polym Sci 35: 1562
193. Kim KJ, Liang W, Choi KY (1989) Ind Eng Chem Res 28: 131
194. Kim KJ, Choi KY (1989) Chem Eng Sci 44: 297
195. Choi KY, Lei GD (1987) AIChE J 33: 2067
196. Kim KJ, Choi KY (1988) Chem Eng Sci 43: 965
197. Wittmer P (1988) Makromol Chem 170: 1
198. Ivanchev SS, Zherebin YI (1974) Polym Sci USSR 16: 956
199. Vuillemenot J, Barbier B, Riess G, Banderet A (1965) J Polym Sci Part A 3: 1969
200. Bock LA, Lewis RN (1979) US Pat 4.146,583
201. Sanchez J (1991) US Pat 5.039754
202. Tsvetkov NS, Zhukovskii VY, Markovskaya RF, Prieto EM (1983) Polymer Science USSR 25: 1708
203. Woodward AE, Smets G (1955) J Polym Sci 17: 51
204. Tsvetkov NS, Markovskaya RF (1974) Vysokomol Soedin Ser A 16(9): 1936
205. Zherebin YL, Ivanchev SS, Galibei VI (1971) Zhurnal Organicheskoi Khimii 7(8): 1660
206. Komai T, Matsushima M (1979) US Pat 4.169,848
207. T. Komai T, Izumi T, Suyama S (1984) US Pat 4.469,862
208. Oshibe Y, Yamamoto T (1978) Kobunski Ronbunshu 44: 73
209. T. Komai T, Matsushima M, Nakayama M (1981) US Pat 4.283,512
210. Sugimura T, Suda I, Minoura Y (1966) Kogyo Kagaku Zasshi 69: 718
211. Korshak VV, Rogozhin SV, Makarova TA (1958) Akad Nauk SSSR Otdel Khim Nauk 1958: 1482
212. Harrison JB, Mageli OL (1964) US Pat 3.117,166
213. Galigei VI, Arkhipova-Kalenchenko EG (1977) Zhurnal Organicheskoi Khimii 13: 227
214. Ivanchev SS, Semenova VA, Matveyentseva MS, Zubareva MM, Zyat'kov IP, Emelin YD, Dovnarovich NA (1981) Zhurnal Organicheskoi Khimii 17: 524
215. Bylina GS, Semenova VA (1977) Zhurnal Organicheskoi Khimii (Engl. transl.) 13: 1842
216. Suyama S, Taura K, Ishigaki H (1990) Kobunshi Ronbunshu 47: 517
217. Ivanchev SS, Uvanova LR, Matveyentseva MS, Zyat'kov IP (1983) Polymer Science USSR 25: 2138
218. Hahn W, Fischer A (1956) Makromol Chem 21: 106
219. Dais VA, Priddy DB (unpublished data)
220. Drumwright RA, Kasl P, Priddy DB, Macromolecules (submitted for publication)
221. Goethals EJ (1989) Telechelic polymers: Synthesis and applications, CRC, Boca Raton, Florida
222. For an excellent review of the preparation of telechelomers using FR chemistry see Boutevin B (1990) Adv in Polym Sci 94: 69
223. Corner T (1984) Adv in Polym Sci 62: 95
224. Konter W, Bomer B, Kohler KH, Heitz W (1981) Makromol Chem 182: 2619
225. Kawabe M, Kimura M (1988) Polym Preprints 27(2): 312
226. Ohishi H, Inaba S, Kawabe M, Klumura M (1991) Polym Preprints 32(3): 152
227. Bamford CH, Jenkins AD, Wayne RP (1960) Trans Faraday Soc 56: 932
228. Haas HC, Schuler NW, Kolesinski HS (1967) J Polym Sci Part A1 5: 2964
229. Tong W, Priddy DB (1991) US Pat 5,034,485

Editor: J. L. Koenig
Received: July 2, 1992

Novel Bis(Naphthalic Anhydrides) and Their Polyheteroarylenes with Improved Processability

Alexandr L. Rusanov
A.N. Nesmeyanov Institute of Organo-Element Compounds,
Russian Academy of Sciences, Russia

Novel bis(naphthalic anhydrides) containing "bridging" groupings and/or bulky substituents are reviewed. Reactions of these compounds with aromatic diamines and bis(*o*-phenylene diamines) led to the formation of polynaphthylimides and polynaphthoylenebenzimidazoles combining solubility in organic solvents with high thermal, heat and chemical resistance.

Advances in Polymer Science, Vol. 111
Springer-Verlag Berlin Heidelberg 1994

1 Introduction

Rapidly developing branches of modern technology put forward increasingly higher requirements from applied polymers. In this connection thermally and chemically resistant polymers have been widely studied [1–11]. Of special interest have been aromatic polymers containing heterocycles in their backbone structures. The interest is accounted for by a complex of valuable properties, sometimes unique physicochemical and mechanical ones, displayed by polymers and related materials [1–23].

The majority of investigations into the synthesis of polyheteroarylenes have been devoted to the polymers based on bis(phthalic anhydrides). These mono- mers are widely employed in the synthesis of polyimides [24–31] and a series of completely or partially ladder-like [32–39] polyheteroarylenes, i.e. polybenzoyl- enebenzimidazoles [22, 35–46], polybenzoylene-s-triazoles [47–50], polyiso- indoloquinazolinediones [51–56], etc. This is illustrated by the following scheme.

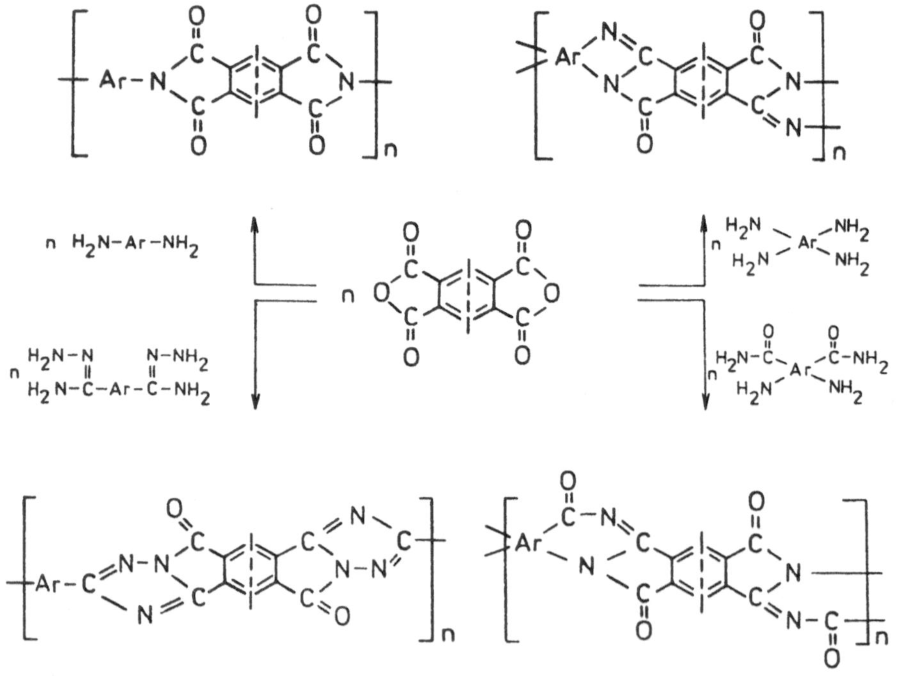

Scheme 1

Almost all polyheteroarylenes are obtained by stepwise processes which makes it possible to terminate the reaction at the following stages: formation of a prepolymer soluble in organic solvents, processing of a prepolymer in products followed by their cyclization to form the desired infusible and insoluble polyheteroarylenes as the final products.

It should be noted that both polyimides and ladder polyheteroarylenes derived from bis(phthalic anhydrides) possess relatively low thermal and chemical stability [37, 39, 57, 58], which to a great extent, is mainly attributed to the strained five-membered imide moieties presented in the formed heterocycles, large positive charges on the carbonylic carbons in the systems, comparatively low orders of the bonds C–C(O)– and N–C(O). This is illustrated by the molecular diagram of 1,2-benzoylenebenzimidazole, i.e. the simplest individual

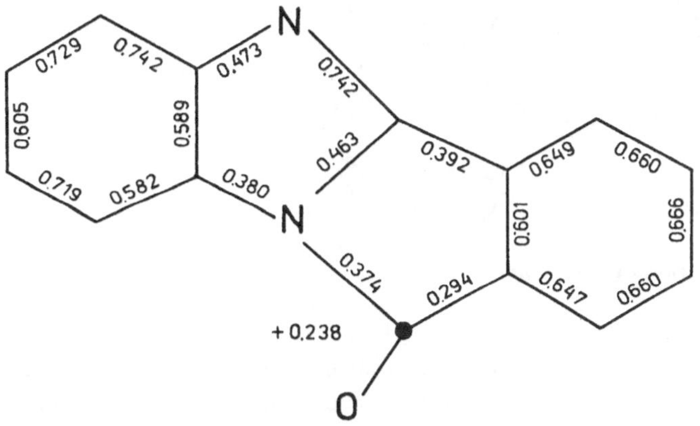

1,2-benzoylenebenzimidazole

Table 1. Results of quantum-chemical calculations of the simplest model compounds by the PPP method and experimental decomposition points

Structural formula of the compounds	Calculation data by the PPP method			Decomposition temp °C	
	Lowest bond order	The highest positive charge q_c	Resonance energy, β	Thermal decomposition	Thermohydrolytic decomposition
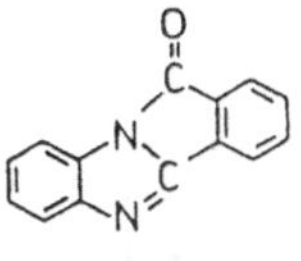	0.294	+ 0.238	5.766	425–450	350–375
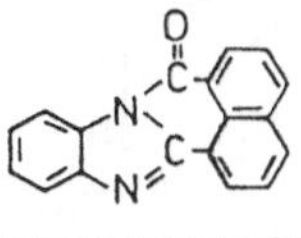	0.323	+ 0.227	5.909	500–525	475–500

compound simulating polybenzoylenebenzimidazoles; as well as by the data on thermal and thermohydrolytical decomposition of the compound (Table. 1).

A far lower strain of the cycle, positive charges on the carbonylic carbon atoms and higher orders of the corresponding bonds are characteristic of 1',8'-naphthoylene-1,2-benzimidazole, the molecular diagram of which is as follows [37, 39, 57, 58]:

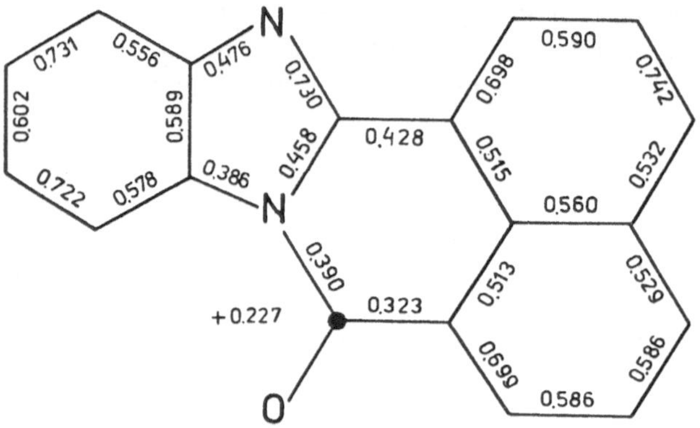

1',8'-naphthoylene-1,2-benzimidazole

Correspondingly, under the conditions of thermal and thermohydrolytical decomposition stability of this compound exceeds that of 1,2-benzoylene-benzimidazole (Table 1).

These data reveal a logical pathway to a possible enhancement in thermal and chemical stability of polyheteroarylenes, i.e. in the above mentioned reactions a replacement of bis(phthalic anhydrides) by bis(naphthalic) ones the structure of which would determine the formation of systems containing six-membered imide rings – either isolated ones or those condensed with other heterocyclic moieties.

Preliminary investigations [59–68] have shown that polynaphthylimides and polynaphthoylenebenzimidazoles, which are formed when bis(naphthalic anhydrides) are employed as monomers, display superior chemical, thermal and

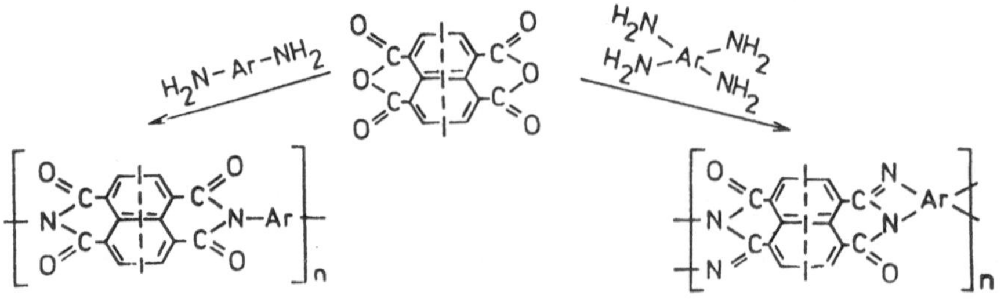

Scheme 2

fire resistance compared with the analogous systems derived from bis(phthalic anhydrides) [69–71] (Scheme 2). As a consequence, in view of design of materials possessing improved properties, polymers based on bis(naphthalic anhydrides) seem to be more perspective than the analogous polymers derived from bis(phthalic anhydrides).

But it should be noted that the majority of polyheteroarylenes based on bis(naphthalic anhydrides) are systems displaying low processability of products: The final polyheteroarylenes are infusible and soluble only in strong acids (sulfuric, methylsulfuric, polyphosphoric) [59–71] and the reactions involving bis(naphthalic anhydrides) are difficult to carry out as stepwise processes due to a reduced electrophilic reactivity of the anhydrides [39, 72–74] and the easy formation of six-membered naphthalimide systems.

Therefore, the design of polyheteroarylenes based on bis(naphthalic anhydrides), to give satisfactory processability of products, is an important and urgent problem in the chemistry of heat resistant polymers. One of the most widely employed pathways of improving processability of polyheteroarylenes relies on improving solubility in organic solvents. The following ways are used:

– synthesis of polyheteroarylenes derived from the most widely used bis(naphthalic anhydrides), i.e. naphthalene-1,4,5,8-tetracarboxylic and perylene-3,4,9,10-tetracarboxylic acid anhydrides

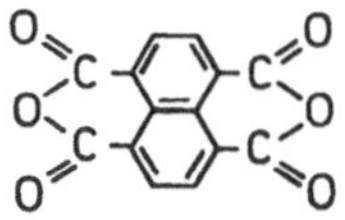

naphthalene-1,4,5,8-tetracarboxylic   perylene-3,4,9,10-tetracarboxylic
acid anhydrides   acid anhydrides

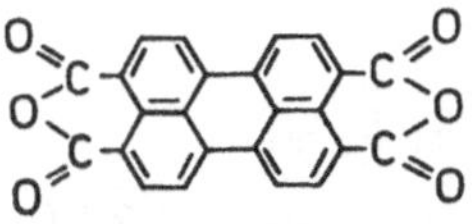

and their nucleophilic comonomers differing by a specific structure, which determines an improved processability of the desired polyheteroarylenes;
– design of bis(naphthalic anhydrides) containing structural fragments which determine an improved processability of related polyheteroarylenes, and employment of bis(naphthalic anhydrides) in reactions with di- and tetra-functional nucleophilic comonomers.

The first approach was discussed earlier [26, 75, 76]; therefore the present review is an attempt to analyse the results obtained in the second direction.

According to Refs. [77, 78], the major approach to imparting to poly-heteroarylenes solubility in organic solvents relies on modification of the starting dianhydrides:

– introduction of the "hinge" bonds into molecules, which facilitate rotation of some fragments in the macromolecules relative to the other ones;
– introduction of bulky substituents into the dianhydride molecules.

The above mentioned approaches and their combinations have been employed in the design of novel bis(naphthalic anhydrides) intended for the preparation of polyheteroarylenes soluble in organic solvents.

2 Bis(Naphthalic Anydrides) with Structural Fragments that Improve Solubility of Polyheteroarylenes

The simplest modification of the most widely used bis(naphthalic anhydrides) to improve processability of polyheteroarylenes is the introduction of single bonds or "hinge" groups between two fragments of naphthalic anhydride. Such bis(naphthalic anhydrides) [76] are mainly prepared from acenaphthene [79] and its simplest derivatives by the following approaches:

– reaction between two molecules of acenaphthene or its derivatives and the corresponding compounds to produce bis-acenaphthyls followed by oxidation [80] and dehydration of the bis(naphthalic acids) obtained:

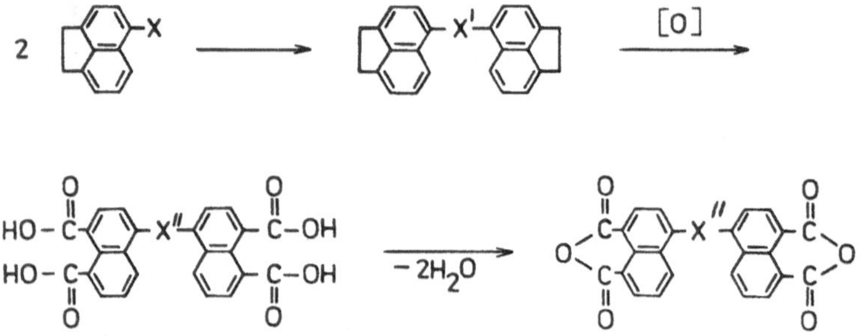

Scheme 3

– reaction between two molecules of acenaphthoquinone (the product of partial oxidation of acenaphthene) [79] and the corresponding compounds, followed by oxidation of the formed bis-acenaphthoquinones to bis(naphthalic anhydrides):

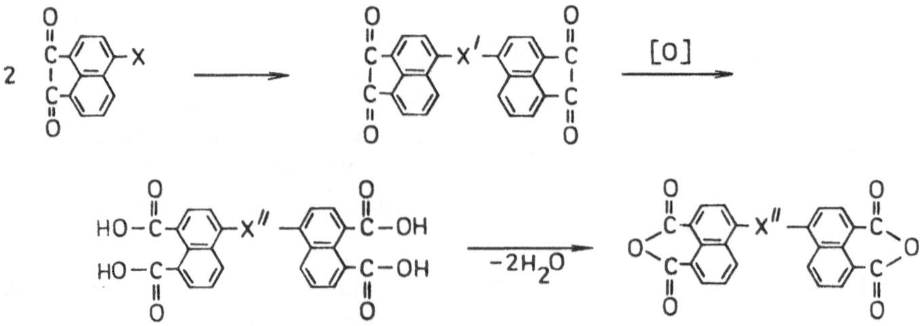

Scheme 4

– reaction between two molecules of derivatives of naphthalic anhydride (a
derivative of acenaphthene [79]) and the corresponding compounds to obtain
bis(naphthalic anhydrides):

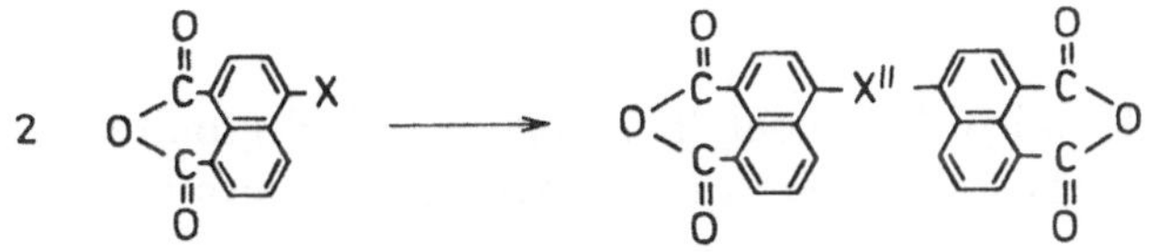

Scheme 5

– reaction between derivatives of naphthalic anhydride and acenaphthene or its
derivatives to get compounds containing both acenaphthene and anhydride
functions, followed by oxidation of the acenaphthyl group and dehydration of
the obtained acidic function:

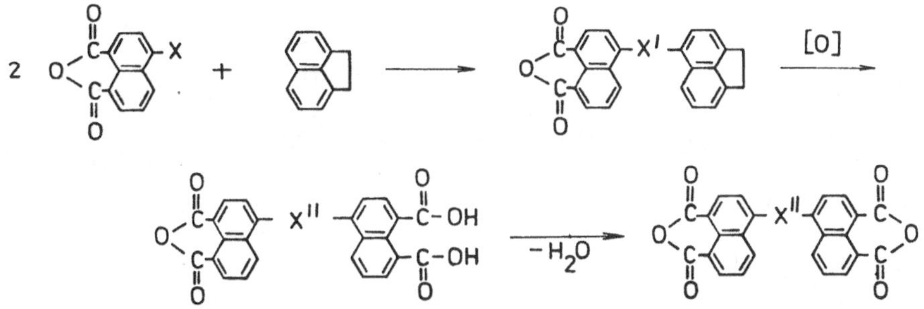

Scheme 6

The simplest representative of such monomers is 4,4'-bis(naphthalic anhydride) which has been prepared from 4-bromoacenaphthene according to the
following scheme [81]:

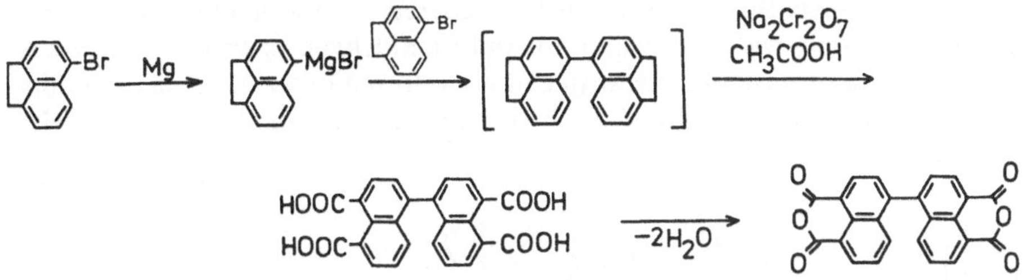

Scheme 7

This dianhydride can also be prepared by the Ulman reaction, i.e. condensation of 4-bromonaphthalic anhydride with copper powder in an inert
solvent [82]:

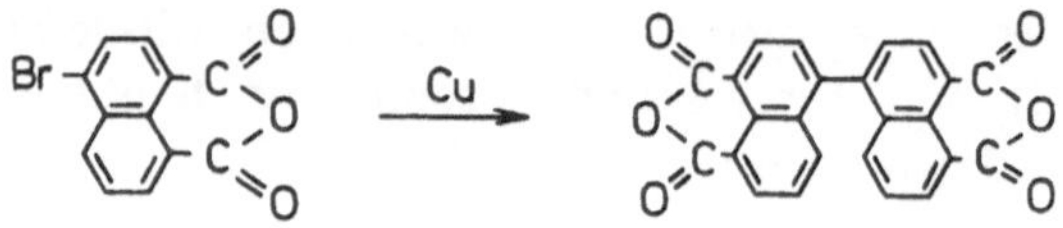

Scheme 8

A series of bis(naphthalic anhydrides) containing one "bridging" group between two fragments of naphthalic anhydride was prepared by the above mentioned methods.

4,4'-Diacenaphthylmethane and 4,4'-diacenaphthylketone were prepared via condensation of acenaphthene with methylal or phosgene in a CS_2 medium [83], respectively. Both compounds are oxidized by sodium bichromate in acetic acid to give keto-4,4'-dinaphthalic acid [83]:

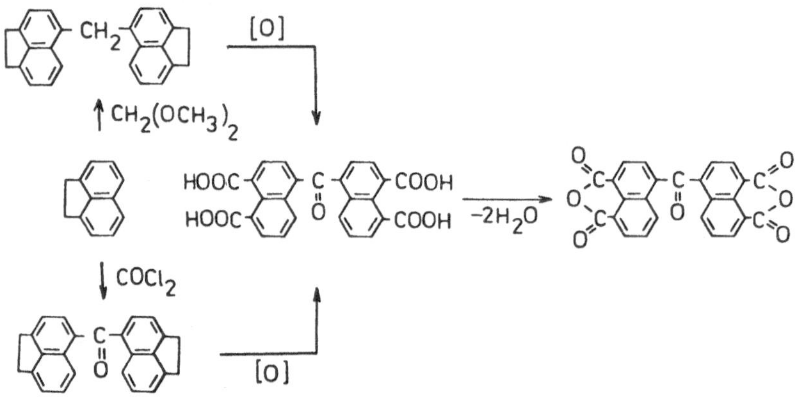

Scheme 9

Significant shortcomings of such a technique for the preparation of 4,4'-keto-bis(naphthalic anhydride) are low yields of bis-acenaphthyls (4,4'-diacenaphthylmethane-15 %; 4,4'-diacenaphthyl ketone-5 %) and work with phosgene and carbon disulfide.

4,4'-Diacenaphthylmethane can be prepared via the interaction of acenaphthene not only with methylal but with formaldehyde as well [84], or via the condensation of acenaphthene with chloral followed by treating diacenaphthyl-trichloroethane formed with alkali hydroxides solutions in aqueous diethylene glycol [85]:

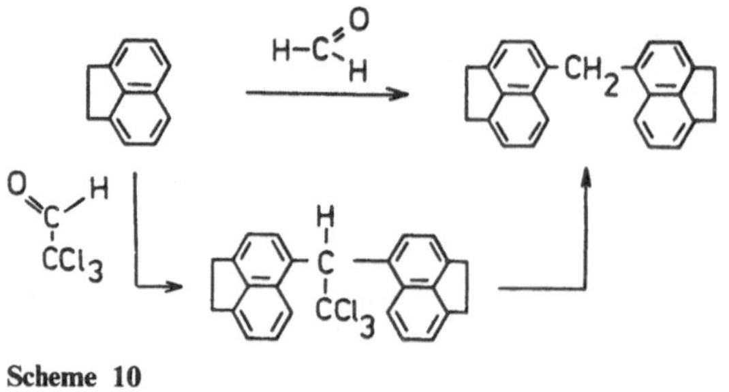

Scheme 10

A series of papers deal with the synthesis of 4,4'-oxy-bis(naphthalic anhydride). It was prepared from 4,4'-diacenaphthyl oxide and simple derivatives of naphthalic anhydride.

In particular 4,4'-diacenaphthyl oxide was prepared by fusing 4-chloroacenaphthene and acenaphthene-4-sulfonic acid with an alkali [86, 87] or from 4-oxyacenaphthene according to the following scheme [88, 89]:

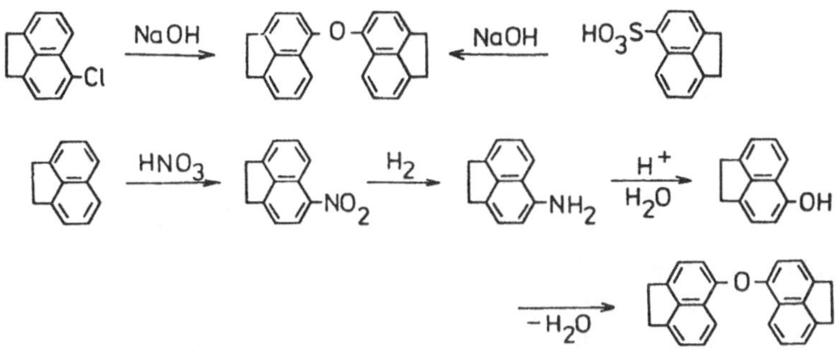

Scheme 11

Oxidation of the product led to 4,4'-oxy-bis(naphthalic anhydride). On the other hand, 4,4'-oxy-bis(naphthalic anhydride)

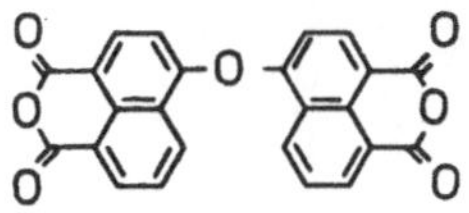

was obtained [90] by heating 4-oxynaphthalic anhydride with CuO in nitrobenzene;

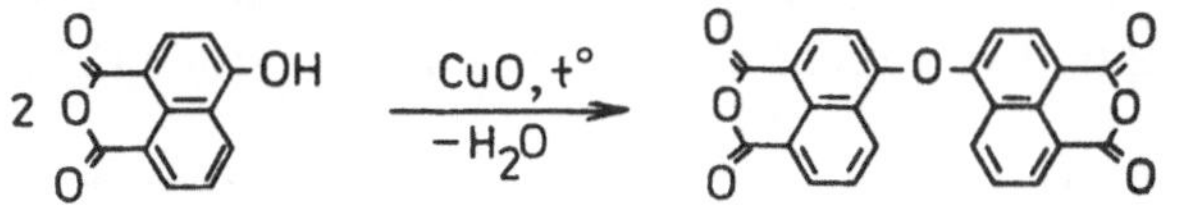

Scheme 12

as well as via alkali fusion of 4-chloronaphthalic anhydride under pressure [87, 91]:

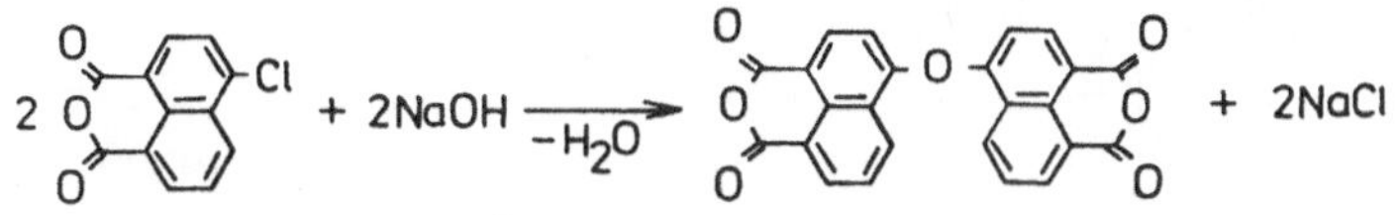

Scheme 13

Similar reactions were also employed for preparation of sulfide- and sulfone-containing bis(naphthalic anhydrides). Thus, the interaction of 4-bromonaphthalic anhydride with sodium sulfide yields 4,4'-sulfide-bis(naphthalic anhydride) [92, 93]:

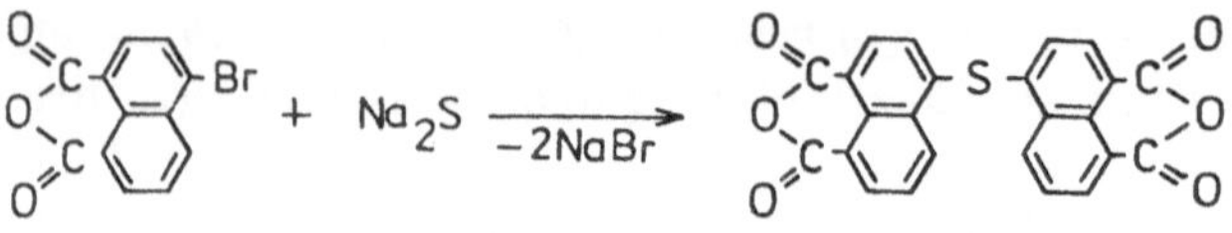

Scheme 14

which was further oxidised to 4,4'-sulfone-bis(naphthalic anhydride) [92, 93]:

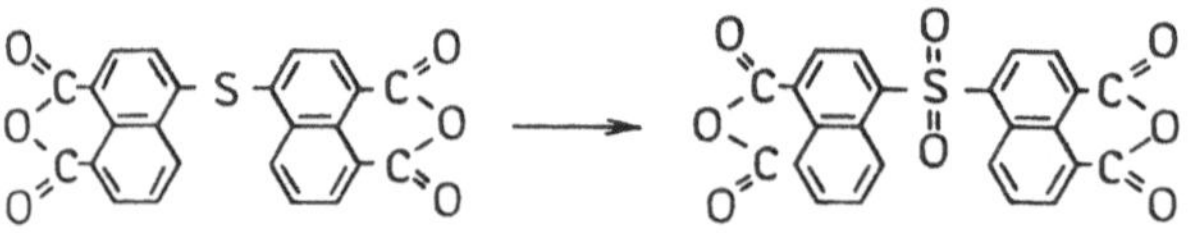

Scheme 15

This bis(naphthalic anhydride) was prepared from acenaphthene derivative-4,4'-diacenaphthyl sulfone, which was synthesized by the interaction of acenaphthene with a mixture of dimethylsulfate and sulfuric anhydride [94] or chlorosulfonic acid [95]. Sulfone-bis-(4,5-dicarboxynaphthyl-1) dianhydride results from oxidation of bis-acenaphthyl with potassium permanganate in pyridine [94]:

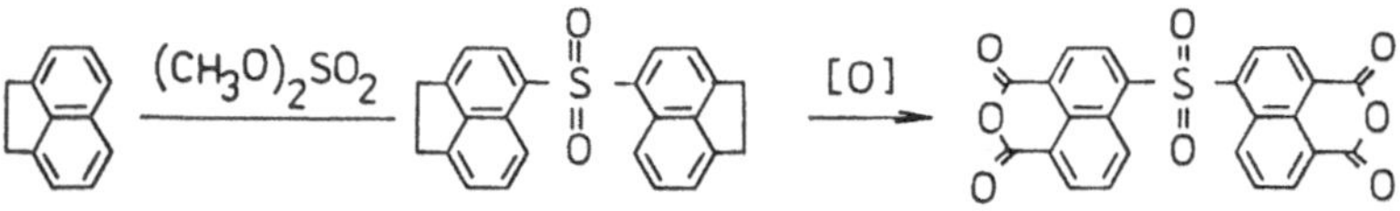

Scheme 16

Its isomer 2,2'-sulfone-bis(naphthalic anhydride) resulted from sulfonation of acenaphthene with chlorosulfonic acid at 125–130 °C and subsequent oxidation of the obtained bis-(acenaphthyl)-sulfone with chromium anhydride [95]:

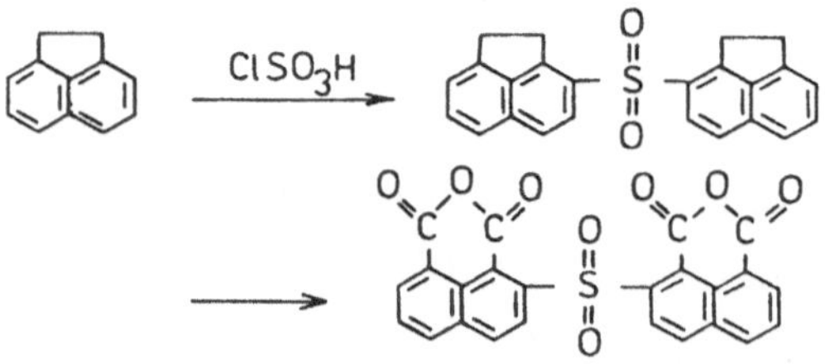

Scheme 17

Some characteristics of simple bis(naphthalic anhydrides) are presented in Table 2.

In a series of bis(naphthalic anhydrides) containing more than one "hinge" grouping, special attention has been drawn to the systems having not less than two ether bonds. These monomers were obtained by the reaction between two-

Table 2. Bis(naphthalic anhydrides) of general formula:

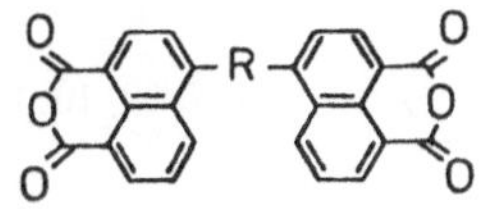

–R–	M.p., °C	Method of preparation	Yield, %	Ref.
—	362	Oxidation of 4,4'-binaphthyl	71	81
>C=O	348	1. Oxidation of 4,4'-diacenaphthylmethane	—	83
		2. Oxidation of 4,4'-diacenaphthylketone	15	
–O–	356	1. Heating of 4-hydroxy (naphthalic anhydride) with CuO	—	90
		2. Alkali fusion of 4-chloro (naphthalic anhydride).	—	91
		3. Oxidation of 4,4'-diacenaphthyl oxide	—	88
–SO$_2$–	355–357	Oxidation of 4,4'-diacenaphthyl sulfone	95	94

fold mole amounts of 4-bromonaphthalic anhydride and alkali metals bis-phenolates in bipolar aprotic solvents according to Scheme 18 [92, 96–102].

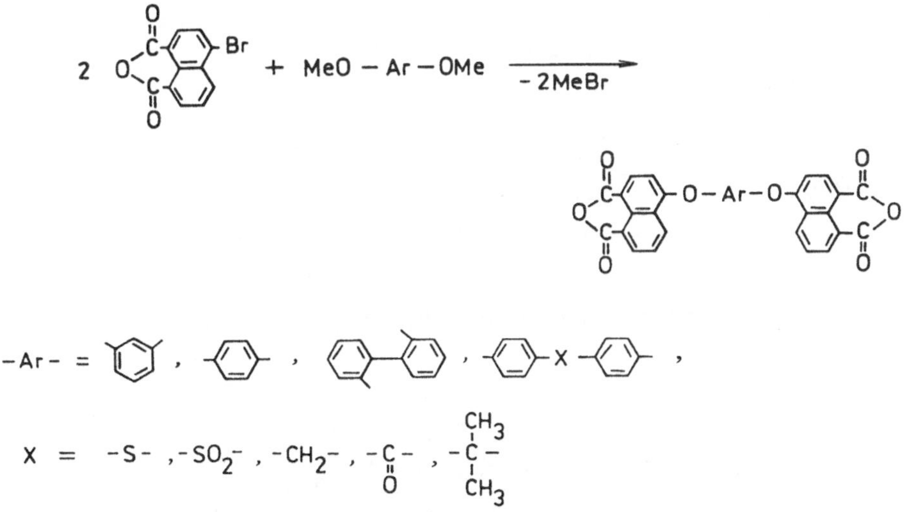

Scheme 18

Some characteristics of several bis(ethernaphthalic anhydrides) are presented in Table 3.

An alternative method for synthesis of similar bis(ethernaphthalic anhydri-des) is the reaction involving 4-nitronaphthalic anhydride instead of 4-bromo-

Table 3. Bis(naphthalic anhydrides) of general formula:

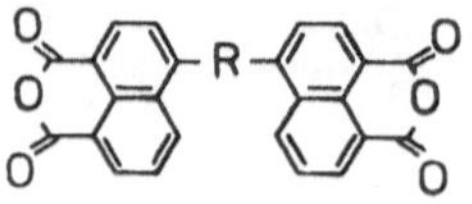

[97, 99, 105]

-R-	M.p., °C	Yield, %	Ref.
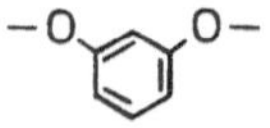	271–2	81.6	97
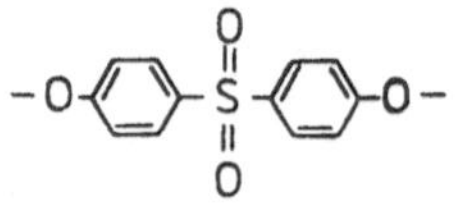	> 300 370–372	93.0 80.0	97 105
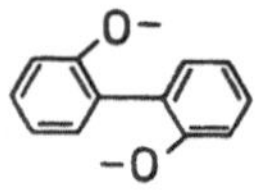	287.5–287.6	95.0	97
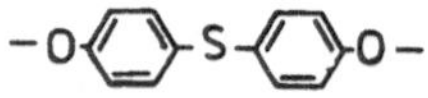	274.4–275.5	96.7	97
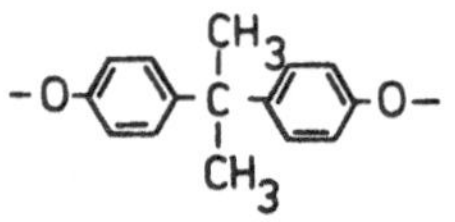	294–295	70.0	99
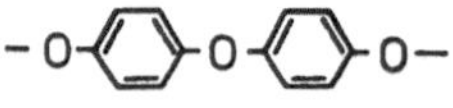	314–315	79.0	105

naphthalic anhydride, Scheme 19 [103], but the process has only been studied rather superficially.

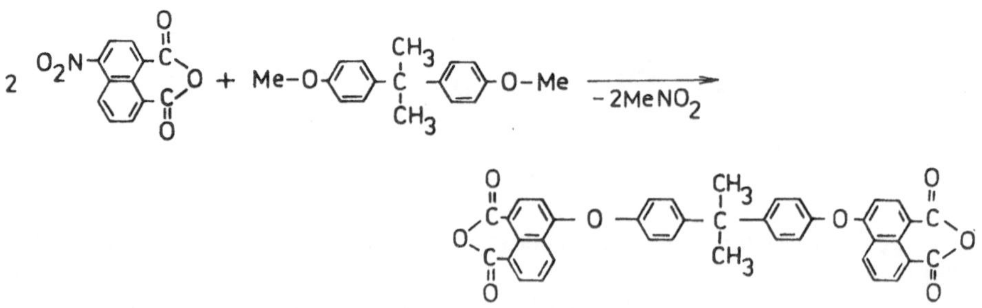

Scheme 19

Both 4-bromo- and 4-nitroderivatives of acenaphthoquinone were also used as coreagents of alkali metals bisphenolates [100, 104–106]. In these cases, bis-acenaphthoquinones result from the first stages of the processes:

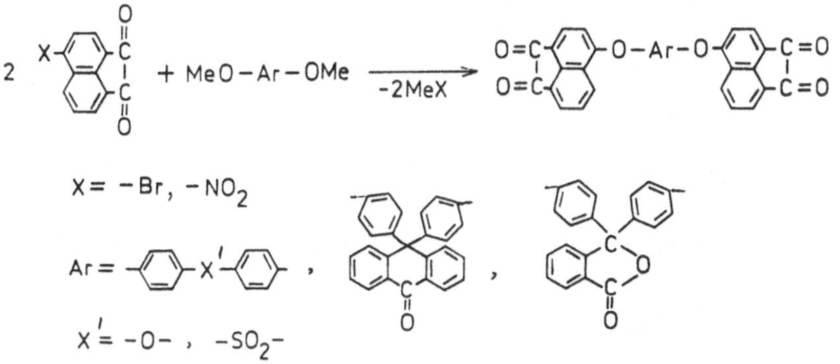

$X = -Br, -NO_2$

$Ar = $

$X' = -O-, -SO_2-$

Scheme 20

Several representatives of the synthesized bis-acenaphthoquinones were further oxidized to bis(naphthalic acids), which were transformed into bis(naphthalic anhydrides) [105, 106]:

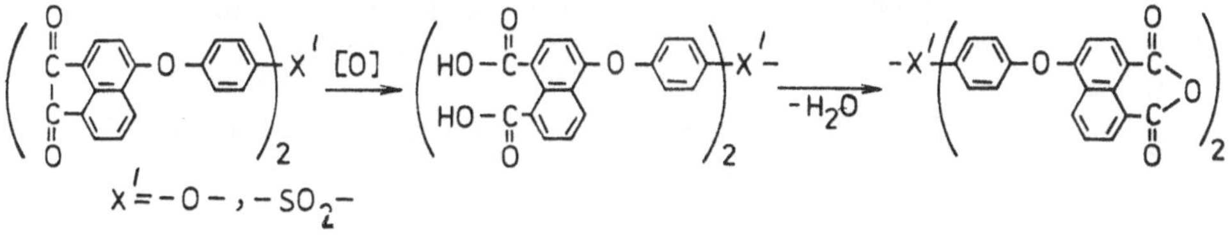

$x' = -O-, -SO_2-$

Scheme 21

Some characteristics of these dianhydrides are given in Table 3.

Practically all the above considered reactions for the synthesis of bis(ethernaphthalic anhydrides) were employed for preparing bis-(thio-ethernaphthalic anhydrides) [92, 93, 107–110]; bis-thiophenolates were used instead of alkali metals bis-phenolates:

$Ar = $

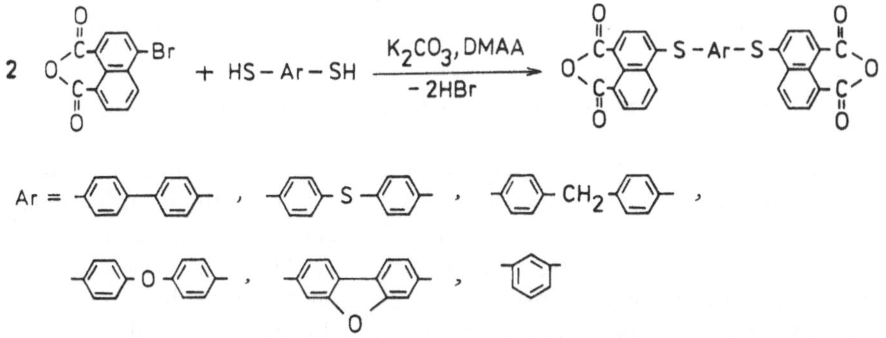

Scheme 22

Table 4. Sulfur-containing bis(naphthalic anhydrides) of general formula:

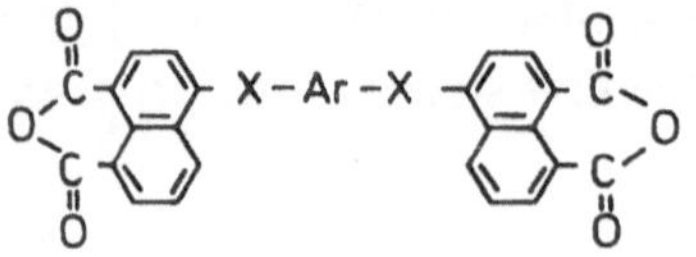

[92, 97, 107, 108]

-X-	-Ar-	Yield, %	M.p., °C	Ref.
- S -	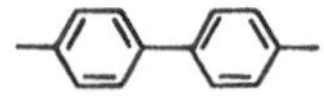	70	370–372	107, 108
— " —	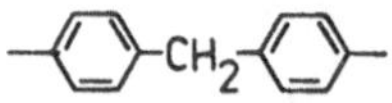	82	294–296	107, 108
— " —	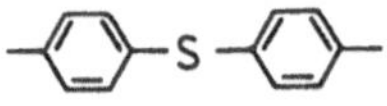	56	250–251	92, 107, 108
— " —	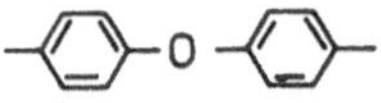	80	289–291	107, 108
— " —	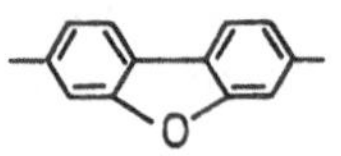	73	372–374	107, 108
— " —		75	222.5–223	97
$-\overset{O}{\underset{O}{\overset{\|}{\underset{\|}{S}}}}-$		92	400–401	107, 108
— " —	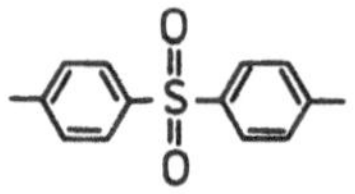	85	395–396	107, 108
— " —	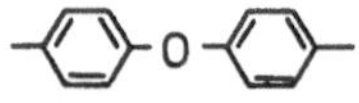	87	358–362	107, 108
— " —	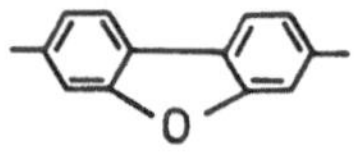	91	341–342	107, 108

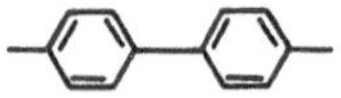

Some characteristics of bis(thioethernaphthalic anhydrides) are given in Table 4.

As mentioned in Sect. 1, bis(naphthalic anhydrides) are characterized by lower electrophilic reactivity compared to that of bis(phthalic anhydrides); introduction of ether and thioether moieties, i.e. electron-donating substituents, leads to a further decrease in electrophilic reactivity. These circumstances have predetermined an expedient transition from electron-donating "hinge" moieties to electron-accepting ones, in particular from dianhydrides with thioether groups to dianhydrides with sulfone fragments. This is carried out according to

Scheme 23 [107, 108]. Some characteristics of a series of bis(sulfone-naphthalic anhydrides) are set out in Table 4.

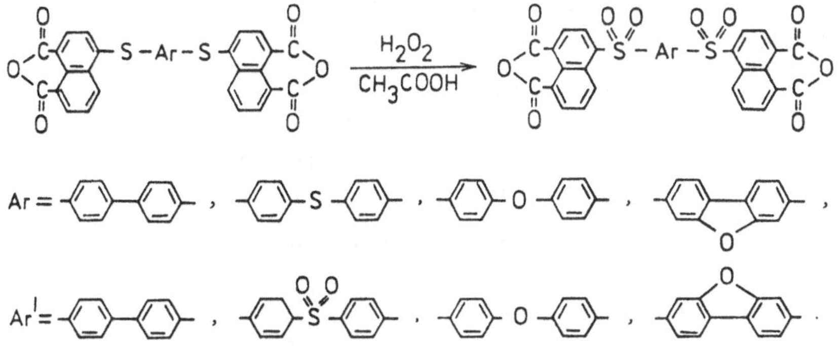

$$Ar = \text{...}, \text{...}, \text{...}, \text{...},$$

$$Ar^{I} = \text{...}, \text{...}, \text{...}, \text{...}.$$

Scheme 23

Another important group of bis(naphthalic anhydrides) – the so called bis(ketonaphthalic anhydrides) – was formed by the Friedel–Crafts reaction between aromatic dicarboxylic acids dichlorides and two-fold molar amounts of acenaphthene and subsequent oxidation of the obtained bis-acenaphthyls to bis(ketonaphthalic anhydrides) [111–115] according to Scheme 24.

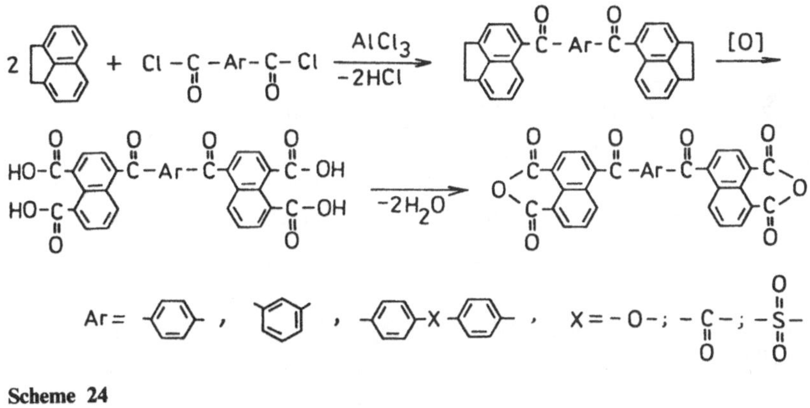

$$Ar = \text{...}, \text{...}, \text{...}, \quad X = -O-; -\overset{O}{\underset{O}{C}}-; -\overset{O}{\underset{O}{S}}-$$

Scheme 24

Some characteristics of the synthesized bis(ketonaphthalic anhydrides) are given in Table 5.

This approach was employed for preparing bis(ketonaphthalic anhydrides) having organoelement central moieties, i.e., hexafluoroisopropylidene, diphenyl-silyl and m-carboranylene [116–118]. Some characteristics of the synthesized dianhydrides are presented in Table 5.

An alternative method for synthesizing the above mentioned bis-(ketonaphthalic anhydrides) and isomeric systems relies on the Friedel–Crafts

Table 5. Bis(naphthalic anhydrides) of general formula:

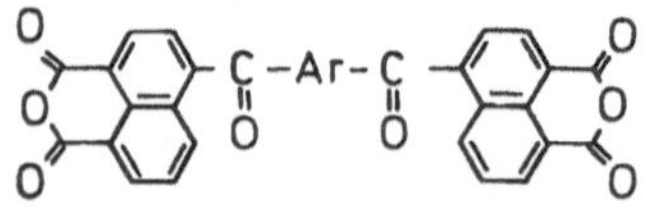

[111, 112, 114, 116–118]

–Ar–	M.p., °C	Yield, %	Ref.
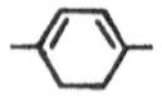	> 450	70.8	111, 112, 114
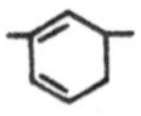	434–436	90.2	111, 114
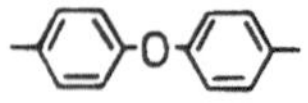	309–311	93.0	111, 114
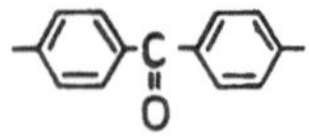	313–314	94.0	80, 111, 114
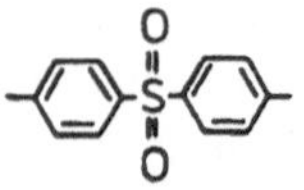	341–342.5	89.1	80, 111, 114
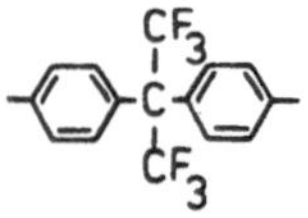	288–290	96.0	84, 111
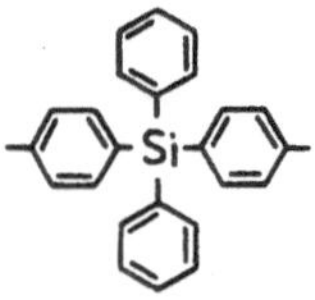	308–309	57.0	116, 118
$-CB_{10}H_{10}C-$	278–280	35.0	117, 118

reaction of 4-chloroformyl(naphthalic anhydride) with aromatic hydrocarbons [119]:

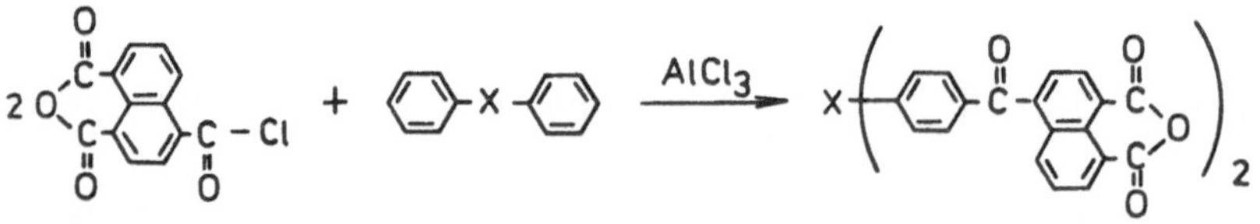

Scheme 25

4-Chloroformyl(naphthalic anhydride), a derivative of naphthalene-1,4,5-tricarboxylic acid, and the acid itself were involved in preparing bis(naphthalic anhydrides) containing ester and amide groups (Scheme 26):

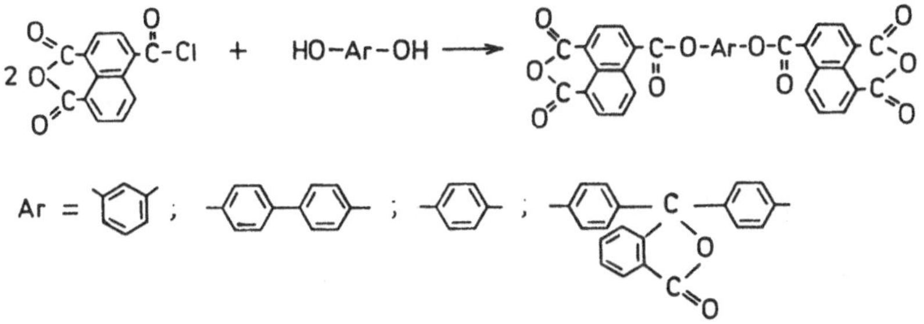

Scheme 26

Thus, condensation of aromatic bis-phenols with two-fold molar amounts of 4-chloroformyl(naphthalic anhydride) in organic solvents in the presence of bases yielded bis(esteranaphthalic anhydrides), Scheme 27 [120].

To obtain a dianhydride containing a diphenyl fragment, the reaction in a melt of naphthalene-1,4,5-tricarboxylic acid with 4,4'-diacetoxydiphenyl was employed [121]. Some characteristics of a series of bis(naphthalic anhydrides) with ester moieties are given in Table 6.

Table 6. Bis(esternaphthalic anhydrides) of general formula:

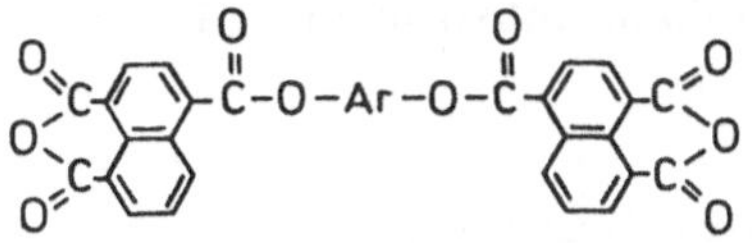

[120, 121]

−Ar−	M.p., °C	Method for preparation	Yield, %	Ref.
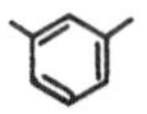	—	Condensation of 4-chloroformyl (naphthalic anhydride) with resorcine	95	120
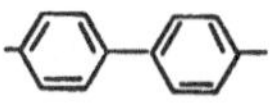	350	1. Condensation of 4-chloroformyl (naphthalic anhydride) with 4,4'-dioxybiphenyl.		120
		2. Condensation of naphthalene-1,4,5-tricarboxylic acid with 4,4'-diacetoxydiphenyl	79.7	121
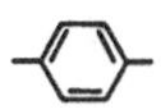	—	Condensation of 4-chloroformyl (naphthalic anhydride) with hydro-quinone	—	120
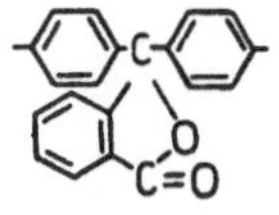	—	Condensation of 4-chloroformyl (naphthalic anhydride) with phenolphthaleine	—	120

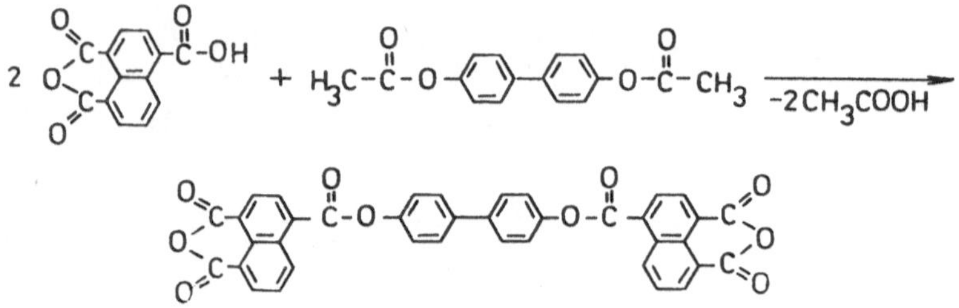

Scheme 27

Melt condensation of 1,4,5-naphthalenetricarboxylic acid and aromatic diacetamides yielded [122] dianhydrides containing amide groups, Scheme 28.

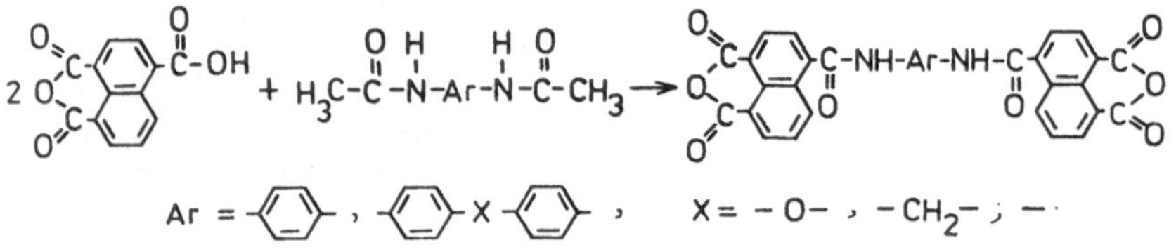

$$Ar = \text{⟨⟩} , \text{⟨⟩–X–⟨⟩} , \qquad X = -O- , -CH_2- ; -$$

Scheme 28

Some characteristics of bis(naphthalic anhydrides) containing amide groups are presented in Table 7.

Table 7. Bis(amidenaphthalic anhydrides) of general formula:

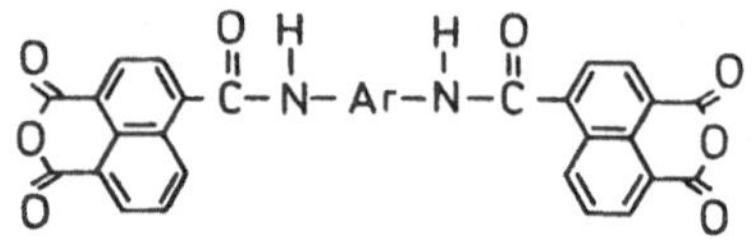

–Ar–	M.p., °C	Yield, %	Ref.
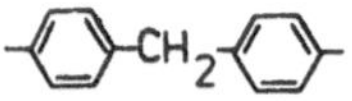 ⟨⟩–CH₂–⟨⟩	320	76.5	122
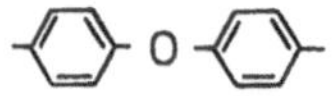 ⟨⟩–O–⟨⟩	335	79.3	122
 ⟨⟩	339–342	80	122
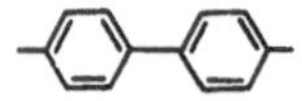 ⟨⟩–⟨⟩	> 350	96.3	122

It should be noted, however, that the introduction of ester and amide groups into bis(naphthalic anhydrides) and thus into the backbone structures of the desired polymers reduces one of the primary advantages which relies on the displacement of bis(naphthalic anhydrides) for bis(naphthalic anhydrides), i.e. an enhancement in chemical stability of the polymeric systems.

In this connection, employment of 4-chloroformyl(naphthalic anhydride) in synthesis of bis(naphthalic anhydrides) containing various aromatic hetero-cycles seems to be more preferable. In particular, bis(naphthalic anhydrides) with 1,3,4-oxadiazole and benzoxazole rings were synthesized within the frame-work of investigation [123], Scheme 29.

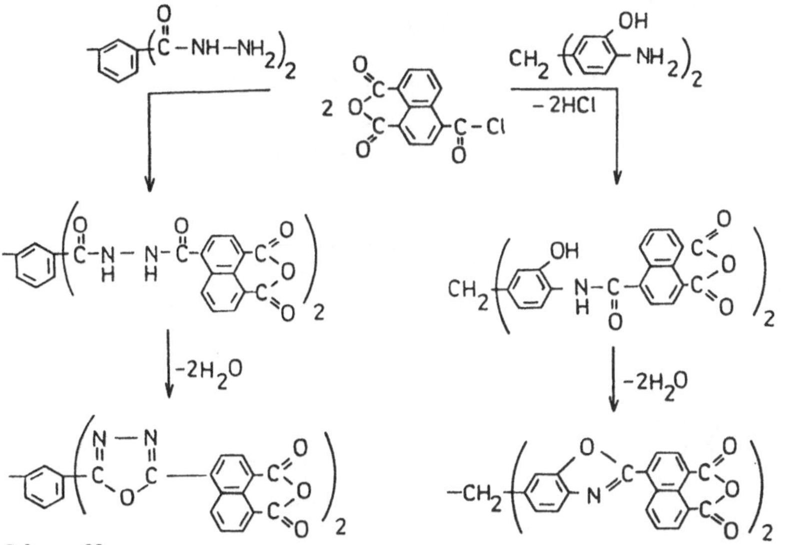

Scheme 29

Some characteristics of the synthesized dianhydrides are presented in Table 8.

Table 8. Heterylcontaining bis(naphthalic anhydrides) of general formula:

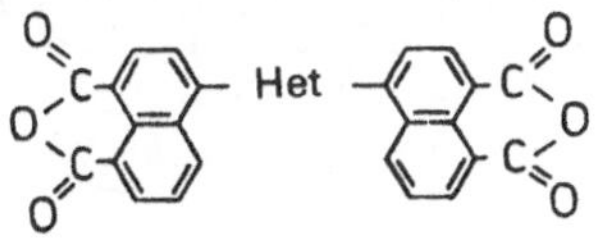

−Het−	M.p., °C	Yield, %	Ref.
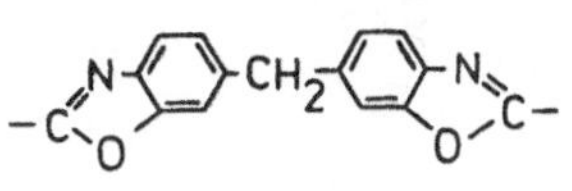	414–416	82	123
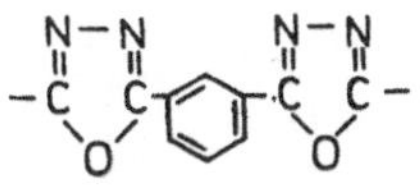	432–434	85	123

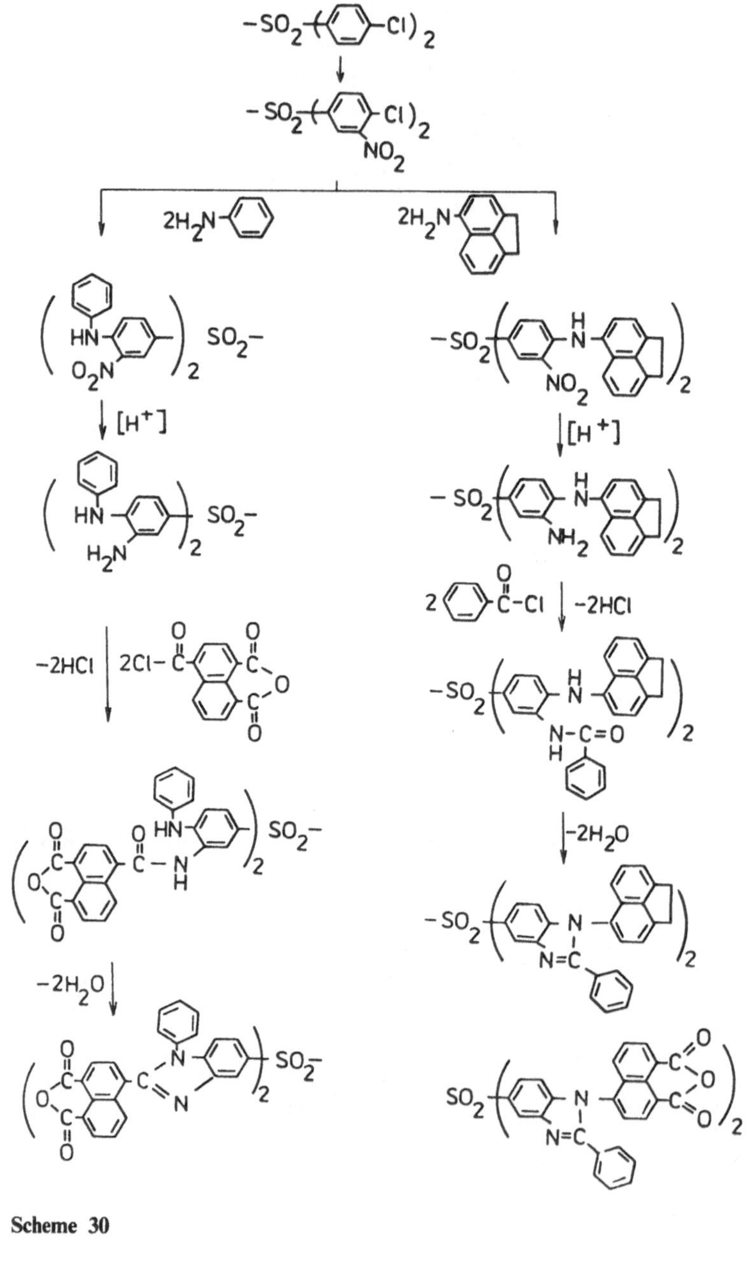

Scheme 30

Involvement of 4-chloroformyl(naphthalic anhydride) in preparing monomers with phenyl-substituted heterocyclic fragments proves to be correct. Thus, bis[1-phenyl-2(4,5-dicarboxynaphthyl) benzimidazole-5-yl]-sulfone dianhydride was obtained by treating bis [(3-amino-4-phenylamino)-phenyl]sulfone – a product based on commercial 4,4'-dichlorodiphenyl-sulfone with a two-fold molar amount of 4-chloroformyl(naphthalic anhydride) and subsequent cyclodehydration of the compound prepared [124, 125].

Its isomer bis[2-phenyl-1(4,5-dicarboxynaphthyl)benzimidazole-5-yl]-sulfone dianhydride – was prepared by treating 3,3'-dinitro-4,4'-dichlorodiphenyl sulfone with 4-aminoacenaphthene, reduction of nitro groups in 3,3'-dinitro-4,4'-di(aminoacenaphthyl) diphenyl sulfone to primary aminogroups, benzoylation of the amino groups, cyclodehydration of o-benzamidoimine fragments into 2-phenyl-1-acenaphthylbenzimidazole cycles and oxidation of acenaphthyl groups to pery-carboxylic ones and anhydridization of the latter [125], Scheme 30.

Several characteristics of the synthesized bis(naphthalic anhydrides) containing different heterocycles are presented in Table 9.

Table 9. Bis(phenylbenzimidazole-naphthalic anhydrides) of general formula:

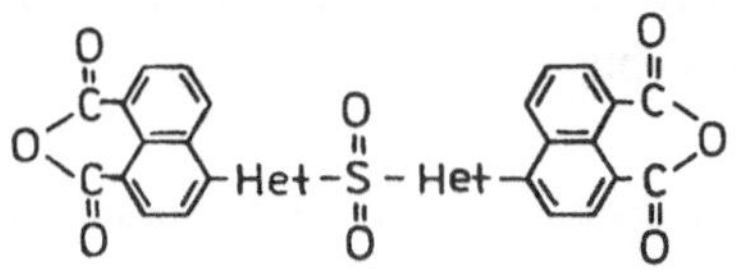

–Het–	M.p., °C	Methol for preparation	Yield, %	Ref.
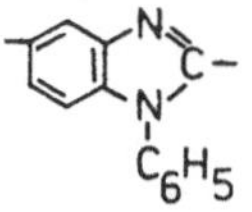	377–379	From 4-chloroformyl (naphthalic anhydride) and 3,3'-diamino-4,4' diphenylamino (diphenyl sulfone)	40	124
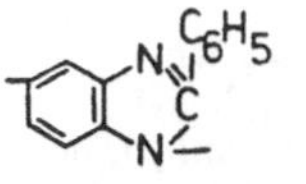	355–356	From 3,3'-dinitro-4,4'-dichloro-diphenyl sulfone and 4-amino-acenaphthene.	42	125

Derivatives of acenaphthene and naphthalic anhydride containing α-diketone moieties in the position 4, were applied for preparing bis(naphthalic anhydrides) having phenylquinoxaline fragments [126–129]. Synthesis of these monomers was carried out according to Scheme 31.

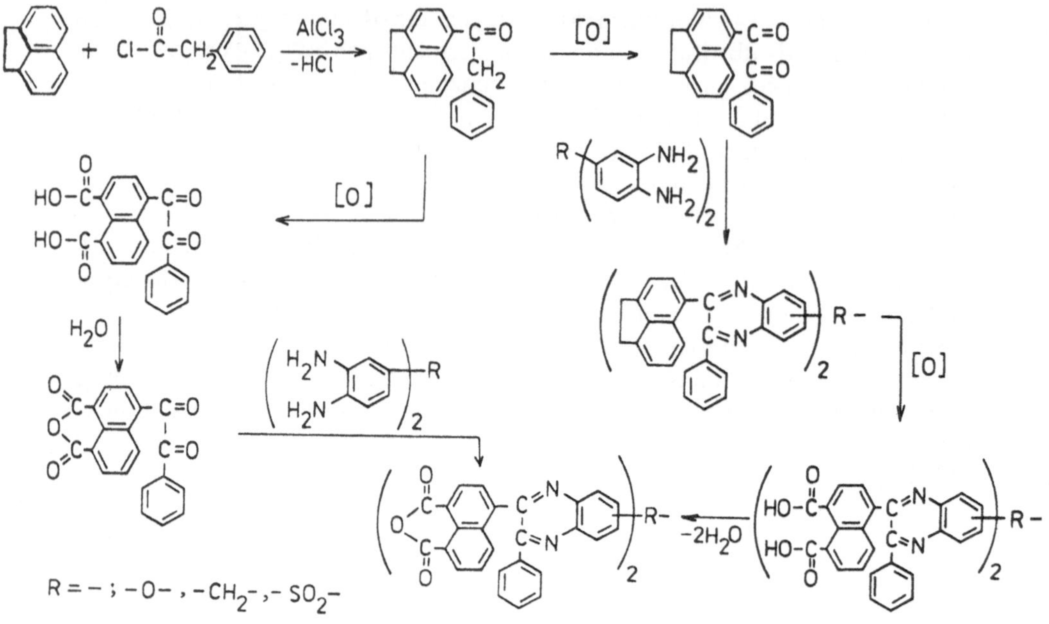

Scheme 31

Several characteristics of the synthesized bis(phenylquinoxalylnaphthalic anhydrides) are tabulated in Table 10.

Table 10. Yields and melting intervals for bis(naphthalic anhydrides) of general formula:

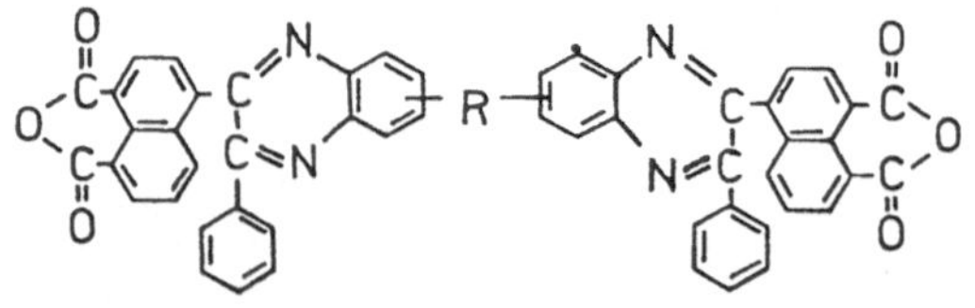

–R–	melt., °C	Yield, %	
		Oxidation of bis-acenaphthyls	Interaction of bis-(o-phenylenediamine) with 4-(phenyl glyoxalyl) naphthalic anhydride
—	380–393	41	77
–O–	315–335	42	68
–CH$_2$–	304–320	51	71
–SO$_2$–	352–364	37	73

Broad melting points of bis(phenylquinoxalylnaphthalic anhydrides) follow from the fact that these products are mixtures of isomers. This is typical for the formation of bis- and polyphenylquinoxaline systems [130–132].

Along with phenyl substituents providing an enhancement in solubility of the desired polyheteroarylenes [133], acetyl- and carboxy- ones were also introduced into bis(naphthalic anhydrides).

Diacetyl and dicarboxy derivatives of 4,4'-bis(naphthalic anhydride) were obtained by acetylation of acenaphthene, nitration of 4-acetylacenaphthene [134] and oxidation of nitroacetylacenaphthene [135]

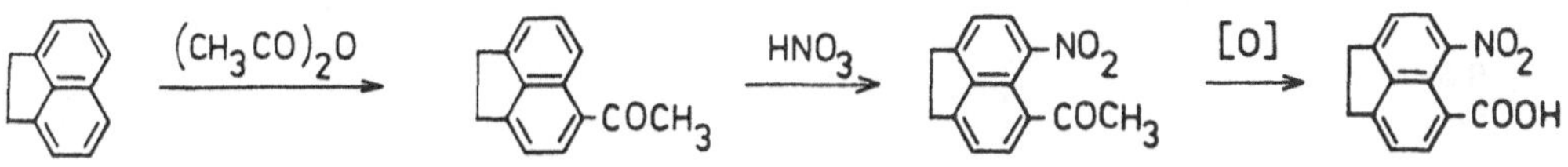

Scheme 32

followed by formation and dimerization of the diazo compound [136–138].

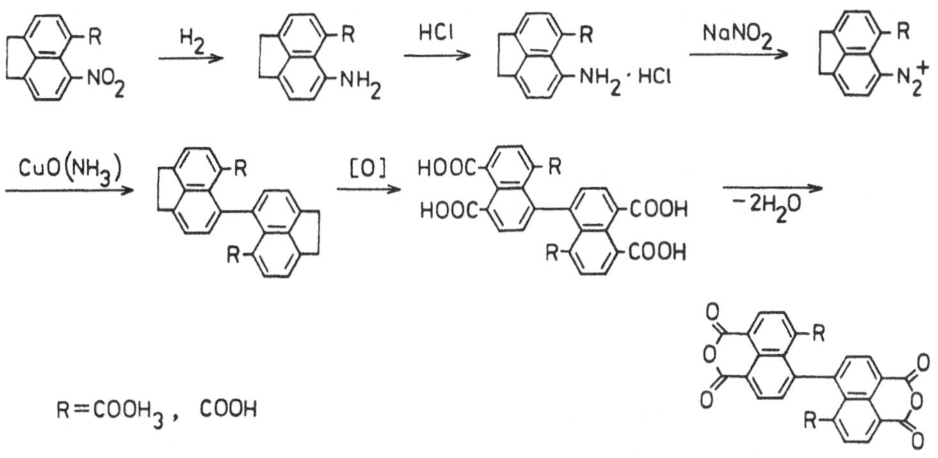

$R = COOH_3$, $COOH$

Scheme 33

1,1'-Bis-naphthyl-4,4',5,5',8,8'-hexacarboxylic acid dianhydride is currently produced on a semi-commercial scale; it is used in the production of cubogenes [139, 140].

It is a well known fact that introduction of 1,1-dihalogenethylene groups macromolecules of different polyheteroarylenes enhances solubility of the latter [141–146]. Therefore, the synthesis of bis(naphthalic anhydrides) containing these fragments was carried out.

On the basis of the chemistry of chloral derivatives, one could suppose that transition from diacenaphthyltrichloroethane to the desired bis(naphthalic anhydride) can be carried out through oxidation of 1,1,1-trichloro-2,2-bis(acenaphth-4-yl), subsequent anhydridization of pery-dicarboxylic groups, and dehydrochlorination of 1,1,1-trichloroethane moieties:

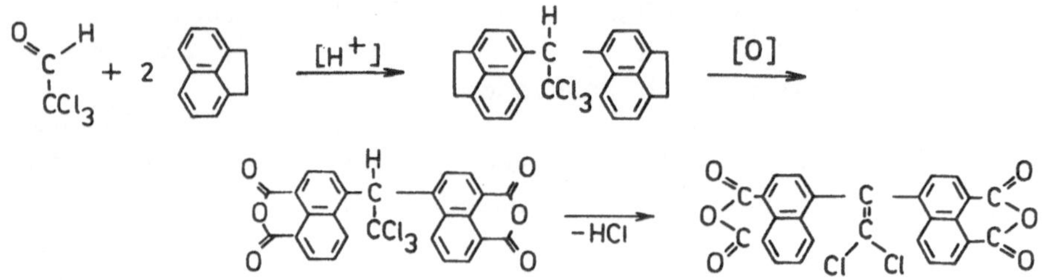

Scheme 34

Contrary to expectations, studies [111, 114, 148–150] have shown that oxidation of diacenaphthyltrichloroethane with sodium or potassium bichromate in acetic acid (glacial) at 90–110 °C was accompanied by dehydrochlorination of the central groups and simultaneous anhydridization of perydicarboxylic groups, Scheme 35. An attempt to dehydrochlorinate 1,1,1-trichloro-2,2-bis(acenaphth-4-yl) failed.

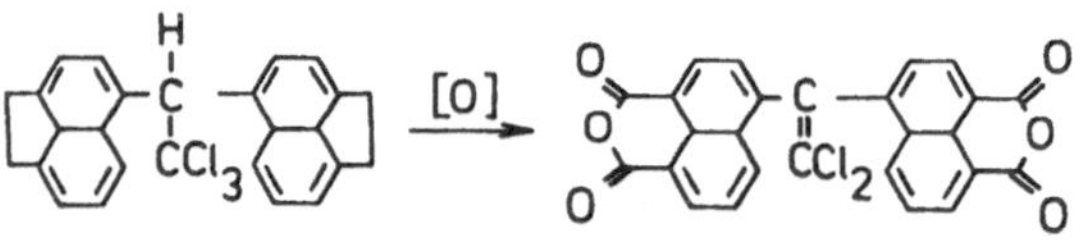

Scheme 35

Of interest is that on replacement of chloral for bromal, the reactions ran in agreement with the expected results of Refs [141–147] and led to 1,1-bis(1,8-dicarboxy-naphthyl-4)-2,2-dibromoethylene [150–151]:

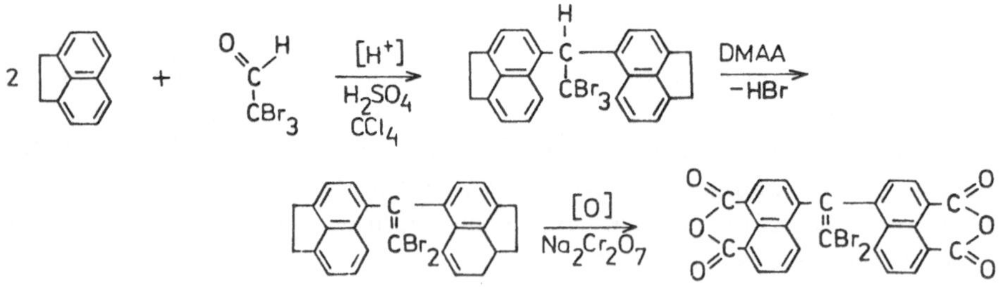

Scheme 36

Some characteristics of the synthesized dianhydrides are given in Table 11.

Table 11. Some properties of bis-(naphthalic anhydrides) of general formula

−Hal	M.p., °C	Yield, %
−Cl	400–402	90
−Br	410–411	93

The literature gives no data on relative reactivity of bis(naphthalic anhydrides). Therefore, quantum-chemical calculations were carried out using the PPP method to obtain some parameters, in particular, the affinity to the electron (A), which determines electrophilic reactivity of dianhydrides towards anions [152].

The calculations results for the values A of various bis(naphthalic anhydrides) are tabulated in Table 12.

Similarly to a series of bis(naphthalic anhydrides) [153], the largest value of A is characteristic for naphthalene-1,4,5,8-tetracarboxylic acid dianhydride; values of A for bis(naphthalic anhydrides) containing electron-accepting "bridging" groups are larger than those of bis(naphthalic anhydrides) with electron-donating groups. Comparison of the absolute A values reveals that in the case of bis(naphthalic anhydrides) the values of A are substantially larger than those of bis(phthalic anhydrides). This is consistent with results in Ref. [153] and is explained by a higher aromaticity of bis(naphthalic anhydrides). But larger values of A cannot be explained by the higher reactivity of bis(naphthalic anhydrides) compared to that of bis(phthalic anhydrides), since it also depends on the strain of anhydride cycle.

The values of the positive charges on the carbonyl carbon atoms (q) were also calculated. They mark the reactive center, which is the primary target of the nucleophilic attack [113]. In bis(naphthalic anhydrides) containing electron-donating "bridging" groups, a larger positive charge is characteristic for the carbonyl carbon atoms located in the positions 5 and 5′, and in bis(naphthalic anhydrides) having electron-accepting "bridging" groups, it is characteristic of the carbonyl carbon atoms in the positions 4 and 4′ with respect to the mentioned "bridging" groups.

The fact that all bis(naphthalic anhydrides) display lower reactivity than naphthalene-1,4,5,8-tetracarboxylic acid dianhydride, has predetermined an expedient synthesis of their polyheteroarylenes. It involves catalytic processes which find wide application in polycondensation and polycyclocondensation reactions [154, 155].

Table 12. Electronic characteristics of naphthalic anhydride and bis(naphthalic anhydrides)

Compound	Electron affinity A, eV		q_1	q_2
	Calculated by Coupman's theory	Calculated from energies difference		
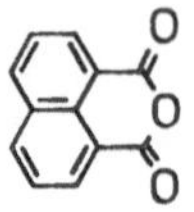	3.1493	1.7426	0.2478	0.2478
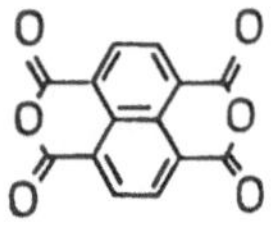	4.0102	2.6044	0.2477	0.2477
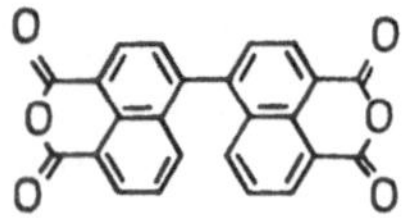	3.7990	2.4770	0.2474	0.2469
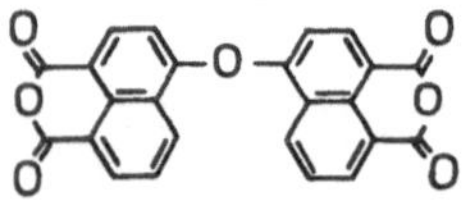	3.4488	2.0819	0.2097	0.2184
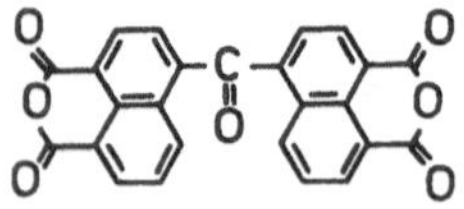	3.8704	2.4896	0.2486	0.2475
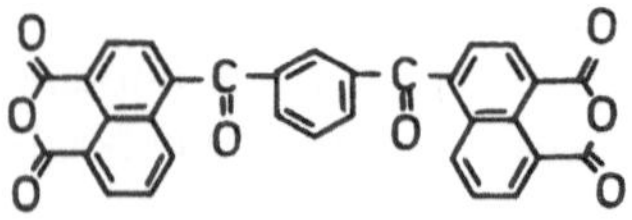	3.6475	2.1878	0.2490	0.2475

3 Polyheteroarylenes Based on Bis(Naphthalic Anhydrides)

3.1 Polynaphthylimides

The majority of polynaphthylimides described in literature results from the interaction of naphthalene-1,4,5,8-tetracarboxylic acid anhydride with various diamines [156]. In addition, a series of bis(naphthalic anhydrides) with "hinge" groupings was employed in synthesis of polynaphthylimides. In particular, the interaction of bis(naphthalic anhydrides) containing ester groups with 4,4′-diaminodiphenyl oxide [120, 121] was carried out in amide solvents according to Scheme 37:

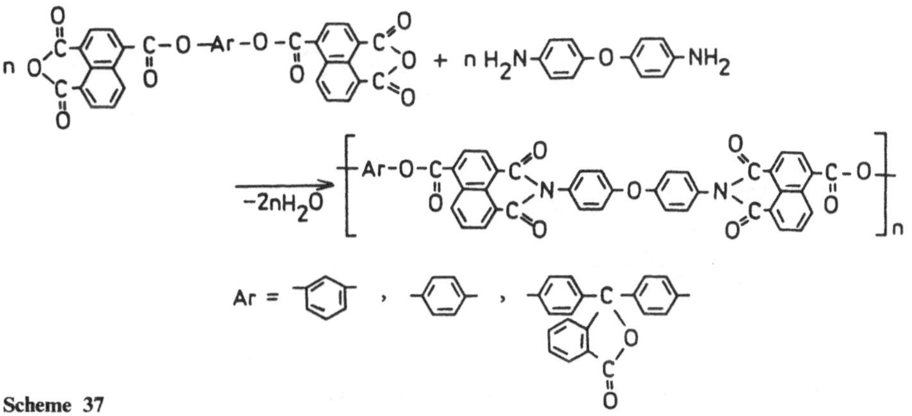

Scheme 37

By this reaction, film-forming prepolymers were obtained which became insoluble in organic solvents after thermal postcyclization.

The bulk reaction between bis(naphthalic anhydrides) having amide bonds and aromatic diamines occurred according to the following scheme [122]:

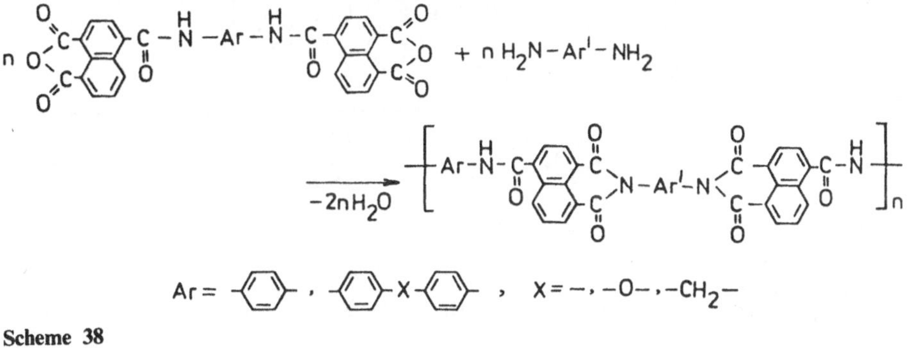

Scheme 38

This resulted in low-molecular polymers (η = 0.12–0.38) which are only soluble in strong acids. These polymers exhibit a rather high thermal stability, they decompose in a nitrogen atmosphere at 400–500 °C. The introduction of ester and amide groups into polyimide macromolecules reduces the primary advantage of such systems, i.e. chemical stability.

Application of bis(naphthalic anhydrides) having ether and thioether bonds seems to be preferable.

Thus, the interaction of dioxo- and dithio-bis(naphthalic anhydrides) with 4,4′-diaminodiphenyl oxide, as shown in Scheme 39 [97, 98] resulted in polynaphthylimides with ether and thioether groups, which were soluble in phenolic solvents. Reactions were carried out in *m*-cresol at 10 % concentration using quinoline or *iso*-quinoline as catalyst.

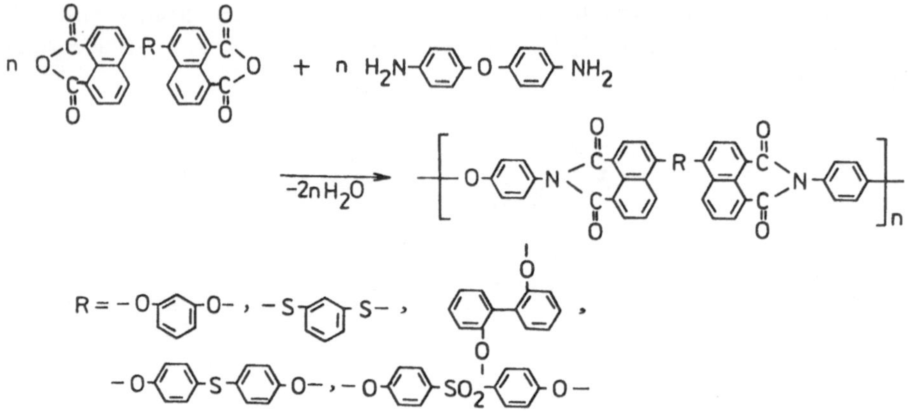

Scheme 39

In contrast to a more rigid polymer derived from naphthalene-1,4,5,8-tetracarboxylic acid dianhydride precipitating from *m*-cresol at this concentration, the obtained polyimides retained solubility in *m*-cresol up to a concentra-

Table 13. Some properties of PNI's of general formula:

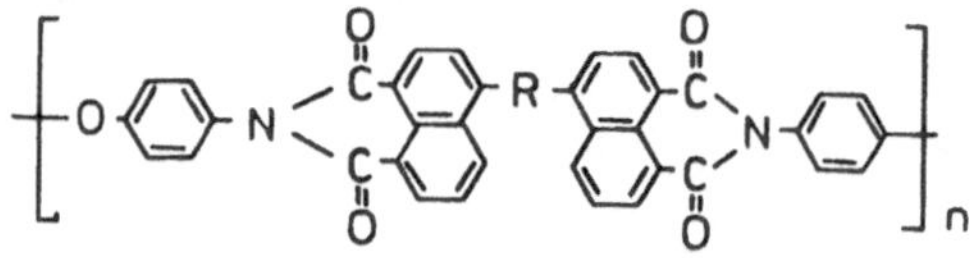

-R-	η_{inh} H_2SO_4 30 °C	T_g, °C	Solubility*			
			H_2SO_4	*m*-cresol	TCE phenol	N-MP
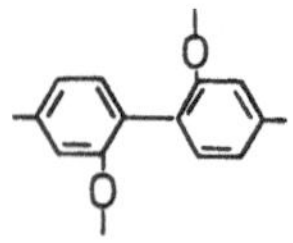	0.15	260	S	S	S	ins
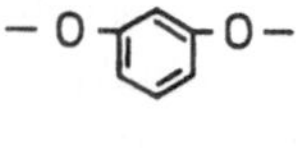 −O−⬡−O−	0.20	260	S	S	S	ins
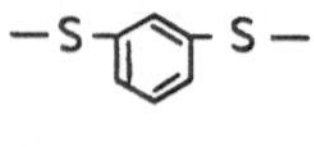 −S−⬡−S−	0.26	253	S	S	S	ins
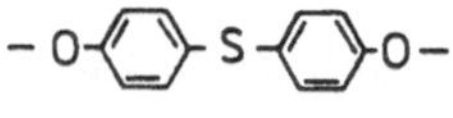 −O−⬡−S−⬡−O−	0.30	269	S	S	S	ins
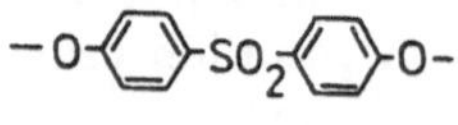 −O−⬡−SO₂−⬡−O−	0.29	317	S	S	S	ins

* Here and in following tables: S-soluble; pS-partially soluble; ins-insoluble

tion of 15 %, but their viscosity characteristics are rather small (η_{inh} = 0.15–0.30 dl/g). TGA of the polymers reveals that the mass loss starts at 450 °C in air, and at 500 °C in nitrogen. The viscosity characteristics and glass transition temperatures of the synthesized polynaphthylimides are listed in Table 13.

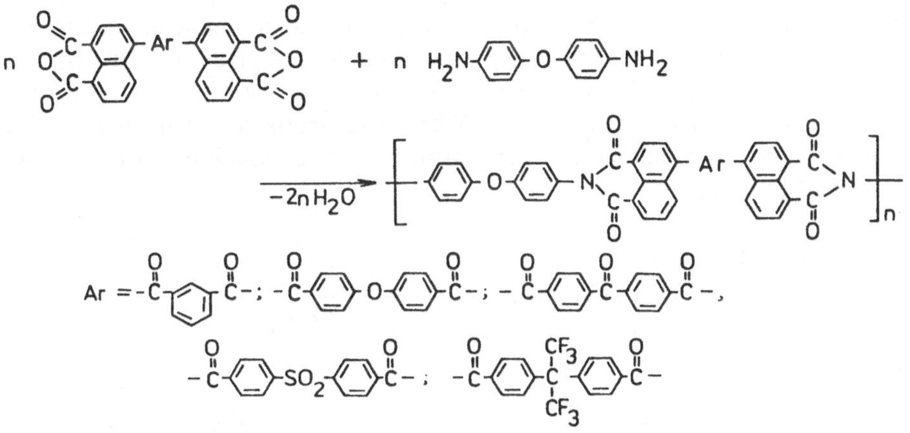

Scheme 40

Table 14. Some properties of polynaphthylimides of general formula:

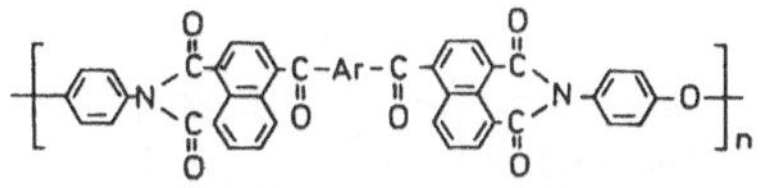

–Ar–	η_{red}, dl/g (0.5% H_2SO_4 25°C)	T_g °C	$T_{10\%}$ °C	Solubility			
				H_2SO_4	m-cresol	TCE:phenol (3:1)	TCE
	0.72	310	500	S	S	S	pS
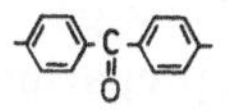	0.53	300	515	S	S	S	pS
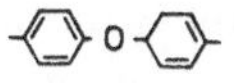	0.78	275	490	S	S	S	S
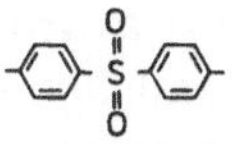	0.78	335	520	S	S	S	pS
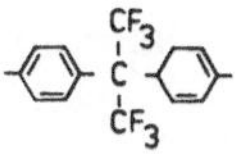	0.32	305	510	S	S	S	S

A substantial enhancement of the viscosity characteristics of the polymers was achieved by applying bis(ketonaphthalic anhydrides) as monomers [157, 158]. Synthesis of polynaphthylimides was carried out in a *m*-cresol medium using benzoic acid as catalyst (Scheme 40). The synthesized polynaphthylimides were soluble in H$_2$SO$_4$, phenolic solvents, tetrachloroethane-phenol mixtures, and the polymers derived from dianhydrides having ether bonds and hexafluoroisopropyl groups as the central fragments were also soluble in tetrachloroethane. Some characteristics of polynaphthylimides are given in Table 14.

Polynaphthylimides having free carboxylic groups were prepared by the interaction of 1,1-bisnaphthyl-4,4',5,5',8,8'-hexacarboxylic acid dianhydride with 4,4'-diaminodiphenyl oxide and benzimidazole-containing aromatic diamines [159–161]

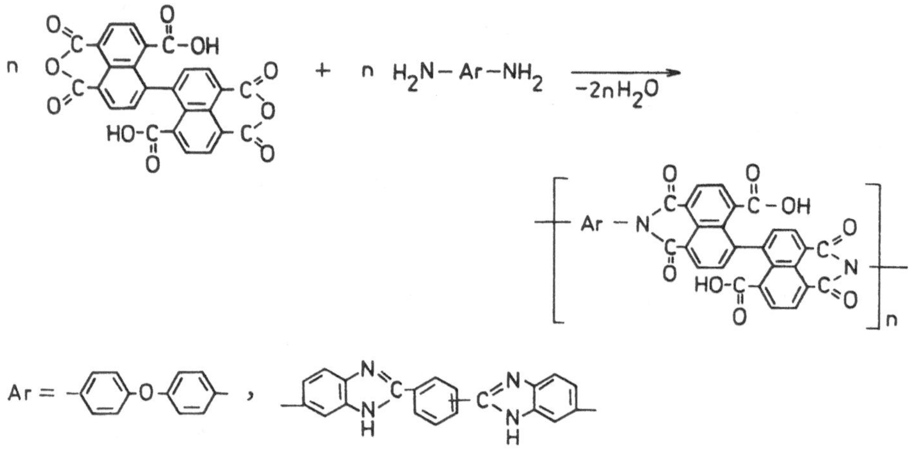

Scheme 41

High-temperature (125–130 °C) polycondensation is carried out in bipolar organic solvents with the addition of bases. Polycondensation occurs best of all in DMSO in the presence of sodium acetate (equimolar amount), which is attributed to the high solubility and stability of monosodium salts formed with strong acidic [138] carboxylic groups of the dianhydride and polynaphthylimide. The rest free carboxylic groups are probably the catalysts of the polycondensation process, and this fact suggests the possibility of preparing polyimides possessing high viscosity characteristics (Table 15).

The high molecular polynaphthylimides obtained have good solubility in the presence of bases not only in organic solvents (DMSO, DMAA) but in water as well. Polynaphthylimides do not soften until decomposition starts [temperature 520–525 °C according to TGA (ΔT = 10 °C/min)]. On spinning into hydrochloric acid as a coagulation bath, polynaphthylimides form films and fibers with strength up to 35 cN/tex and elongation up to 20% at breaking point.

High solubility in organic solvents is typical for polynaphthylimides containing *N*-naphthylimide or *N*(*p*-phenoxy) naphthylimide substituents. The compounds are synthesized according to Scheme 42 [162, 163].

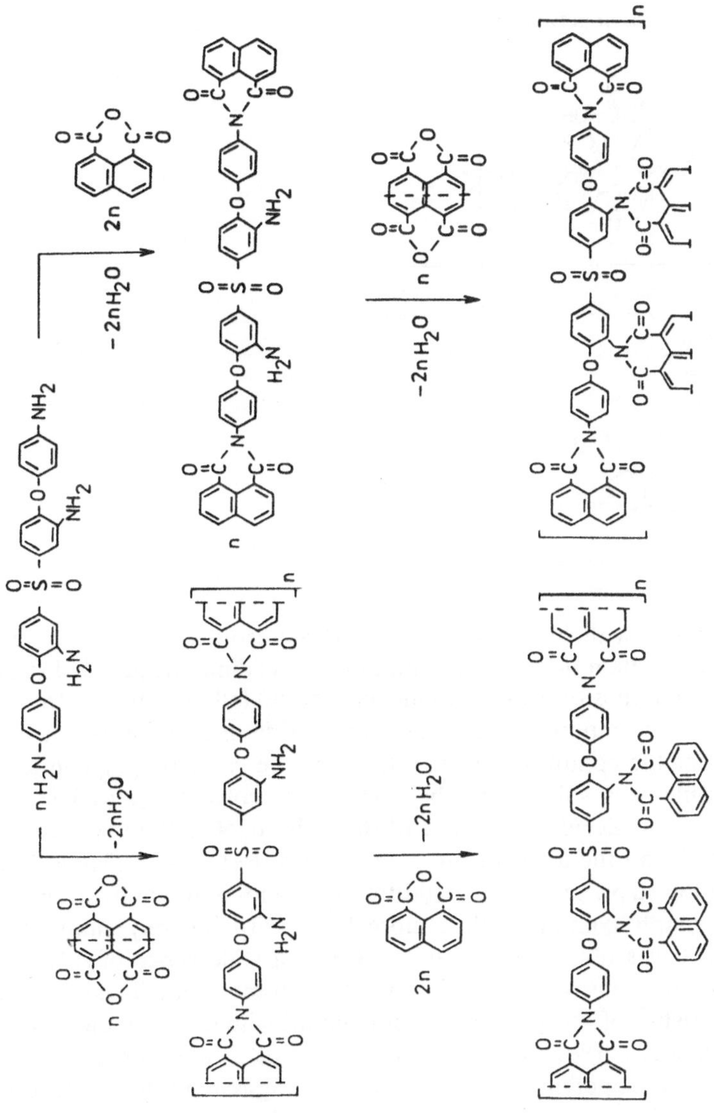

Scheme 42

Table 15. Intrinsic viscosities of polynaphthylimides of general formula:

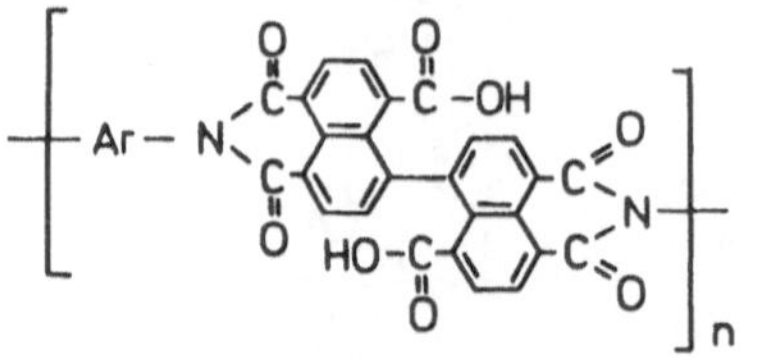

–Ar–	[η]
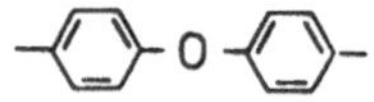	1.4
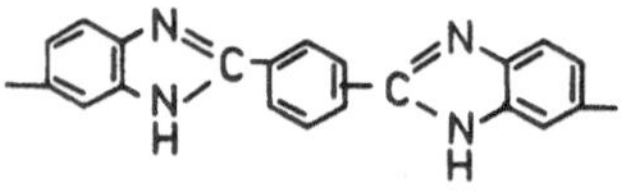	2.0

Polynaphthylimide with *N*-naphthylimide *ortho*-substituents was prepared by the reaction of 3,3′-diamino-4,4′-di(*p*-aminophenoxy) diphenyl sulfone [164] with an equimolar amount of bis(naphthalic anhydride) followed by treatment with a two-fold molar amount of naphthalic anhydride in a *m*-cresol-benzoic acid medium under the conditions of high-temperature catalytic polycondensation. The polymer was soluble in phenolic, amide solvents as well as in tetrachloroethane; its molecular mass was of the order of 60 000, the softening temperature 340 °C and the temperature of 10 % mass loss 480 °C (Table 16).

In the case of the stepwise alternate addition of mono- and dianhydrides to the reaction, polynaphthylimide with a more bulky *N*(*p*-phenoxy)-naphthylimide *o*-substituent was obtained. It was soluble in a wide range of solvents including chloroform along with the compounds (solvents) mentioned above. Thermal characteristics of the synthesized polynaphthylimides evaluated from the thermomechanical curves and dynamical TGA data, are similar to the thermal characteristics of polynaphthylimides containing *N*-naphthylimide substituents of polynaphthylimides containing *N*-naphthylimide substituents (Table 16). The molecular mass of the synthesized polymer being of the order of 25 000, impedes preparing films from it which display satisfactory characteristics.

The high solubility of the systems synthesized by both the first and the second approach is probably attributed to the asymmetry of the polymer molecules, the presence of bulky side substituents which dilute the polymer structure, the positive effect of the SO_2 group, and the mild conditions of the cyclization.

Summarizing the current studies in the field of polynaphthylimides based on the bis(naphthalic anhydrides) considered, it should be noted that at present

Table 16. Some properties of polynaphthylimides having the general formula:

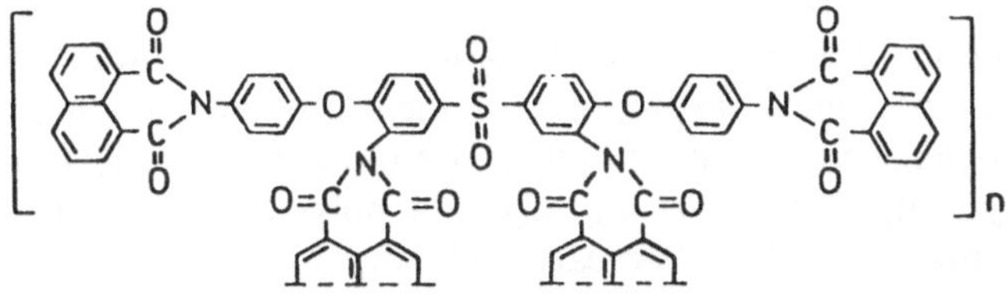

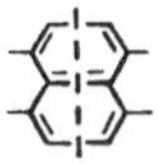	Molecular weight $\bar{M}_w$	T_g °C	$T_{10\%}$ °C	Solubility			
				MP	m-Cr	TCE	CHCl$_3$
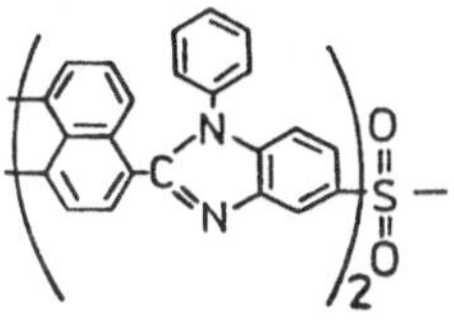	25 000	340	480	S	S	S	S

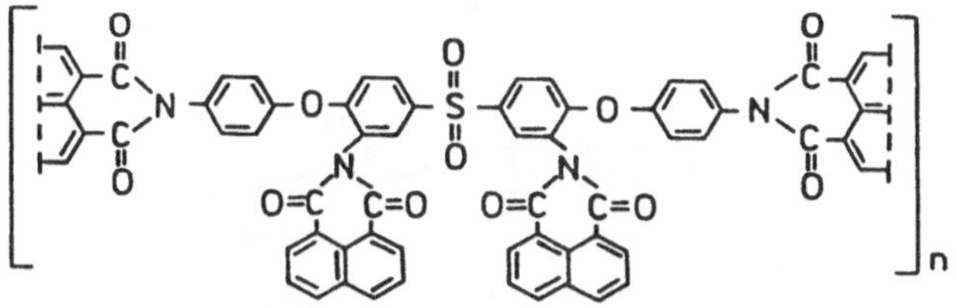

	Molecular weight $\bar{M}_w$	T_g °C	$T_{10\%}$ °C	Solubility			
				MP	m-Cr	TCE	CHCl$_3$
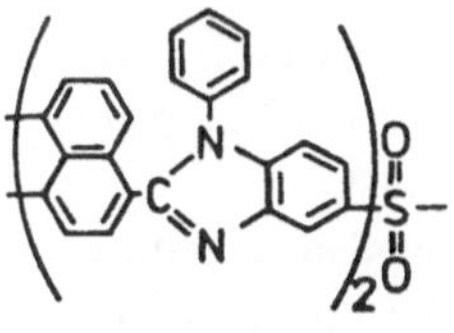	60 000	340	480	S	S	S	ins

progress is far from being rapid. However, the availability of the starting compounds, especially of bis(ethernaphthalic anhydrides) and bis(keto-naphthalic anhydrides) and a possible variation of the structure open an important perspective to the design of polyimides which combine high thermal, heat and chemical stability, and good processability of the final products.

3.2 Polynaphthoylenebenzimidazoles

Compared to polynaphthylimides, more attention has been paid to poly-naphthoylenebenzimidazoles derived from the bis(naphthalic anhydrides) mentioned above. It can easily be explained despite the fact that to prepare polynaphthoylenebenzimidazoles, bis-(o-phenylenediamines) are introduced into the reaction. (They are not so available and therefore more expensive than aromatic diamines which are employed in the synthesis of polynaphthylimides). The naphthoylenebenzimidazole cycle is one of the most stable (both thermally and hydrolytically) heterocycles [165], and polynaphthoylenebenzimidazoles based on napthalene-1,4,5,8-tetracarboxylic dianhydride were successfully applied in the production of fibers possessing a unique complex of properties [166–176].

The successful application of simple bis(naphthalic anhydrides) containing oxy-, keto- and sulfo-groups, i.e. oxy-bis(4,5-dicarboxynaphthyl-1), keto-bis(4,5-dicarboxynaphthyl-1) and sulfone-bis(4,5-dicarboxynaphthyl-1), was reported in Refs. [177–180]; 3,3',4,4'-tetraaminodiphenyl oxide and 2,2-bis(3,4-diaminophenyl) hexafluoropropane were used as monomers:

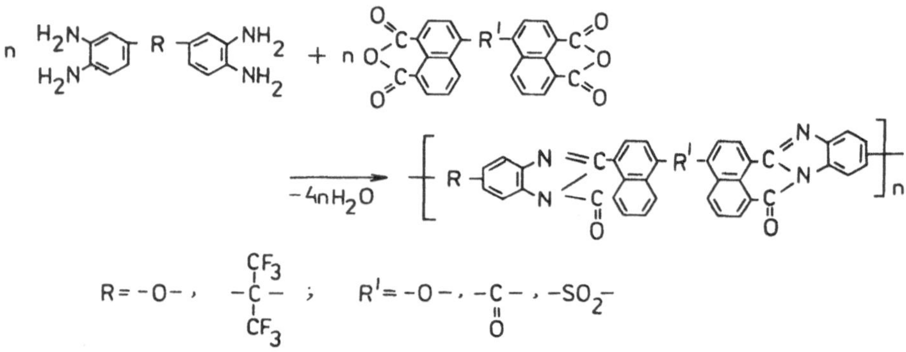

$$R = -O-, \quad -\overset{\overset{\displaystyle CF_3}{|}}{\underset{\underset{\displaystyle CF_3}{|}}{C}}- \; ; \quad R^I = -O-, \; -\overset{\overset{\displaystyle O}{\|}}{C}-, \; -SO_2-$$

Scheme 43

Analogously to the procedure described in Refs. [181, 182], polynaphthoylenebenzimidazoles were synthesized by high-temperature catalytic polycondensation in a m-cresol medium using benzoic acid as catalyst.

Synthesis of polynaphthoylenebenzimidazoles based on 3,3',4,4'-tetraaminodiphenyl oxide was a heterogeneous process, irrespective of the nature of the applied bis(naphthalic anhydrides), while synthesis of polynaphthoylenebenzimidazoles based on 2,2-bis(3,4-diaminophenyl)-hexafluoropropane was a homogeneous one. Irrespective of homogeneity or heterogeneity of the polycyclocondensation processes, the desired polymers were obtained in quantitative yields and were characterized by high degrees of cyclization. Some characteristics of the synthesized polynaphthoylenebenzimidazoles are set out in Table 17.

Analysis of the data on solubility of the synthesized polynaphthoylenebenzimidazoles (Table 17) demonstrates that polymers derived from 3,3',4,4'-tetraaminodiphenyl oxide are only soluble in strong acids of the H_2SO_4 type, and in this context, are comparable with polynaphthoylenebenzimidazoles based on the same bis(o-phenylenediamine) and naphthalene-1,4,5,8-tetra-

Table 17. Some properties of polynaphthoylenebenzimidazoles of general formula:

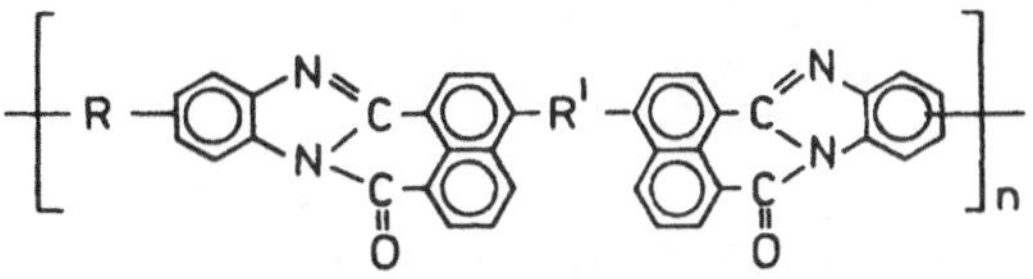

-R-	-R¹-	η_{red}, dl/g (H$_2$SO$_4$, 25 °C)	Solubility			
			m-cresol	TCE:phenol (3:1)	N-MP	H$_2$SO$_4$
-O-	-C- (=O)	1.60	ins	ins	ins	S
-O-	-SO$_2$-	0.80	ins	ins	ins	S
-O-	-O-	0.90	ins	ins	ins	S
-C(CF$_3$)(CF$_3$)-	-C- (=O)	0.35	S	S	S	S
-C(CF$_3$)(CF$_3$)-	-SO$_2$-	0.35	S	S	S	S
-C(CF$_3$)(CF$_3$)-	-O-	0.22	S	S	S	S

carboxylic acid dianhydride [60–64]; thus in this series of polymers introduction of one "hinge" group into bis(naphthalic anhydrides) leads to no significant change in solubility.

In contrast, polynaphthoylenebenzimidazoles based on 2,2-bis(3,4-diaminophenyl) hexafluoropropane are soluble not only in H$_2$SO$_4$, m-cresol and a tetrachloroethane-phenol mixture (3:1) (this is characteristic for polymers derived from bis(o-phenylenediamine) and naphthalene-1,4,5,8-tetracarboxylic acid dianhydride [183, 184]) but also in N-MP (N-methylpyrolidane). An enhanced solubility of polynaphthoylenebenzimidazoles of this series is explained not only by the introduction of a "hinge" group between naphthalic anhydride moieties, but also by low viscosity characteristics of the systems caused by low basicity of 2,2-bis(3,4-diaminophenyl)hexafluoropropane [185].

According to thermomechanical analysis data, several of the polymers studied have softening temperatures lower than their thermal decomposition temperature. Thus the softening temperature of the polymer is about 410 °C, that is 80–90 °C lower than that of the polymer derived from 1,4,5,8-naphthalenetetracarboxylic acid dianhydride [182], and its decomposition temperature.

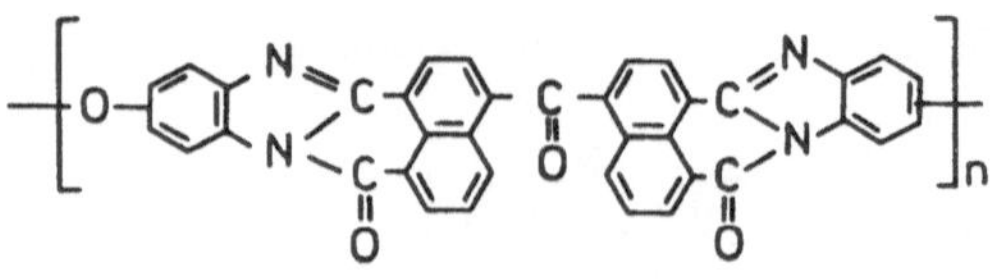

However, the introduction of "bridges" has no significant effect on the thermal stability of the synthesized polymers. The temperatures at which decomposition starts is consistent with the corresponding literature data for the polynaphthoylenebenzimidazoles synthesized earlier [182], i.e. about 500 °C in air, and about 520 °C in an inert atmosphere.

Introduction of additional "hinge" groups was considered to be the simplest structural variation of bis(naphthalic anhydrides) for the preparation of the desired polyheteroarylenes soluble in organic solvents.

Among polynaphthoylenebenzimidazoles based on bis(naphthalic anhydrides) having more than one "hinge" group, one of the first synthesized polymers were the sulfide-containing ones [92, 107–110]. The polymers were prepared by high-temperature polycondensation in m-cresol using benzoic acid as catalyst; the polymers were synthesized according to the following scheme:

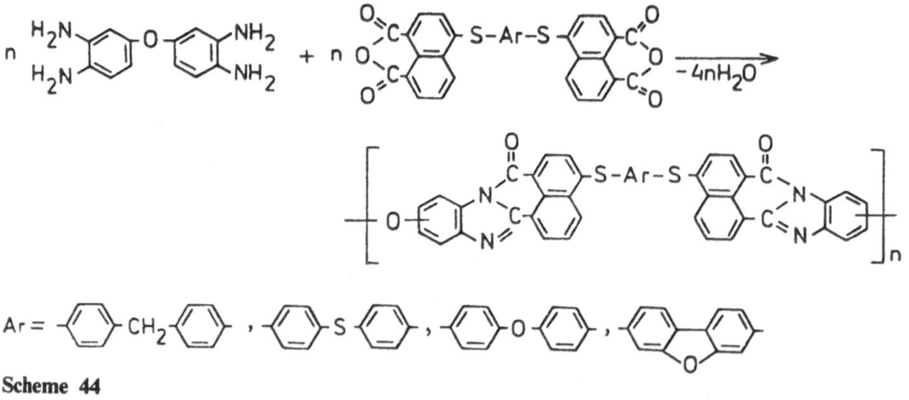

Scheme 44

All polycondensation reactions proceeded in solutions (homogeneously) and yielded well-cyclized polymers, some characteristics of which are given in Table 18.

The synthesized sulfide-containing polynaphthoylenebenzimidazoles are soluble not only in H_2SO_4 but in phenolic solvents (m-cresol, phenol : tetrachloroethane mixtures), which in combination with high softening (310–380 °C) and decomposition (450–470 °C) temperatures stimulate interest in the systems.

All sulfide-containing polynaphthoylenebenzimidazoles are characterized by far from high viscosity ($\eta_{red} = 0.4$–0.8 dl/g) so that it is difficult to prepare materials with suitable mechanical properties.

Table 18. Some properties of polynaphthoylenebenzimidazoles of general formula:

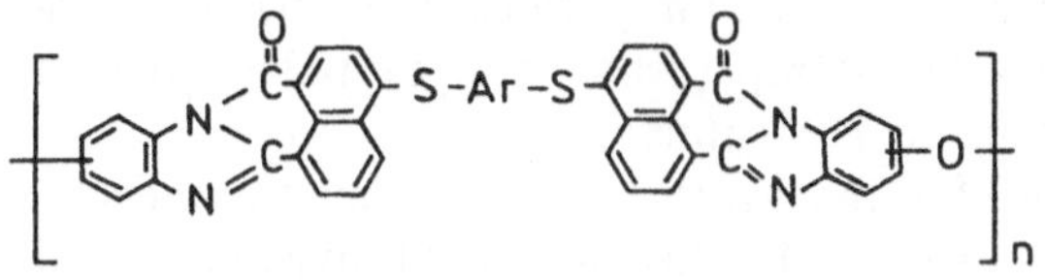

-Ar-	T_g, °C	$T_{10\%}$, °C	η_{red}, dl/g	
			H_2SO_4	TCE:phenol (3:1)
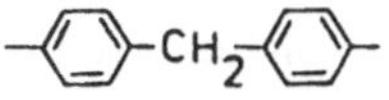	330	450	0.4	0.5
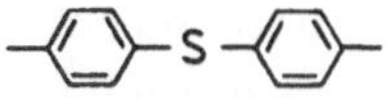	310	470	0.5	0.5
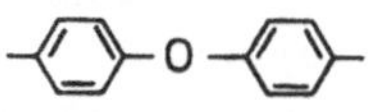	330	460	0.8	0.7
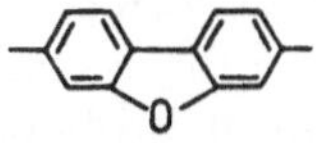	380	460	—	0.4

This viscosity and the corresponding molecular-mass characteristics of sulfide-containing polynaphthoylenebenzimidazoles is probably brought about by electron-donating sulfide groups in the starting bis(naphthalic anhydrides) which determine a reduced electrophilic reactivity of the monomers. On preparing polynaphthoylenebenzimidazoles to enhance the electrophilic reactivity of the monomers, an attempt has been made to use bis(naphthalic anhydrides) having two or more sulfone groups [92, 107–110]:

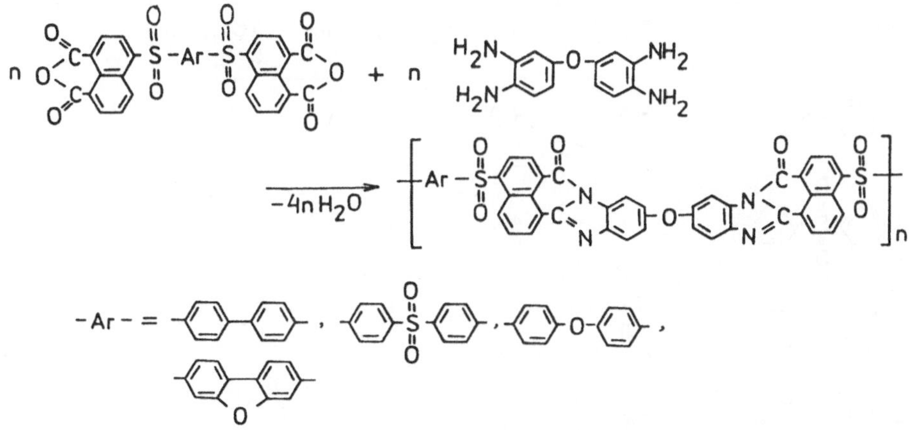

Scheme 45

Similarly to the preparation of polynaphthoylenebenzimidazoles from sulfide-containing bis(naphthalic anhydrides) [92, 107–110], the synthesis has been carried out by a high-temperature catalytic polycondensation method in *m*-cresol using benzoic acid as catalyst.

All the reactions for the synthesis of polyheteroarylenes except for the processes involving 4,4′-bis(1,8-dicarboxynaphthylsulfone-4)-diphenylsulfone dianhydride occurred under homogeneous conditions and led to the desired polyheteroarylenes in yields close to quantitative.

Some characteristics of the synthesized polynaphthoylenebenzimidazoles are given in Table 19.

The synthesized polynaphthoylenebenzimidazoles display better solubility than the polymers of this class derived earlier from 4,4′-sulfone-bis(naphthalic anhydride) [92]. This demonstrates the positive effect of the second sulfone group introduced into dianhydride and into the main chains of macromolecules. An exception is the polymer with three sulfone groups in the chain; insolubility of this polynaphthoylenebenzimidazole in the above mentioned solvents is accounted for by its crystallinity – 20 to 30 % according to X-ray data.

However, the polynaphthoylenebenzimidazoles containing two or more sulfone groups are less soluble than analogous systems with sulfide groups.

Low viscosity characteristics of polymers are quite unexpected for systems derived from dianhydrides activated by sulfone groups. The most probable explanation of this phenomenon are the cleavage reactions of naphthoylene-

Table 19. Some characteristics of polynaphthoylenebenzimidazoles of the general formula:

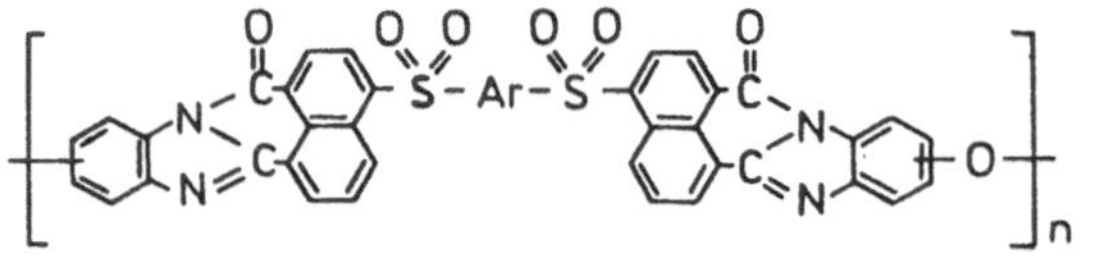

–Ar–	T_g, °C	$T_{10\%}$, °C	η_{red}, dl/g H_2SO_4	TCE:phenol (3:1)
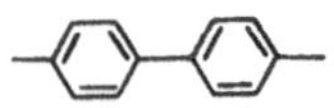	380	430	0.4	—
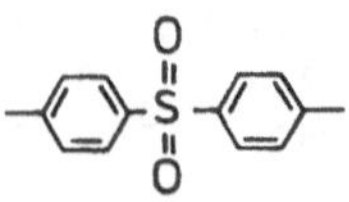	—	400	0.4	—
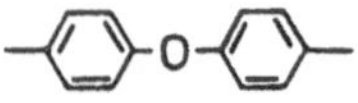	350	420	—	0.3
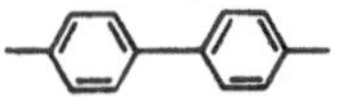	370	410	0.4	—

benzimidazole rings by *o*-phenylenediamine moieties of monomers or terminal groups of the growing macromolecular chains:

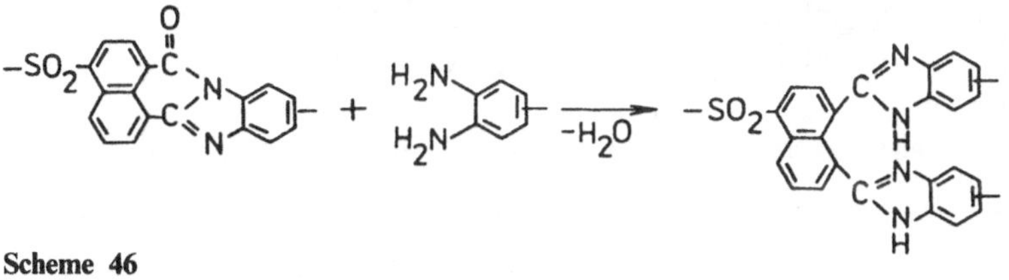

Scheme 46

The reactions are analogous to those described earlier for mono- and bis-benzoylenebenzimidazoles [186–189].

This supposition is consistent with the idea assuming the correlations between electrophilic reactivity of the starting anhydrides and susceptibility of the desired carbonyl-containing heterocycles to hydrolysis and action of nucleophilic reagents [39]. This fact is responsible for the rather low thermal stability of the synthesized polynaphthoylenebenzimidazoles determined by dynamic TGA in air (Table 19); the 10 % mass loss temperature is 400–430 °C. Unsatisfactory properties of the synthesized polyheteroarylenes make applications of bis(naphthalic anhydride) systems containing sulfone groups directly bounded by anhydride cycles rather difficult.

Far better results have been obtained in the synthesis of polynaphthoylene-benzimidazoles based on bis(naphthalic anhydrides) having ether bonds [99–102]:

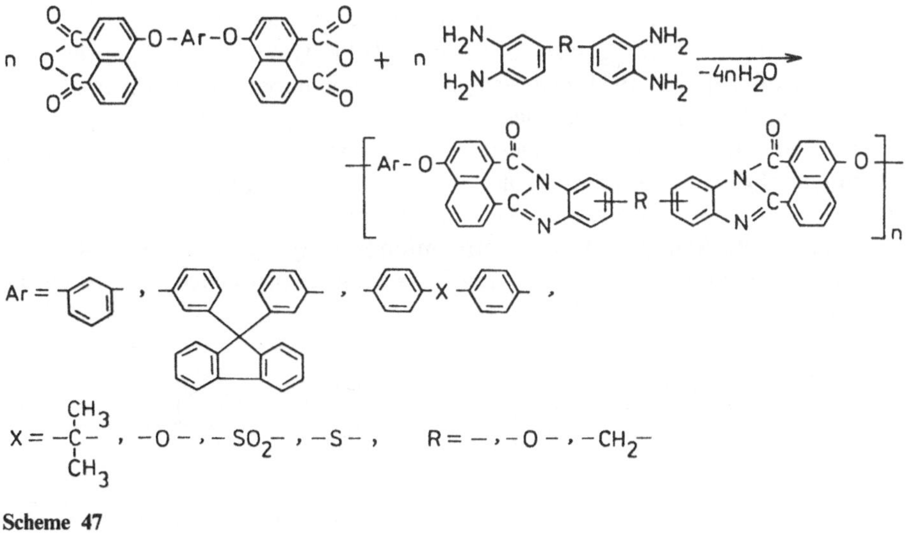

Scheme 47

These reactions have been carried out both under traditional conditions, i.e. high-temperature polycondensation in *m*-cresol and application of benzoic acid as catalyst [99]; and in a series of other solvents [190] – mixtures of *m*-cresol and polyphosphoric acid, *m*-cresol and di-*m*-cresyl phosphate, *m*-cresol

Table 20. Reduced viscosities of polynaphthoylenebenzimidazoles of general formula:

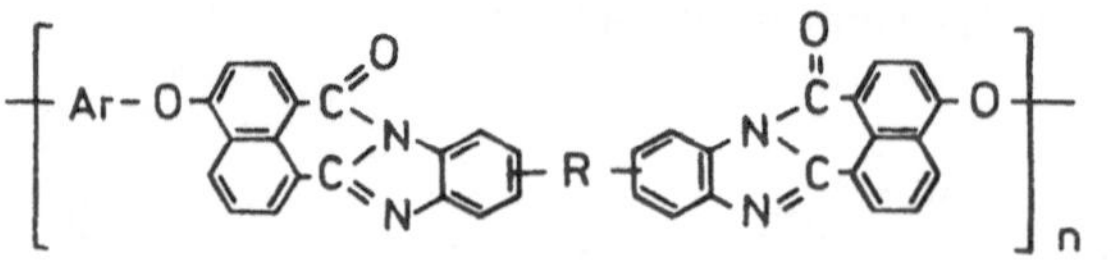

[190]

synthesized in the mixtures of m-cresol with P_2O_5 (method "a") and with benzoic acid (method "B") (reaction time 30 h)

–Ar–	–R–	η_{red}, dl/g	
		Method "a"	Method "b"
$-p\text{-}C_6H_4\text{-}C(CH_3)_2\text{-}p\text{-}C_6H_4-$	—	1.10	1.00
–"–	–O–	0.94	0.75
$-m\text{-}C_6H_4-$	—	3.95	1.10
–"–	–O–	1.18	0.68
$-p\text{-}C_6H_4\text{-}SO_2\text{-}p\text{-}C_6H_4-$	—	1.00	0.49
–"–	–O–	0.97	0.30
$-p\text{-}C_6H_4\text{-}S\text{-}p\text{-}C_6H_4-$	—	1.89	1.03
–"–	–O–	1.03	0.62

η_{red} measured in TCE:phenol (3:1) mixture at 25 °C

and P_2O_5, phenol and P_2O_5, and p-chlorophenol and P_2O_5. Of the series of the studied series, the m-cresol-P_2O_5 mixture has shown itself to be the best (Table 20 [190]). Polynaphthoylenebenzimidazoles prepared in other solvents, in particular in m-cresol-benzoic acid mixtures, exhibit good viscosity characteristics (Table 21) and molecular masses for obtaining stable and elastic films (Table 22) which have interesting electrophysical properties.

Comprehensive investigations of thermal [191], dynamical, mechanical [191] and relaxation [192, 193] characteristics of polynaphthoylenebenzimidazoles of this series have shown that these systems are of significant practical interest.

Logical extension of these investigations is connected with the development of polynaphthoylenebenzimidazoles based on 4,4' oxy-bis(p-phenylenoxy)-dinaphthalene-1,8,1',8'-tetracarboxylic dianhydride [105]. Its reaction with bis-(o-phenylenediamines)-3,3',4,4'-tetraaminodiphenyl oxide,   3,3',4,4'-tetra-aminodiphenylmethane and 3,3'-diaminobenzidine proceeded according to Scheme 48

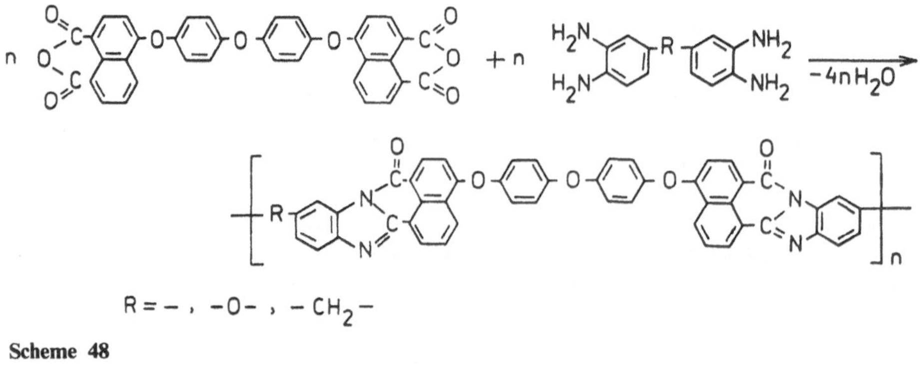

$R = -\ ,\ -O-\ ,\ -CH_2-$

Scheme 48

Table 21. Some properties of polynaphthoylenebenzimidazoles of general formula:

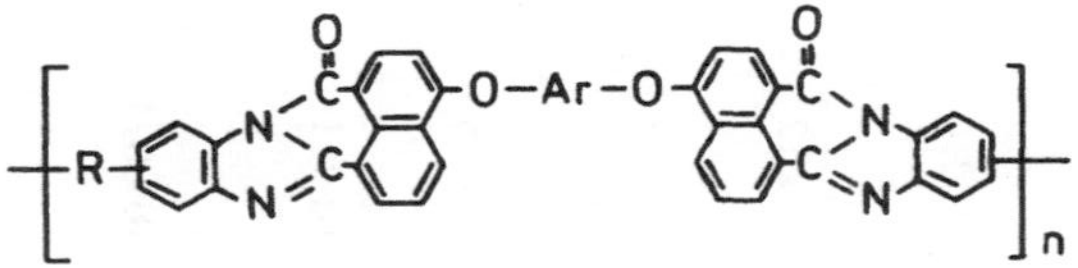

R-	-Ar-	η_{red}, dl/g	T_g, °C	T_{degr} Ar	T_{degr} Air	CF$_3$COOH	m-cresol	TCE:phenol
—	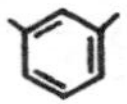 (phenylene)	1.10	395	490	470	s	ins	ins
-O-	"	0.68	368	490	490	s	pS	pS
-CH$_2$-	"	0.29	—	460	440	s	s	s
—	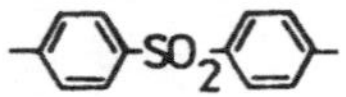 (–SO$_2$– diphenyl)	0.49	404	450	430	s	s	s
-O-	"	0.30	360	460	430	s	s	s
-CH$_2$-	"	0.20	380	450	430	s	s	s
—	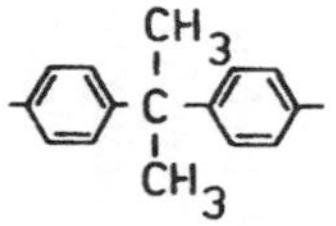 (–C(CH$_3$)$_2$– diphenyl)	1.00	373	460	450	s	s	s
-O-	"	0.75	312	460	450	s	s	s
-CH$_2$-	"	0.37	355	460	440	s	s	s
—	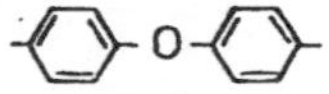 (–O– diphenyl ether)	0.40	340	—	520	ins	s	s
-O-	"	0.50	320	—	500	ins	s	s
-CH$_2$-	"	0.70	320	—	380	ins	s	s

(Solubility columns: CF$_3$COOH, m-cresol, TCE:phenol)

Table 22. Properties of the films based on polynaphthoylenebenzimidazoles of general formula:

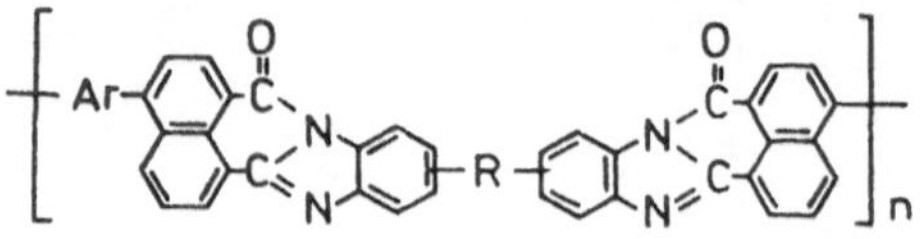

-Ar-	-R-	Mechanical properties at 25 °C		Electrical properties	
		σ, MPa	ε, %	Volume resistivity (ohm·cm)	$+ g \delta$ a + 50 Hz
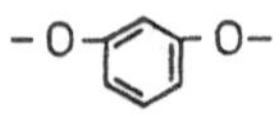 $-O\langle\bigcirc\rangle O-$	—"—	112	85	$2 \cdot 10^{16}$	0.0017
— " —	-O-	108	85	$7 \cdot 10^{16}$	0.0014
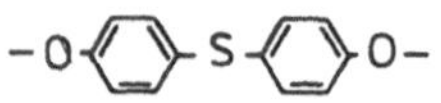 $-O\langle\bigcirc\rangle S\langle\bigcirc\rangle O-$	—"—	135	70	$3 \cdot 10^{16}$	0.0012
— " —	-O-	91	75	$4 \cdot 10^{16}$	0.0014
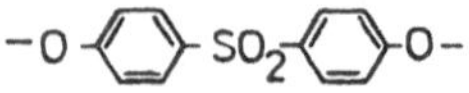 $-O\langle\bigcirc\rangle SO_2\langle\bigcirc\rangle O-$	—"—	92	60	$1 \cdot 10^{16}$	0.0015
— " —	-O-	111	80	$3 \cdot 10^{16}$	0.0014
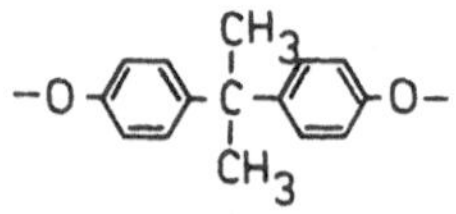 $-O\langle\bigcirc\rangle \overset{CH_3}{\underset{CH_3}{C}}\langle\bigcirc\rangle O-$	—"—	95	70	$6 \cdot 10^{16}$	0.00094
— " —	-O-	98	113	$1 \cdot 10^{16}$	0.0016

under conditions of high-temperature catalytic polycyclization in phenolic solvents. All the reactions for the synthesis of polynaphthoylenebenzimidazoles proceeded from beginning to end under homogeneous conditions and led to polymers displaying moderate viscosity characteristics (Table 21).

Introduction of an additional oxygen atom does not lead to any enhancement in solubility of the polymers: they are soluble in acidic and phenolic solvents but insoluble in DMSO and N-MP.

A comparison of the softening temperature and the region of the noticeable thermal decomposition gives evidence to the fact that the polymers can successfully be processed by molding.

Of significant interest are polynaphthoylenebenzimidazoles derived from aroylene-bis(naphthalic anhydrides) according to Scheme 49 [111, 114, 115, 194].

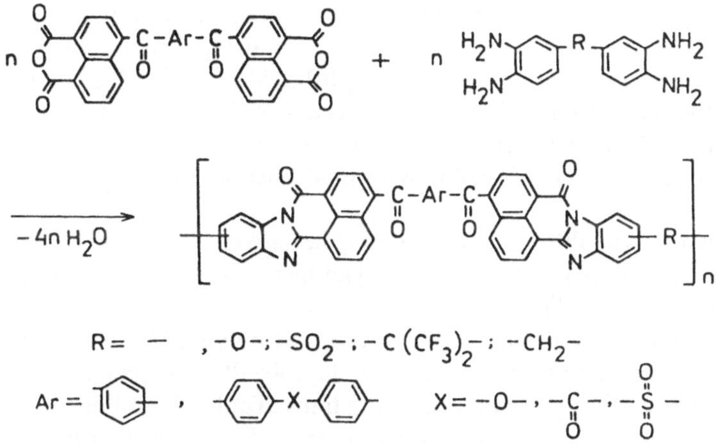

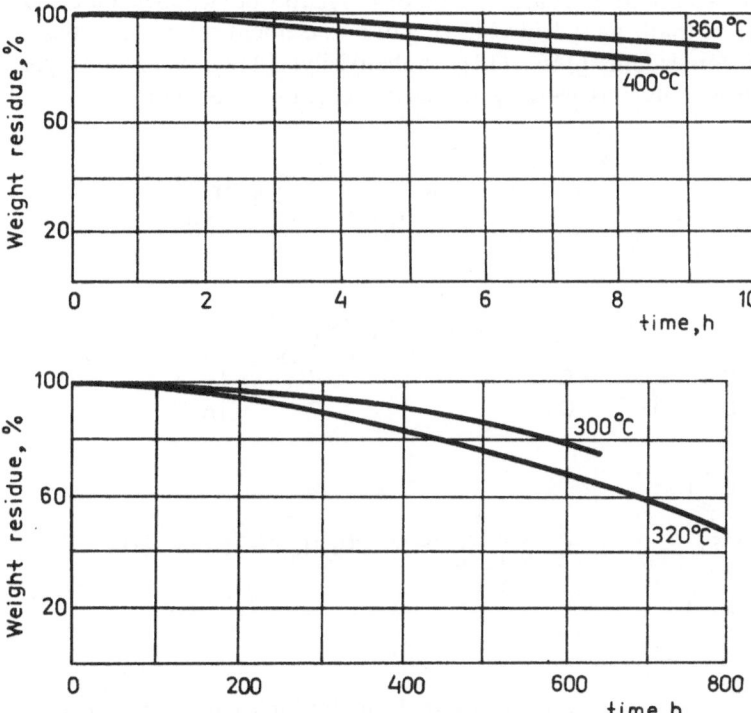

Scheme 49

Fig. 1. Isothermal aging in air of polynaphthoylenebenzimidazole

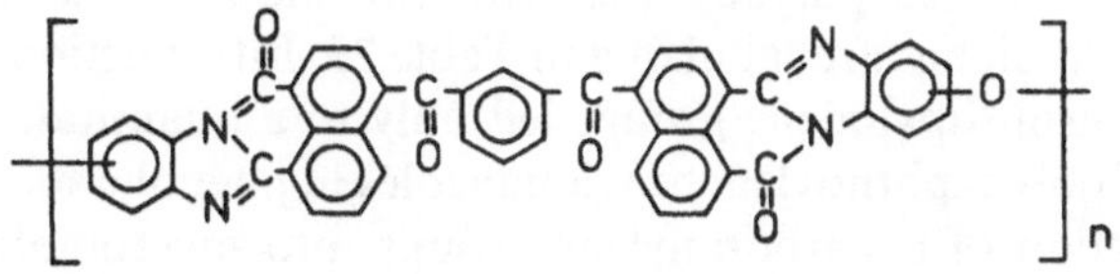

Synthesis of these polynaphthoylenebenzimidazoles was carried out by the high temperature polycondensation technique in *m*-cresol using benzoic acid as catalyst. All reactions for synthesis of polynaphthoylenebenzimidazoles ran homogeneously and led to well-cyclized high molecular polymers. Some characteristics of polynaphthoylenebenzimidazoles based on a series of aroylene-bis(naphthalic anhydrides) and 3,3′,4,4′-tetraaminodiphenyl oxide are shown in Table 23.

The synthesized polynaphthoylenebenzimidazoles are characterized by high thermal stability both under conditions of dynamical TGA (Table 23) and isothermal aging in air (Fig. 1).

As seen from the data in Table 23, the synthesized polynaphthoylenebenzimidazoles are soluble both in H_2SO_4 and phenolic solvents. Molecular masses of the polymers are high enough to obtain strong films (Table 23) and fibers; the difference between the softening temperatures (345–375 °C) and thermal decomposition temperatures (490–550 °C) is large enough to process the polymers by pressing.

A far better solubility is characteristic of polynaphthoylenebenzimidazoles based on bis(naphthalic anhydrides) having organoelement central moieties, i.e. hexafluoroisopropyledene [116, 118], diphenylsilyl [118], *m*-carboranylene [117, 118]. These polymers were synthesized according to Scheme 50:

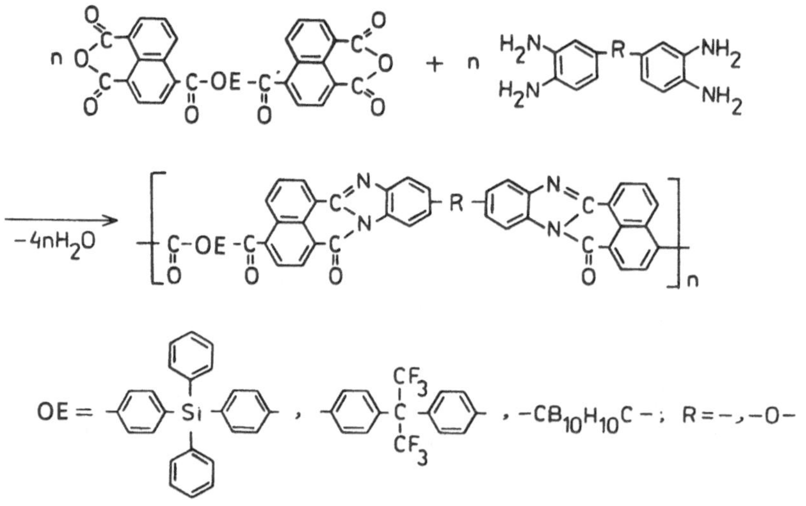

Scheme 50

Synthesis of these polymers was carried out by high-temperature catalytic polycondensation in *m*-cresol using benzoic acid as catalyst. Some characteristics of the synthesized polynaphthoylenebenzimidazoles containing organoelement central moieties are tabulated in Table 23. Introduction of diphenylsilyl and hexafluoroisopropylodene groups led only to a quantitative enhancement in solubility of polynaphthoylenebenzimidazoles in phenolic solvents [116, 118], while introduction of *m*-carboranylene groups into macromolecules imparted solubility in N-MP [117, 118].

Table 23. Some properties of polynaphthoylenebenzimidazoles of general formula:

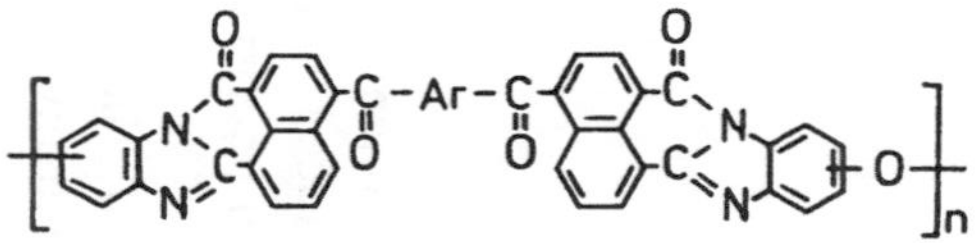

-Ar-	η_{red} H$_2$SO$_4$ dl/g	T$_{soft}$, °C	T$_{10\%}$, °C	Solubility				Film properties	
				m-cresol	TCE:phenol	H$_2$SO$_4$	N-MP	σ, Mpa	ε, %
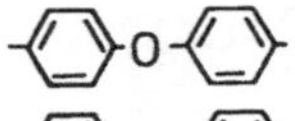 (diphenyl ether)	1.00	345	550	S	S	S	ins	126	15
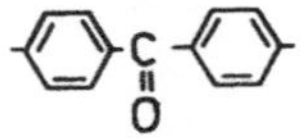 (C=O)	1.40	355	520	S	S	S	ins	110	13
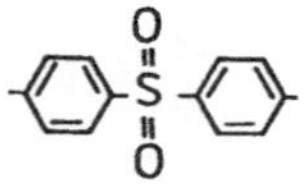 (O=S=O)	1.35	375	525	S	S	S	ins	100	11
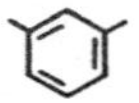 (methyl)	3.90	370	530	S	S	S	ins	104	11
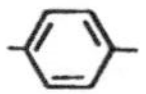 (CF$_3$–C–CF$_3$)	0.53	—	490	pS	S	S	ins	—	—
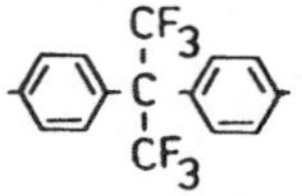	1.80	360	545	S	S	S	ins	108	13
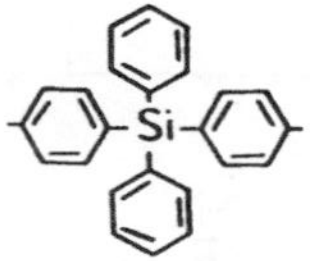 (Si)	1.57	375	550	S	S	S	ins	92	10
–CB$_{10}$H$_{10}$C–	0.40	—	550	S	S	S	S	—	—

Significant attention has been paid to the synthesis of polynaphthoylene-benzimidazoles based on bis(naphthalic anhydrides) containing various aromatic heterocycles.

Polynaphthoylenebenzimidazoles containing 1,3,4-oxadiazole and benzoxazole cycles in the backbone structures were prepared [123] by the interaction of bis(naphthalic anhydrides) containing the previously mentioned heterocycles with aromatic bis(o-phenylenediamines) according to Scheme 51.

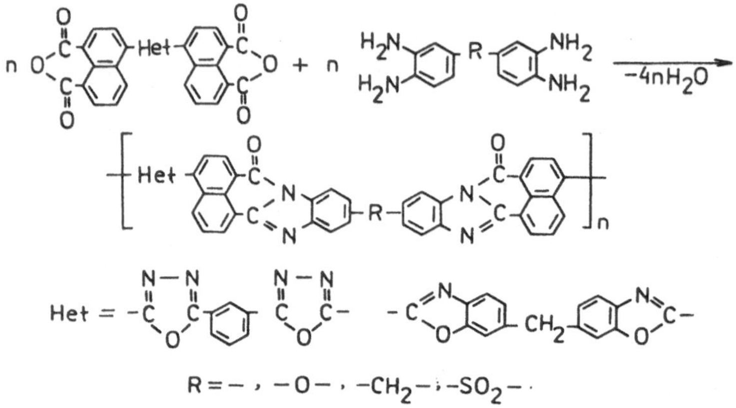

Scheme 51

Synthesis of polynaphthoylenebenzimidazoles was carried out in phenol at 160–180 °C for 7–9 h. The polynaphthoylenebenzimidazoles produced were characterized by comparatively high viscosity and thermal characteristics (Table 24), but they were only soluble in H_2SO_4 and CF_3COOH.

Polynaphthoylenebenzimidazoles displaying fair viscosity and thermal characteristics but far better solubility were prepared by reacting various bis-(o-phenylenediamines) with bis(naphthalic anhydrides) containing phenyl-quinoxaline cycles [126, 128] according to Scheme 52.

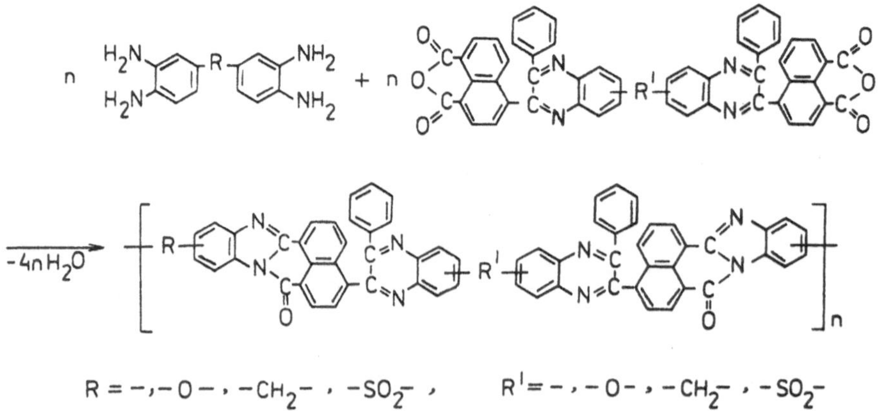

Scheme 52

Table 24. Some properties of the heteryl-containing polynaphthoylenebenzimidazoles of general formula:

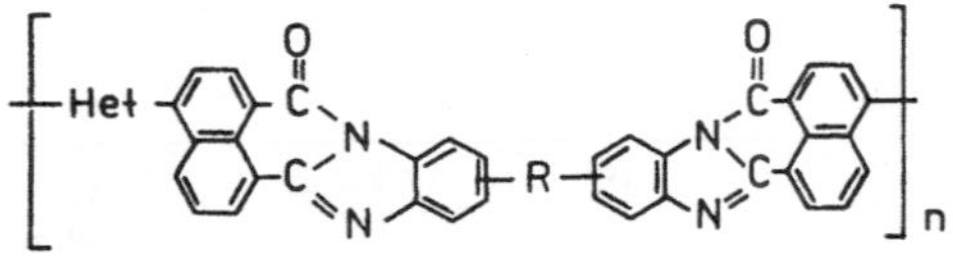

-Het-	-R-	η_{red} dl/g H_2SO_4 25°C	T_g, °C	$T_{10\%}$, °C	Solubility		
					H_2SO_4	m-cresol	TCE:phenol
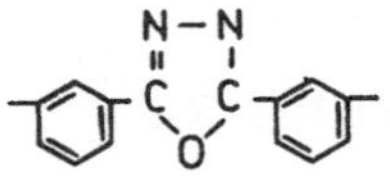	—	0.90	400	525	s	ins	ins
-"-	-O-	1.20	380	510	s	ins	ins
-"-	-CH$_2$-	1.00	385	500	s	ins	ins
-"-	-SO$_2$-	0.80	390	520	s	ins	ins
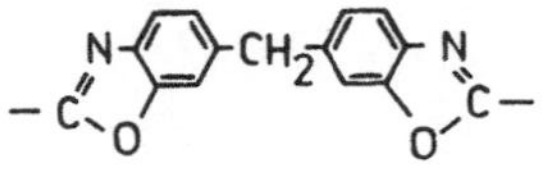	-O-	1.40	500	500	s	ins	ins

Table 25. Some properties of phenylqunoxaline-containing polynaphthoylenebenzimidazoles of general formula:

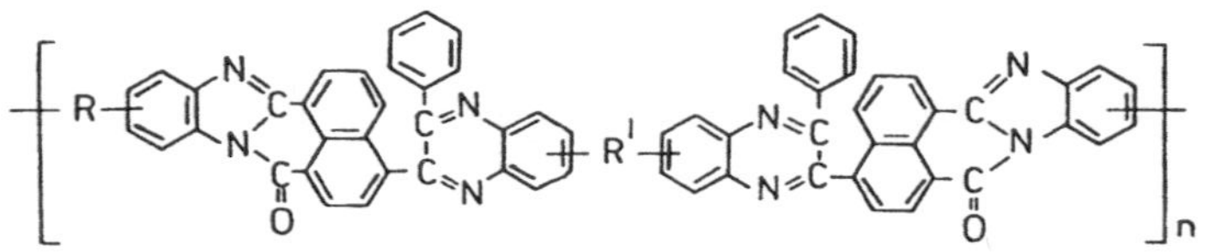

-R-	-R'-	η_{red} (H$_2$SO$_4$, 25°C) dl/g	T$_g$, °C	T$_{10\%}$, °C	Solubility			Film properties 25°C	
					m-cresol	TCE:phenol	Phenol	σ, Mpa	ε, %
—O—	—O—	1.4	340	455	S	S	pS	102	18
—"—	—"—	2.1	320	470	S	pS	pS	95	24
—CH$_2$—	—"—	1.0	320	455	S	S	pS	108	20
—SO$_2$—	—"—	0.8	340	460	S	S	pS	99	20
—O—	—SO$_2$—	1.6	360	480	S	S	pS	110	20
—"—	—"—	2.3	340	470	S	S	pS	130	25
—CH$_2$—	—"—	1.8	340	460	S	S	pS	105	23
—SO$_2$—	—"—	0.4	350	490	S	S	pS	98	23

Table 26. Some properties of phenylqunoxaline-containing polynaphthoylenebenzimidazoles of general formula:

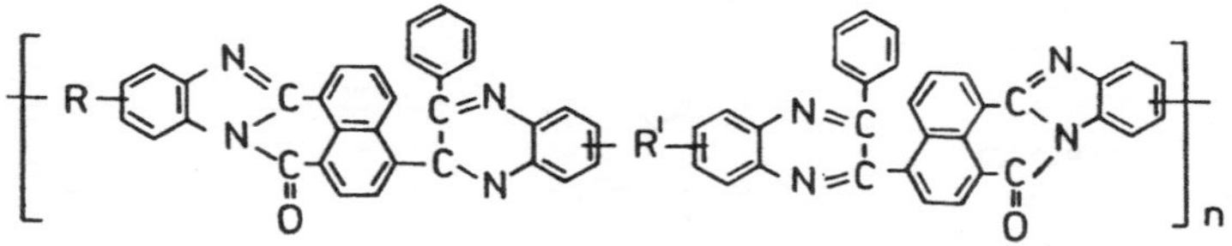

-R-	-R'-	η_{red} dl/g (H$_2$SO$_4$) 25°C	T$_g$, °C	T$_{10\%}$, °C	Solubility			Film properties at 25°C	
					m-cresol	TCE:phenol	Phenol	σ, Mpa	ε, %
—	—	1.8	380	470	S	S	pS	1150	20
-O-	-"-	2.5	350	490	S	S	pS	1070	20
-CH$_2$-	-"-	1.7	350	460	S	S	pS	1230	15
-SO$_2$-	-"-	0.6	360	485	S	S	pS	—	—
—	-CH$_2$-	1.3	330	460	S	S	pS	1120	16
-O-	-"-	1.7	310	450	S	S	pS	1280	18
-CH$_2$-	-"-	1.1	310	440	S	S	pS	1200	17
-SO$_2$-	-"-	0.5	320	460	S	S	pS	—	—

Synthesis of these polynaphthoylenebenzimidazoles was carried out in *m*-cresol using benzoic acid as catalyst.

Polycondensation reactions occurred homogeneously and afforded high molecular polymers (Tables 25, 26) with high degrees of cyclization.

The systems combine high (350 ± 20 °C) glass transition temperatures, high (enough for reliable film formation) viscosity characteristics and solubility in organic solvents. It should also be noted that these polymers only have good solubility in phenolic solvents and mixtures of the latter with halogenated hydrocarbons. In this respect, polynaphthoylenebenzimidazoles derived from bis(*o*-phenylenediamines) and bis-[1-phenyl-2(1,8-dicarboxynaphth-4-yl) benzimidasol-5-yl] sulfone [195] according to Scheme 53

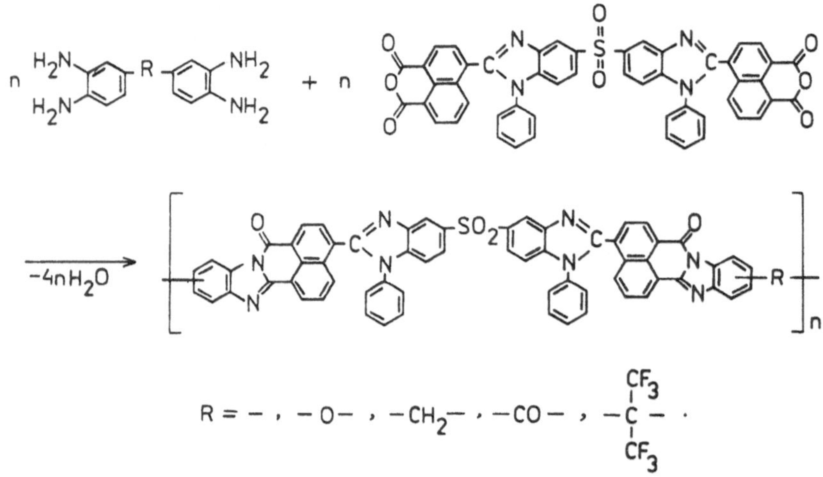

Scheme 53

are of great interest. Synthesis of polynaphthoylenebenzimidazoles was carried out in *m*-cresol using benzoic acid as catalyst at 160–190 °C. The polycondensation process occurred under homogenous conditions and led to polymers possessing the desired high viscosity characteristics (Table 27) and degree of cyclization.

A peculiarity of these polynaphthoylenebenzimidazoles is their solubility in N-MP i.e. a solvent suitable for technology.

The properties of polynaphthoylenebenzimidazoles can be affected by the introduction of 1,1-dichloroethylene and 1,1-dibromoethylene groupings into the macromolecules using the application of the corresponding bis(naphthalic anhydrides) [148–151, 196]. Synthesis of polynaphthoylenebenzimidazoles containing 1,1-dichloroethylene and 1,1-dibromoethylene "bridging" groups [114, 148–151, 196] was carried out in *m*-cresol using benzoic acid as a catalyst:

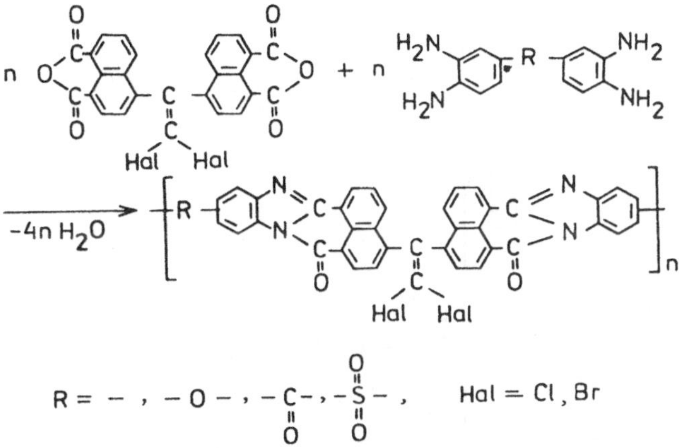

Scheme 54

In all cases, the synthesis proceeded homogeneously and led to well-cyclized and quite high molecular weight systems.

A viscosity-molecular mass relationship [197] for one of the polynaphthoylenebenzimidazoles is presented in Table 28. Some characteristics of the synthesized polynaphthoylenebenzimidazoles containing 1,1-dihalogenethylene groups are tabulated in Table 29. From analysis of the data, all the synthesized polynaphthoylenebenzimidazoles are soluble not only in sulfuric acid but in phenolic solvents as well. The majority of the systems are characterized by the high softening temperatures and high decomposition temperatures under the conditions of dynamic TGA.

However under the conditions of isothermal ageing, at temperatures above 300 °C, all the polymers considered undergo mass losses (Fig. 2, see p. 168) and hydrohalogens are mainly liberated. In a series of alternative dehydrohalogenation reactions – intermolecular and intramolecular – the latter seem more probable because the polymers retain solubility in H_2SO_4 during thermal treatment. Taking into account the high mobility of the pery-protons [198], it may be supposed that the most probable direction of dehydrohalogenation is the process presented by the Scheme 55:

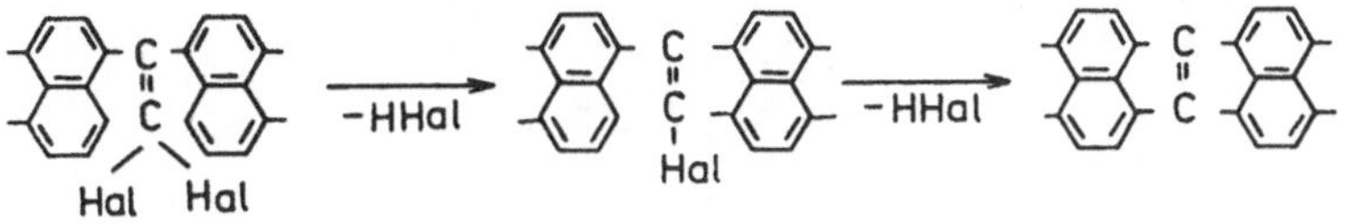

Scheme 55

Similar acenaphth-(9,10)-acenaphthylene structures were described in the literature [199]. The formation of polymers may lead to novel highly condensed

Table 27. Some properties of N-phenylbenzimidazole-containing polynaphthoylenebenzimidazoles of general formula:

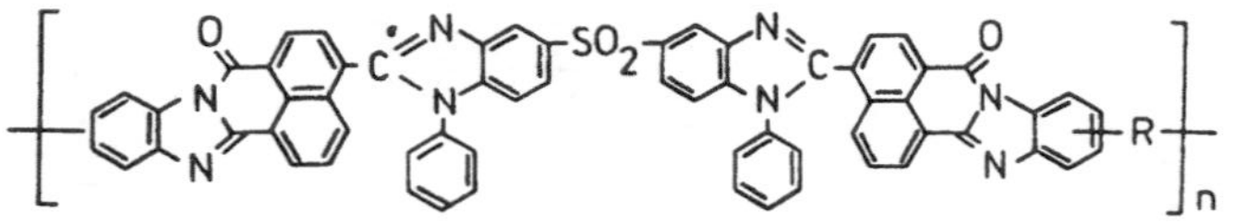

| –R– | η_{red}, dl/g H_2SO_4 25°C | T_g, °C | $T_{10\%}$, °C | Solubility | | | | Film properties 25°C | |
				m-cresol	TCE:phenol	N-MP	TCE	σ, MPa	ε, %
–	2.50	460	550	S	S	pS	ins	100	7
–O–	1.20	420	500	S	S	pS	pS	102	7
–CH₂–	0.90	430	500	S	S	pS	pS	108	65
–C–	—	430	500	S	S	pS	pS	100	7
	0.60	420	520	S	S	pS	S	—	—

Table 28. Viscosities and molecular weights of polynaphthoyl-enebenzimidazoles of general formula:

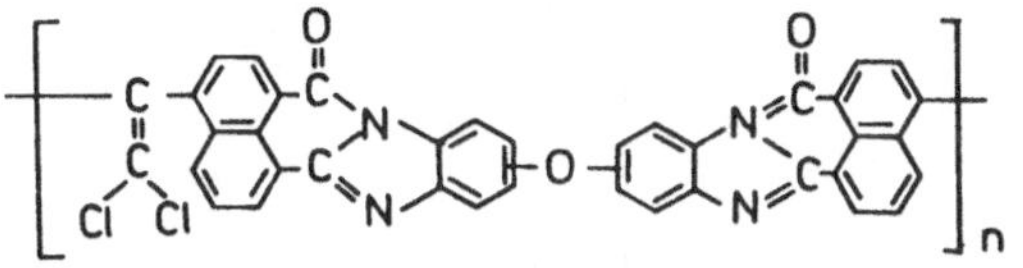

[197]

η_{intr}	$M \times 10^{-3}$
0.39	33
0.73	80
0.98	108
1.14	117

Table 29. Some properties of polynaphthoylenebenzimidazoles of general formula:

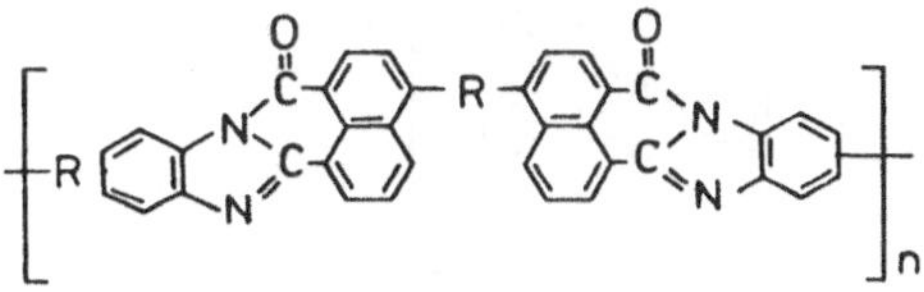

–R–	–R^1–	η_{red}, dl/g	T_g, °C	$T_{10\%}$, °C	Solubility		
					m-cresol	TCE:phenol (1:3)	H_2SO_4
—		0.80	450	500	S	S	S
–O–	–"–	2.00	380	500	S	S	S
–SO$_2$–	–"–	0.50	410	490	S	S	S
–CH$_2$–	–"–	0.70	370	490	S	S	S
–C(CF$_3$)$_2$–	–"–	0.80	380	500	S	S	S
–O–		0.30	380	490	S	S	S

systems of the general formula:

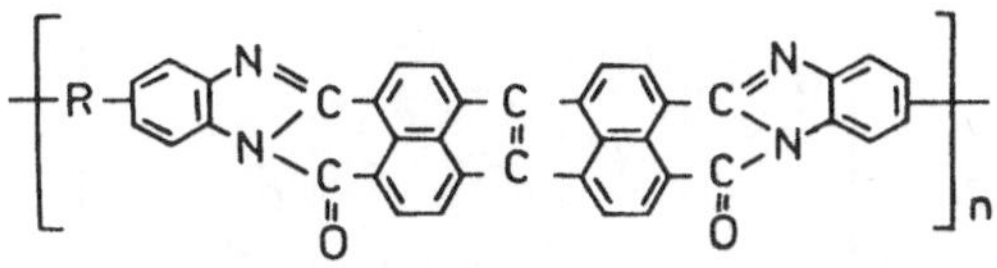

It allows us to consider all the results obtained as a novel approach to the synthesis of "ladder" ("double strand") polymers.

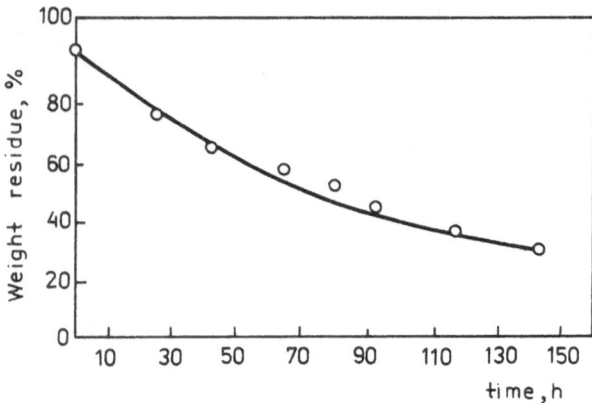

Fig. 2. Curve of the isothermal aging (at 320 °C in air)

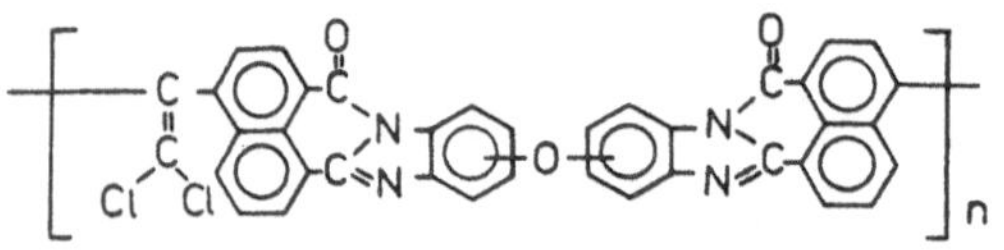

In this respect, polymers derived from 1,1-bis-naphthyl-4,4′,5,5′,8,8′-hexacarboxylic acid seem to be quite close to polynaphthoylenebenzimidazoles containing 1,1-dihalogenethylene groupings [200–203].

Polynaphthoylenebenzimidazole based on 3,3′,4,4′-tetraaminodiphenyl oxide and 1,1-binaphthyl-4,4′,5,5′,8,8′-hexacarboxylic acid dianhydride was studied in more detail; the optimal method for synthesis of the polymer is the reaction of monosodium salt of the acidic reaction component with bis(o-phenylenediamine) in a DMSO medium [201]:

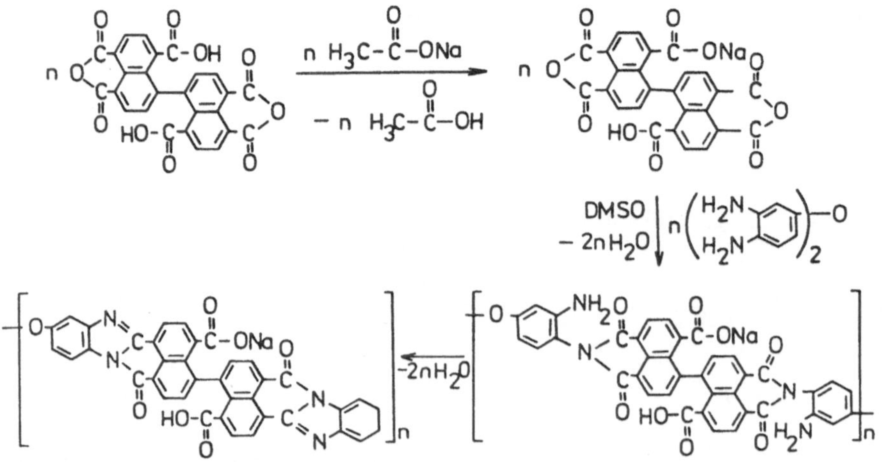

Scheme 56

Due to an increased basicity of the tetraamine compared to that of the diamine [185], the reaction for the preparation of polynaphthoylenebenzimidazole proceeded at 115–120 °C for 3–5 h, i.e. quicker than the synthesis of polynaphthylimides yielding a polymer with $\eta_{int} = 1.3$ dl/g. The synthesis of polynaphthoylenebenzimidazole at higher temperatures led to gel formation.

Polynaphthoylenebenzimidazoles were soluble in organic solvents only when a strong alkali solution (10 % KOH) was added. After precipitation into an acidic bath, the polymer undergoes 10–15 % decomposition.

Strong films were obtained through drying the reaction solutions of monosodium salts of polynaphthoylenebenzimidazoles: spinning of the fibers was carried out a bath composed of DMSO, water, and concentrated HCl in the ratio of 25:25:1.

Attention has been paid to the products of transformation of polynaphthoylenebenzimidazole into perylene [202] and anthanthrone-containing systems [203], Scheme 57. To transform the starting polynaphthoylenebenzimidazole

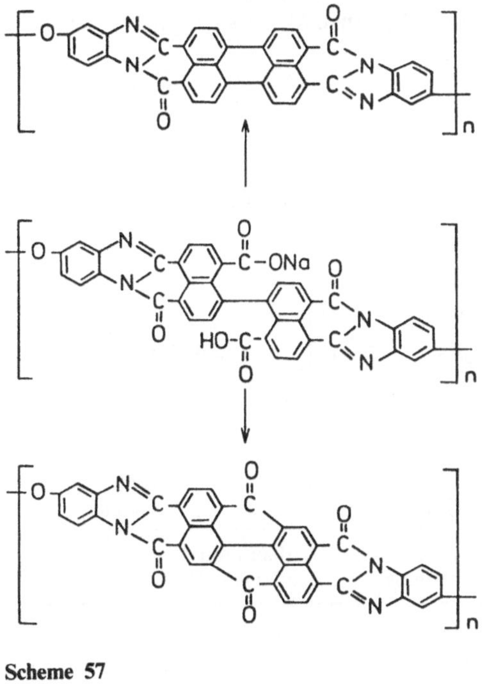

Scheme 57

into a structure containing perylene, decarboxylation was carried out in 10 % solution of KOH: attempts failed to prepare a 1% solution of the polymer in a less alkali medium. The degree of transformation of dicarboxynaphthyl fragments into the perylene fragments was 70 % and [η] of the polymer solutions attained the value 0.8 dl/g, which significantly exceeds the viscosity characteristics of polynaphthoylenebenzimidazoles based on perylene-3,4,9,10-tetracarboxylic dianhydride [204].

Cyclization reactions induced by H_2SO_4 were employed in the transformation of the starting polynaphthoylenebenzimidazole into the anthanthron-
containing structure. An efficient cyclization process is observed in 2 % oleum at
40 °C; under these conditions, cyclization proceeds up to 80 % over 2 h, and a
decrease in $[\eta]$ of the polymer does not exceed 0.1 dl/g.

Along with cyclization effected by alkali and acidic reagents, high-condensed
systems may be obtained by thermal treatment of the starting polynaphthoylenebenzimidazoles in an inert medium. Both cyclodehydration and cyclode-
carboxylation processes are observed. As a result, the thermally treated systems
contain both perylene and anthanthrone cycles. The thermal treatment is the
optimal method of cyclization of polynaphthoylenebenzimidazole fibers and
this is confirmed by the data in Table 30.

Products of the thermal treatment of the starting polynaphthoylenebenzimidazole have a higher LOI value than the threads "Arimid-PM" and "LOLA";
they are not inferior to "LOLA" with regard to the thermal decomposition
temperature but they have a worse thermal stability characterized by thermo-
ageing, and they are inferior to "Arimid-PM" threads in strength.

Table 30. Properties of fibers based on polynaphthoylenebenzimidazole

[201–203]

and products of its chemical and thermal cyclization

	Starting polymer	Products of chemical cyclization		Product of thermal treatment
		Perylene-containing	Anthanthrone-containing	
Tenacity, cN/teks	20–25	7	12	20–25
Elongation at break, %	20–25	13	15–27	15–30
Maximal mechanical modulus, GPa	5.7	—	—	6.9
$T_{init, degr}$, °C				
in air	530	540	550	550
in inert atm.	600	610	620	610
Zero strength temperature, °C	580	—	—	630
LOI, %	—	—	—	65

4 Conclusions

The data presented in this review testify to the appearance of a large series of novel bis(naphthalic anhydrides). These monomers are characterized by structural moieties which determine an enhanced solubility (and processability) of the derived polyheteroarylenes. Bis(naphthalic anhydrides) with ether groups may be employed in the preparation of systems similar to polyetheramides but having higher glass transition temperatures. The majority of novel bis(naphthalic anhydrides) may be used in the preparation of novel polynaphthoylenebenzimidazoles which combine satisfactory processability with very high thermal and chemical stability.

5 References

1. Segal CL (ed) (1967) High-temperature polymers. Marcel Dekker, New York
2. Korshak VV (1969) Thermally stable polymers. Nauka, Moscow
3. Korshak VV (1970) Chemical structure and thermal properties of polymers. Nauka, Moscow
4. Frazer AH (1968) High-temperature resistant polymers. Interscience, New York
5. Lee H, Stoffey D, Neville K (1967) New linear polymers. McGraw-Hill, New York
6. Cotter RJ, Matzner M (1972) Ring-forming polymerization Part B-1. Academic, New York
7. Cotter RJ, Matzner M (1972) Ring-forming polymerization Part B-2. Academic, New York
8. Bühler K-U (1978) Spezialplaste Akademie Verlag, Berlin
9. Cassidy PE (1980) Thermally stable polymers. Marcel Dekker, New York
10. Critchley JP, Wridth WW (1983) Heat-resistant polymers. Plenum, New York
11. Rusanov AL, Tugushi DS, Korshak VV (1988) Progress in polyheteroarylenes chemistry. TGU, Tbilissi
12. Kardash IE, Teleshov EN (1971) In: Results of science. Chemistry and technology of high-molecular compounds, vol 6, VINITI, Moscow p 3
13. Jones II (1968) J Macromolec Sci-Revs C2: 303
14. Marvel CS (1968) Pure Appl Chem 16: 351
15. Marvel CS (1973) Appl Polymer Symp 22: 47
16. Marvel CS (1975) J Macromolec Sci Revs Macromolec Chem 13: 219
17. Krongauz ES (1973) Uspekhi Khim 42: 1584
18. Izyneev AA, Teplyakov MM, Samsonova VG, Maximov AD (1967) Uspekhi Khim 36: 1090
19. Korshak VV, Teplyakov MM (1968) In: Progress in polymer chemistry. Nauka, Moscow 198
20. Korshak VV, Teplyakov MM (1971) J Macromolec Sci-Revs Macromolec Chem C5: 409
21. Mandric G (1972) St cerc chim 20: 1001
22. Korsak VV, Rusanov AL, Plieva L Kh (1977) Faserforsch Textiltechn 28: 371
23. Neuse EW (1982) Adv Polymer Sci 47: 1
24. Adrova NA, Bessonov MI, Layus LA, Rudakov AP (1968) Polyimides – a new class of thermally stable polymers. Nauka, Leningrad
25. Bessonov MI, Koton MM, Kudryavtsev VV, Layus LA (1983) Polyimides – class of thermally stable polymers. Nauka, Leningrad
26. Korshak VV, Rusanov AL, Batirov I (1986) New Trends in thermally stable polyimides. Donish, Dushanbe
27. Wilson D, Stenzenberger HD, Hergenrother PM (eds) (1990) Polyimides. Blackie, New York
28. Mittal K (ed) V1 (1984) Polyimides: synthesis, characterization and applications. Plenum, New York
29. Mittal K (ed) V2 (1984) Polyimides: synthesis, characterization and applications. London Plenum, New York
30. Sroog CE (1976) J Polymer Sci Macromolec Revs 11: 161

31. Sroog CE (1991) Progr Polymer Sci
32. Tessler MM (1967) ACS Polymer Prepr 8: 152
33. Tessler MM (1966) J Polymer Sci A1 4: 2521
34. Overberger CG, Moore JA (1970) Adv Polym Sci 7: 113
35. Berlin AA, Liogon'kiy BI, Shamraev GM (1971) Uspekhi Khim 40: 513
36. Nartsissov B (1974) Macromolec Sci-Revs C11(1): 143
37. Rusanov AL, Leont'eva SN, Iremashvili Ts G (1975) Uspekhi Khim 44: 151
38. Rusanov AL (1979) Uspekhi Khim 48: 115
39. Korshak VV, Rusanov AL (1981–2) J Macromol Sci-Revs C21(2): 275
40. Dawans F, Marvel CS (1965) J Polymer Sci A3: 3549
41. Sillion B, Reboult A (1966) Compt Rend 262C: 471
42. Bell VL (1967) J Polymer Sci B5: 941
43. Bell VL, Jewell RA (1967) J Polymer Sci A5: 3043
44. Colson JG, Michel RH, Paufler RM (1966) J Polymer Sci A4: 59
45. Korshak VV, Rusanov AL, Katsarava RD (1968) Dokl Acad Nauk SSSR 178: 105
46. Korshak VV, Rusanov AL, Katsarawa RD (1969) Vysokomolek Soyed A11: 2090
47. Korshak VV, Rusanov AL (1968) Izv Acad Nauk SSSR, ser khim 2261
48. Kersten H, Meyer G (1970) Makromol Chem 138: 265
49. Korshak VV, Rusanov AL, Leont'eva SN, Jashiashvili TK (1975) Vysokomolek Soyed Á17: 228
50. Korshak VV, Rusanov AL, Leont'eva SN, Jashiashvili TK (1975) Macromolecules 8: 582
51. Rabilloud G, Sillion B, de Gaudemaris G (1966) Compt Rend 263C: 862
52. Rabilloud G, Sillion B, de Gaudemaris G (1967) Makromol Chem 108: 18
53. Kurihara M, Yoda N (1967) Bull Chem Soc Japan 40: 2429
54. Kurihara M, Yoda N (1968) J Polymer Sci B6: 875
55. Kurihara M (1970) Macromolecules 3: 722
56. Korshak VV, Doroshenko Yu E, Kharitonova NK (1971) Dokl Acad Nauk SSSR 200: 114
57. Korshak VV, Rusanov AL, Pavlova SA, Gribkova PN, Mikadze LA, Bochvar DA, Stankevich IV, Tomilin OB (1969) Dokl Acad Nauk SSSR 184: 95
58. Korshak VV, Pavlova SA, Gribkova PN, Mikadze LA, Rusanov AL, Plieva L Kh, Lekae TV (1977) Izv Acad Nauk SSSR, ser khim 1381
59. Plonka Z Yu, Albrecht VM (1965) Vysokomolek Soyed A7: 2117
60. Van Deusen RL, Goins OE, Sicree AJ (1966) ACS Polymer Prepr 7: 528
61. Van Deusen RL (1966) J Polymer Sci B4: 211
62. Van Deusen RL, Goins OK, Sicree AJ (1968) J Polymer Sci A6: 1777
63. Arnold FE, Van Deusen RL (1969) Macromolecules 2: 497
64. Arnold FE, Van Deusen RL (1971) J Appl Polymer Sci 15: 2035
65. Gerber AH (1973) J Polymer Sci Polymer Chem Ed 11: 1703
66. Berlin AA, Liogon'kiy BI, Shamraev GM, Belova GV (1966) Izv Acad Nauk SSSR, ser khim 945
67. Shamraev GM, Dula AA, Liogon'kiy BI, Berlin AA (1970) Vysokomolek Soyed A12: 401
68. Korshak VV, Rusanov AL, Pavlova SA et al. (1975) Dokl Acad Nauk SSSR 221: 1334
69. Belyakov VK, Belyakova IV, Medved SS et al. (1971) Vysokomolek Soyed A13: 1739
70. Krasnov EP, Aksenova VA, Belyaev AA et al. (1973) Vysokomolek Soyed A15: 1606
71. Krasnov EP, Aksenova VA, Khar'ka SN (1973) Vysokomolek Soyed A15: 2093
72. Imai Y, Ueda M, Sei-i gakkaishi (1975) J Soc Fiber Sci Technol Japan 31(5): 135
73. Elberson L, Landstrom L (1972) Acta Chem Scand 26: 239
74. Koton MM, Kudryavtsev VV, Adrova NA, Kalnin'sh KK, Dubkova AM, Svetlychniy VM (1974) Vysokomolek Soyed A16: 2081
75. Vinogradova SV, Vygodskiy Ya S, Korsak VV, Spirina TN (1979) Acta Polymerica 30: 3
76. Zapadinskiy BI, Liogon'kiy BI, Berlin AA (1973) Uspekhi Khim 42: 2037
77. Harris FW, Feld WA, Lanier LN (1976) ACS Polymer Prepr 17(2) : 353
78. Harris FW, Lanier LN (1977) In: Structure-solubility relationships in Polymers. Academic, New York, p 183
79. Dashevskiy MM (1966) Acenaphthene. Khima, Moscow
80. Suvorov BV, Sembaev D Kh, Tkacheva GD et al. (1981) In: Chemistry and physics of high-molecular compounds 14(55), Alma-Ata, Nauka, 3
81. Jones LA, Watson R (1972) Can J Chem 51: 1833
82. Gurdeska K, Kabas G, Pugin A et al. (1975) US Pat 3,898,234
83. Dziewonski K, Kahl W, Koczorowska W, Wulfson A (1933) Roczn Chem 13: 154
84. Robinson JG, Brain SA (1982) US Pat 4,334,055

85. Titov VI, Moskvichev Yu A, Timoshenko GN, Mironov GS (1980) USSR Pat 891,612
86. Morgan G, Harrison H (1928) J Soc Chem Ind 47: 16
87. Morgan G, Harrison H (1930) J Soc Chem Ind 49: 413
88. Jedlinski Z, Gaik U, Mzyk Z, Fudal M, Kowalski B (1979) ACS Polymer Prepr 20(2): 559
89. Jedlinski Z, Kowalski B, Gaik U (1992) Polish J Chem (in press)
90. Germ Pat 545 212 (1933) Frdl 18: 1510
91. Jedlinski Z, Kowalski B, Fabjansca-Sweca G, Gaik U (1992) Angew Makromol Chem (in press)
92. Kravchenko TV, Dvalishvili TI, Romanova TA, Tkacheva GD (1982) Vysokomolek Soyed B24: 852
93. Rudaya LI, Smolenkova LA, Kvitko I Ya, El'tsov AV, Kravchenko TV, Dvalishvili TI (1982) USSR Pat 977,457. Bull Izobr 44
94. Rusanov AL, Berlin AM, Mironov GS, Moskvichev Yu A, Kolobov GV, Korshak VV (1981) Vysokomolek Soyed A23: 1586
95. Dzewonski K, Grünberg B, Shoena I (1930) Bull Int Acad Polon Sciences Letters A: 518
96. Loughran GA, Arnold FE, US Pat 3,987,003
97. Loughran GA, Arnold FE (1977) ACS Polymer Prepr 18(1): 831
98. Loughran GA, Arnold FE (1976) Org Prepr Proced Int 8: 98
99. Jedlinski Z, Kowalski B, Gaik U (1983) Macromolecules 16: 522
100. Jedlinski Z (1984) Macromol Chem Suppl 7: 17
101. Jedlinski Z, Gaik U, Kowalski B (1986) Pol Pat 131,685; (1987) R Zh Khim 18S554P
102. Jedlinski Z, Kowalski B (1986) Pol Pat 133,225; (1987) R Zh Khim 8S588P
103. Williams FJ (1974) US Pat 3,850,964
104. Jedlinski Z (1981) Pol Pat 34,517
105. Korshak VV, Rusanov AL, Berlin AM et al. (1989) Vysokomolek Soyed A31: 51
106. Ustinov VA, Perepechkina EP, Plakhtinskiy VV et al. (1979) USSR Pat 653,249; (1979) Bull Izobr 11
107. Korshak VV, Rusanov AL, Berlin AM et al. (1984) Vysokomolek Soyed B26: 713
108. Korshak VV, Rusanov AL, Berlin AM et al. (1986) Chemia Stosovana 30: 171
109. Korshak VV, Bulycheva EG, Berlin AM et al. (1987) Abstracts of communications at XI International Microsymposium on polycondensation. Prague, A-7
110. Rusanov AL, Korshak VV, Shalikiani MO et al. (1987) Abstracts of communications at 31st IUPAC Macromolecular Symposium. Merseburg, GDR, 187
111. Korshak VV, Rusanov AL, Berlin AM et al. (1980) USSR Pat 784,258; (1988) Switz Pat 666,032; (1988) Fr Pat 2,584,407; (1989) Brit Pat 2,176,778; (1989) Ind Pat 164,541; (1987) Ger Offen DE 3,526,155
112. Brit Pat 1,069,337 (1965). To Farbenfabriken Bayer Aktiengesellschaft.
113. Korshak VV, Rusanov AL, Berlin AM et al. (1988) Dokl Acad Nauk Tadzh SSR 31,8,526
114. Korshak VV, Bulycheva EG, Shifrina ZB et al. (1988) Acta Polymerica 39: 460
115. Korshak VV, Rusanov AL, Jedlinski Z et al. (1991) Acta Polymerica 42: 53
116. Korshak VV, Rusanov AL, Berlin AM et al. (1988) Dokl Acad Nauk SSSR 301: 115
117. Korshak VV, Rusanov AL, Prigozhina MP et al. (1990) Plastmassy 3: 5
118. Bulycheva EG, Prigozhina MP, Ponomarev II et al. (1991) Acta Polymerica 42: 63
119. Sonnenberg J (1977) US Pat 4,002,645; (1977) CA 140666x
120. Mognonov DM, Korshak VV, Izyneev AA, Batotsyrenova AI, Batlayev KE (1978) USSR Pat 659,571; (1979) Bull Izobr 16
121. Yamazaki Y, Suzuki T, Ohkiho Y (1973) Nippon Kagaku Kaishi 1073
122. Yamazaki Y, Suzuki T (1973) Nippon Kagaku Kaishi 1071
123. Korshak VV, Rusanov AL, Berlin AM, Fidler S Kh, Jashiashvili TK, Tugushi DS (1984) USSR Pat 1,119,322
124. Shifrina ZB, Rusanov AL (1992) Vysokomolek Soyed (in press)
125. Rusanov AL, Korshak VV, Batirov I, Shifrina ZB, Tabidze RS, Bocharov SS (1986) I Regional Conference of Middle Asia Republics and Kazakhstan on chemical reactives Abstracts Dushanbe, 77
126. Korshak VV, Rusanov AL, Lekae TV (1983) USSR Pat 1,037,646
127. Korshak VV, Rusanov AL, Lekae TV (1984) USSR Pat 1,100,871
128. Korshak VV, Rusanov AL, Lekae TV (1988) Vysokomolek Soyed B30: 439
129. Rusanov AL, Bocharov SS, Lekae TV, Bulycheva EG, Batirov I, Kolontarov I Ya (1989) Dokl Acad Nauk Tadj SSR 32: 463
130. Hergenrother PM (1971) J Macromol Sci-Revs C6: 1
131. Hergenrother PM (1976) Polymer Eng Sci 16: 303

132. Krongauz ES (1977) Uspekhi Khim 46: 112
133. Korshak VV, Rusanov AL (1983) Uspekhi Khim 52: 812
134. Eskert W (1931) Germ Pat 535,069; (1933) Frdl 13: 614
135. Dokunikhin NS, Vorozhtsov GN, (1966) Zhurn Org Khim 2: 148
136. Dokunikhin NS, Vorozhtosov GN, Kichina FI (1968) Zhurn Org Khim 4: 2000
137. Vorozhtsov GN, Dokunikhin NS, Fel'dblum NB (1975) Zhurn Org Khim 11: 1517
138. Vorozhtsov GN, Dokunikhin NS, Vorozhtsova NI (1975) Zhurn Org Khim 11: 1499
139. Stepanov BI (1977) Introduction to the chemistry and technology of organic dyestuffs. Khimia, Moscow
140. Dokunikhin NS, Vorozhtsov GN, Alekseev VI (1981) Khim prom, 592
141. Korshak VV, Rusanov AL (1989) Uspekhi Khim 56: 1006
142. Rusanov AL (1990) In: Results of science Chemistry and technology of high-molecular compounds. VINITI, Moscow 26: 3
143. Korshak VV, Rusanov AL Fidler S Kh et al. (1984) Plastmassy (10): 28
144. Kekharsayeva ER, Shustov EB, Mikitayev AK, Dorofeev VT (1985) Plastmassy (2): 9
145. Lesiak T, Nowakowski (1980) J Polimery 25: 81
146. Lesiak T, Nowakowski J (1981) Polimery 26: 1
147. Luknitskii FI (1975) Chem Revs 75: 259
148. Korshak VV, Rusanov AL, Berlin AM, Mironov GS et al. (1980) USSR Pat 866,999
149. Korshak VV, Rusanov AL, Berlin AM, Bulycheva EG, Shalikiani MO, Mironov GS, Moskvichev Yu A, Timoshenko GN, Titov VI (1988) Dokl Acad Nauk SSR 299: 131
150. Korshak VV, Bulycheva EG, Shifrina ZB, Berlin AM et al. (1992) Vysokomolek Soyed (in press)
151. Korshak VV, Rusanov AL, Berlin AM, Shalikiani MO, Bakhtadze IG, Putkaradze NV (1986) Soobshch Acad Nauk Gruz SSR 121: 537
152. Korshak VV, Kosobutskiy VA, Bolduzev AI, Rusanov AL et al. (1980) Izv Acad Nauk SSSR, ser khim 1553
153. Pebalk DV, Kotov BV, Neiland O Ya et al. (1974) Dokl Acad Nauk SSSR 236: 1379
154. Korshak VV (1982) Uspekhi Khim 51: 2096
155. Korshak VV, Kazakova GV, Rusanov AL (1989) Vysokomolek Soyed A31: 5
156. Korshak VV, Rusanov AL, Batirov I (1982) Plastmassy 8: 14
157. Rusanov AL, Bulycheva EG (1991) Abstracts of Posters. STEPI-2. Montpellier, France, PI-12
158. Sek D, Piget P, Wanic A (1991) Abstracts of Posters. STEPI-2. Montpellier, France, PI-9
159. Kalashnikov BO, Efros LS, Vorozhtsov GN et al. (1986) Vysokomolek Soyed B28: 232
160. Kuznetsova GB, Silinskaya IG, Kallistov OV et al. (1988) Vysokomolek Soyed A30: 586
161. Kuznetsova GB, Kalashnikov BO, Lazareva MA et al. (1989) Vysokomolek Soyed B31: 763
162. Shifrina ZB, Rusanov AL, Urman Ya G (1992) Izv Acad Nauk SSSR, ser khim (in press)
163. Rusanov AL, Shifrina ZB (1990) All-Union Conf "Fundamental problems of modern science on polymers". Leningrad, Nov 1990 Abstracts P1.P2.-L, 41
164. Korshak VV, Rusanov AL, Shifrina ZB et al. (1986) Dokl Acad Nauk SSSR 289: 367
165. Korshak VV, Pavlova SA, Gribkova PN et al. (1977) Izv Acad Nauk SSSR, ser khim N6: 1381
166. Steinberg JM (1970) US Pat 3,539,677
167. Dunay M, Parker JA (1971) US Pat 3,562,380
168. Chem Eng News (1968) Sept 23: 40
169. Gloor WH (1967) In: Fiber spinning and drawing. Coplan A (ed). Interscience Publishers, New York – London – Sidney – Toronto, p 151
170. Chem Week (1966) Sept 19: 121
171. Gloor WH (1966) ACS Polymer Prepr 7(2): 819
172. Gloor WH (1967) Appl Polymer Symp 6: 151
173. Gloor WH (1969) Appl Polymer Symp 9: 159
174. Kudryavtsev GI, Shchetinin AM (1968) Khim Volok 6: 2
175. Khim Volok (1975) 3: 68
176. Kudryavtsev GI, Askadskiy AA, Khudoshev IF (1978) Vysokomolek Soyed A20: 1879
177. Gaik U, Kowalski B, Jedlinski Z, Rusanov AL, Berlin AM, Fidler S Kh, Korshak VV (1981) Izv Acad Nauk Kaz SSR, ser khim 5: 19
178. Jedlinski Z, Gaik U, Kowalski B, Korshak VV, Rusanov AL, Fidler S Kh (1982) Makromol Chem 183: 1615
179. Gaik U, Kowalski B, Rusanov AL, Berlin AM, Fidler S Kh, Korshak VV, Jedlinski Z (1981) VIII International Microsymposium on polycondensation. Alma-Ata, April. Absts 1

180. Rusanov AL, Berlin AM, Fidler S Kh, Mironov GS, Moskvichev Yu A, Kolobov GV, Korshak VV (1981) Vysokomolek Soyed A23: 1586
181. Korshak VV, Rusanov AL, Berlin AM, Fidler S Kh (1979) USSR Pat 652,194; (1979) Bull Izobr 10: 108
182. Korshak VV, Rusanov AL, Berlin AM, Fidler S Kh, Adyrkhaeva FI (1979) Vysokomolek Soyed A21: 68
183. Korshak VV, Rusanov AL, Berlin AM, Fidler S Kh, Livshits BR, Dymshits T Kh, Avakova NA (1978) USSR Pat 587,139; Bull Izobr 1: 70
184. Korshak VV, Rusanov AL, Berlin AM, Fidler S Kh, Livshits BR, Dymshits T Kh, Silyutina LI, Blinov VF (1979) Vysokomolek Soyed A21: 657
185. Balyatinskaya LN, Milyaev Yu F, Korshak VV et al. (1978) Dokl Acad Nauk SSSR 238: 862
186. Korshak VV, Krongauz ES, Rusanov AL, Travnikova AP, Gitis SS, Aleksandrov VN, Pugacheva SA (1973) USSR Pat 398,579; (1973) Bull Izobr 38: 77
187. Korshak VV, Krongauz ES, Travnikova AP, Rusanov AL, Katsarava RD (1971) Dokl Acad Nauk SSSR 196: 106
188. Korshak VV, Krongauz ES, Rusanov AL, Travnikova AP (1974) Vysokomolek Soyed A16: 35
189. Korshak VV, Krongauz ES, Travnikova AP, Rusanov AL (1974) Macromolecules 7: 589
190. Jedlinski Z, Kowalski B, Gaik U (1985) Makromol Chem 186: 457
191. Gaik U, Kowalski B, Jedlinski Z et al. (1985) Thermochim Acta 83: 309
192. Smigasiewicz S, Kowalski B (1984) Mater Sci (PRL) 10: 491
193. Smigasiewicz S, Kowalski B (1986) J Polym Sci Phys Ed 24: 1961
194. Rusanov AL, Bulycheva EG, Bocharov SS (1990) J Thermal Analysis 36: 1685
195. Korshak VV, Rusanov AL, Berlin AM, Polina TV (1980) USSR Pat 788,689
196. Korshak VV, Rusanov AL, Berlin AM et al. (1981) USSR Pat 866,999
197. Tsvetkov VN, Nowakowski VB, Strelina IA et al. (1989) Vysokomolek Soyed A31: 40
198. Reinhardt BA, Tsai TT, Arnold FE (1984) In: New monomers and polymers. Plenum Press, New York – London, p 41
199. Letsinger RL, Gilpin GA, Vulvo WJ (1962) J Org Chem 27: 672
200. Korshak VV, Rusanov AL, Berlin AM et al. (1981) Vysokomolek Soyed B23: 42
201. Kalashnikov BO, Efros LS, Strelets B Kh et al. (1986) Vysokomolek Soyed B28: 671
202. Kalashnikov BO, Efros LS, Strelets B Kh et al. (1987) Vysokomolek Soyed A29: 2415
203. Sadekova RA, Pronichkina IK, Perepechkina EP et al. (1987) Khim Volok 1: 29
204. Berlin AA, Liogon'kiy BI, Zapadinskiy BI, Shamraev GM (1969) TUPAC International Symposium on Macromolecular Chemistry, August. Budapest. Abstracts I: 18

Editor: Prof. K. Dušek
Received January 29, 1992

Recent Developments in the Synthesis, Thermostability and Liquid Crystal Properties of Aromatic Polyamides

Jian Lin[1] and David C Sherrington*
Department of Pure and Applied Chemistry, University of Strathclyde,
Thomas Graham Building, 295 Cathedral Street, Glasgow G1 1XL, UK

Wholly aromatic polyamides were developed and commercialised some decades ago, and since then a great deal of effort has been expended to try to improve in particular the processability of the early materials. Work has focussed on attempting to improve polymer solubility in organic solvents, and to reduce the melting point of materials, by synthesizing a variety of macromolecular backbones with many different types of substituents. The challenge has been to modify these polymers, while not causing adverse changes in the unique and useful physical properties of this group of polymers. This has led to the development of entirely novel synthetic strategies to allow, for example, the introduction of potentially reactive groups, and also to an interest in the potential lyotropic and thermotropic liquid crystalline properties of these polymers. This review describes the various synthetic methodologies which have emerged in recent years, and then summarises and categorises progress in the understanding of the structure-property relationships of the polymers produced. Properties such as solubility, thermostability, lytropism and thermotropism are highlighted.

*To whom all correspondences should be addressed.
[1] Present address: Shell Research Limited, Sittingbourne Research Centre, Sittingbourne, Kent ME9 8AG, UK.

Advances in Polymer Science, Vol. 111
© Springer-Verlag Berlin Heidelberg 1994

1 Introduction

Some three decades ago, scientists from the Du Pont company developed polycondensation methods which allowed the preparation of high molecular weight wholly aromatic polyamides. The first commercially produced wholly aromatic polyamide fibre was poly(m-phenyleneisophthalamide) (Nomex, Du Pont, 1967) [1a, c]. Some years later, development of the preparation and processing of poly(p-phenyleneterephthalamide) (PPTA) led to the commercialization of the para product Kevlar (Du Pont) in the early seventies [1b, c]. While Nomex shows excellent thermal stability and flame-retardance, and indeed is referred to as a heat and flame resistant aramid fibre, Kevlar fibre also has similar properties, but in addition it has exceptional tensile strength and modulus, and is referred to as an ultra-high strength, high modulus aramid fibre.

Although these products have become of great commercial importance, the fabrication of unsubstituted aromatic polyamides has in general proved to be difficult because they show a tendency to decompose during, or even before melting and are insoluble in most common organic solvents [1d]. There has been therefore an increased interest in the preparation of polyamides with different substituents or structural irregularities in order to improve their processability. The introduction of substituents also makes it possible to incorporate functional groups into aromatic polyamides, which might allow further modification of the reactive centres. This is particularly intriguing in terms of curing of an initially easily processable polymer [2].

The relationship between the primary structure of polyamides and their properties has been studied extensively and reported in the literature. The polymers used in these studies were usually made by solution polycondensation methods. While the number of novel aromatic polyamides is constantly increasing, new synthetic methods have also been developed, along with optimisation of more conventional methods.

It seems timely therefore to review the most used successful synthetic methods for laboratory scale preparation of aromatic polyamides, and the structure-property relationships of the resultant polyamides. We intend to do so by illustrating with some typical examples from the literature, rather than by providing an exhaustive catalogue of polyamides made to date. The necessity for novel methods stems from the fact that in making polyamides with specific structures, functionalised monomers may be needed, and these can have quite different properties such as solubility and reactivity. Alternatively it might be easier to synthesize and purify monomers in another functional form, instead of diacid chloride or diamine. Overall of course, the development of novel polyamides depends on what is understood already in terms of structure-property relationships. It is hoped that this review will provide some general guidelines for the choice of an appropriate method as well as for the design of new structures.

The first part of this work deals with the syntheses of polyamides. For each method, some important factors governing the reaction are discussed and examples given to illustrate a typical preparation under appropriate experimental conditions.

In the second part, examples of polyamides with different structures are presented in order to provide an insight into the relationship between structure and such properties as thermal stability, solubility, liquid crystallinity.

2 Synthetic Methods for the Preparation of Aromatic Polyamides

2.1 Solution Polycondensation of Diamine and Diacid Chloride

The high temperature melt polycondensation method involving a diamine and a diacid has been successfully used in the synthesis of aliphatic polyamides, especially in the fabrication of nylon [1e, f]. Nevertheless this method is not applicable to the preparation of wholly aromatic polyamides because the potential polyamides either do not melt or melt at high temperature with decomposition. Many aromatic diacids will decarboxylate and the aromatic diamines are readily oxidized and have a tendency to sublime [1d]. Scientists from the Du Pont company then developed low temperature polycondensation methods either in solution or at the interface of two solvents. Solution polycondensation involves a diamine and a diacid chloride reacting in an amide solvent such as N-methylpyrrolidone (NMP), hexamethylphosphoramide (HMPA), or dimethylacetamide (DMAc) (Scheme 1) [3].

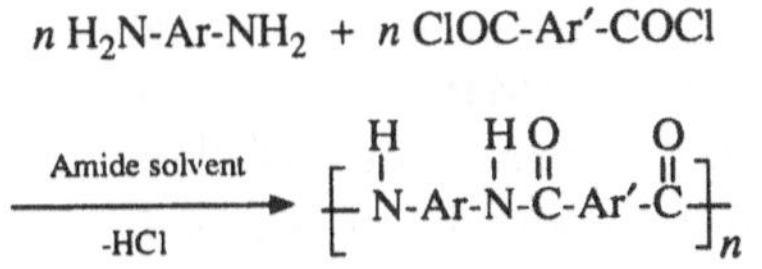

$$n \text{ H}_2\text{N-Ar-NH}_2 + n \text{ ClOC-Ar'-COCl}$$

$$\xrightarrow[\text{-HCl}]{\text{Amide solvent}} \left[\begin{matrix} \text{H} & \text{H} & \text{O} & & \text{O} \\ | & | & \| & & \| \\ \text{N-Ar-N-C-Ar'-C} \end{matrix} \right]_n$$

Scheme 1

The amide solvent serves also as an acid acceptor for the hydrogen chloride produced in the reaction. Other polar aprotic solvents such as dimethylformamide (DMF) and dimethylsulfoxide (DMSO) cannot be used because they react significantly with acid chlorides [1d]. Amide solvents can interact efficiently with the aromatic amide polymers of moderately high molecular weight formed during the condensation reaction, and therefore provide maximum swelling. Consequently this allows the macromolecular chain to continue to grow until completion. Many of the key factors required in polycondensation for obtaining high molecular weight polymers have been studied [4]. These are:

a) stoichiometry of the monomers
b) purity and concentration of monomers
c) temperature of the reaction medium and reaction time
d) nature of the solvent(s) and addition of salt
e) solubility (swellability) of the polymer
f) speed of stirring.

While the importance of purity and stoichiometry of monomers seem to be obvious, the concentration of monomers can also influence considerably the molecular weight of the final polymer. High concentration leads to an highly viscous reaction medium, therefore the mobility of macromolecular chains can be restricted. Also at high concentration, heat generation favours side reactions. At low concentration, competitive side reactions described by Herlinger et al. [5] such as reaction of diacid chloride with the solvent can become significant because of the lower rate of the condensation reaction. The temperature should be adjusted to ensure that the condensation reaction is much faster than the side reactions. In general low temperatures are favoured in this method. The solvent should allow maximum solubility (swellability) of the polymer formed at the early stage of polycondensation, and the solvation properties of amide solvents can usually be increased by addition of salts such as LiCl or $CaCl_2$. Fast precipitation of the polymer needs to be avoided, because it prevents further growth of the macromolecular chains.

Using a solution method, PPTA with $\eta_{inh} = 6.93$ dl/g (C = 0.5 g/dl in conc. H_2SO_4 at 30 °C) was obtained under the following conditions [4c]: A solution of p-phenylenediamine (0.25 M) in HMPA/NMP (2:1) was cooled and stirred at − 15 °C. To this solution, powdered terephthaloyl chloride (1 equiv) was added in one portion with rapid stirring. After 5 min at this temperature and 12 h at room temperature, the polymer was isolated.

The PPTA can be also successfully prepared in a modified solvent mixture, DMAc/HMPA (1 : 1.4) [2b]. Subsequently as HMPA was proved to be carcinogenic in rats [6], scientists from Teijin (Japan) and Akzo (Holland) developed a method involving NMP as the solvent with added $CaCl_2$ [7].

2.2 Polycondensation of Diamine and Diacid via Phosphorylation or with Phosphorus-Containing Activating Agents

The low temperature solution method described above has been used successfully in the preparation of numerous polyamides by reacting different diamines with diacid chlorides. Nevertheless, in this method, the diacid needs to be converted to its diacid chloride. Although the synthesis itself does not seem to be difficult, and usually involves treatment with thionyl chloride, this is an additional reaction step and produces environmentally undesirable SO_2 and HCl. Also the stability of diacid chloride towards hydrolysis is another problem, in terms of purification and storage. In addition when other functional groups are present on the same molecule, they need to be protected before treating with $SOCl_2$. These problems have stimulated the search for other procedures.

In 1974, Higashi et al. described a novel procedure to prepare aromatic polyamides. This reaction involved the complex 1 of an acid with triphenylphosphite in NMP and pyridine [8] (Scheme 2). $CaCl_2$ and LiCl were used along with NMP to improve the molecular weight of the polymer obtained in the synthesis of para wholly aromatic polyamides.

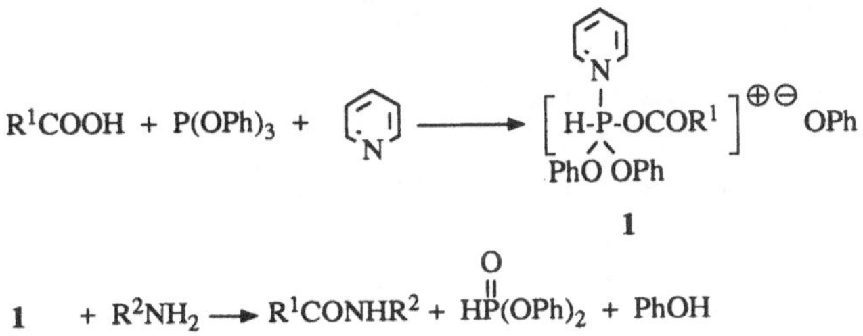

$$R^1COOH + P(OPh)_3 + \underset{N}{\bigcirc} \longrightarrow \left[\underset{PhO\ OPh}{\overset{\overset{\displaystyle\bigcirc}{\underset{N}{|}}}{H\text{-}P\text{-}OCOR^1}} \right]^{\oplus\ominus} OPh$$

1

$$\mathbf{1} \quad + R^2NH_2 \longrightarrow R^1CONHR^2 + \overset{O}{\overset{\|}{HP}}(OPh)_2 + PhOH$$

Scheme 2

The initial study was followed by an extensive investigation of the reaction conditions by the group [9]. Eventually, Krigbaum et al. optimized these and successfully obtained PPTA with $\eta_{inh} = 6.2$ dl/g (C = 0.1 g/dl in conc. H_2SO_4 at 25 °C) [9], which was comparable with the polymer obtained from p-phenylenediamine and terephthaloyl chloride [9].

In this synthetic method involving phosphorylation, several key factors can considerably influence the molecular weight of the final polymer:

a) concentration of monomers
b) ratio of triphenylphosphite to monomer
c) reaction temperature and time
d) concentration of LiCl and $CaCl_2$
e) solvent and amount of pyridine relative to the metal salt.

Gelation often occurs in this reaction but the polycondensation does not cease. The role of the $CaCl_2$ and LiCl salts is quite complicated. They can form complexes with pyridine, i.e. LiCl-2Py and $CaCl_2$-nPy (with up to 8 ligands). These are more soluble in NMP than the salts alone, and NMP with a higher content of metal salt can solubilize more polyamide formed in the reaction medium, leading in due course to higher molecular weight products.

Optimum conditions for the synthesis of a PPTA with $\eta_{inh} = 6.2$ dl/g are indicated in the following example: A solution of NMP (65 ml) containing dissolved $CaCl_2$, LiCl and pyridine (10 ml) was added at room temperature to a mixture of terephthalic acid, p-phenylenediamine (PPD) and triphenylphosphite (TPP). The mixture was heated at 115 °C for 100 min; monomer concentration: 0.083 M; ratio TPP/monomer = 2; Py/($CaCl_2$ + LiCl) = 2.5 mol/mol, $CaCl_2$/LiCl = 1.2 mol/mol).

Since the successful use of phenylphosphine dichloride 2 [11], hexachloro-cyclotriphosphatriazene 3 [12], triphenylphosphine 4 [13] and diphenylchloro-phosphate 5 [14] in the synthesis of polyesters, Ogata et al. studied direct polycondensation of diacid and diamine in the presence of triphenylphosphine dichloride (TPPCl$_2$) [15b] which can form in situ when triphenylphosphine is treated with hexachloroethane or when triphenylphosphine oxide interacts with phosgene or oxalyl chloride (Scheme 3) [15a].

$$Ph_3P + C_2Cl_6 \longrightarrow Ph_3PCl_2 + C_2Cl_4$$

$$Ph_3PO + COCl_2 \longrightarrow Ph_3PCl_2 + CO_2$$

$$Ph_3PO + (COCl)_2 \longrightarrow Ph_3PCl_2 + CO + CO_2$$

$$PhPCl_2 \qquad\qquad 2 \qquad\qquad 3 \qquad\qquad Ph_3P \quad 4 \qquad (PhO)_2PCl \quad 5$$

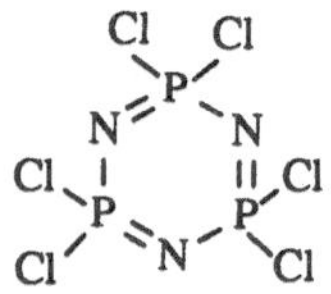

Scheme 3

One of the advantages of this approach is the possibility of reconverting the by-product triphenylphosphine oxide to TPPCl$_2$ by treating it with more oxalyl chloride or phosgene. During the polycondensation reaction, the TPPCl$_2$, which exists in form of an ion pair, reacts with diacid to form the activated inter-mediate 6 in the presence of pyridine, which in turn reacts with the diamine to form the polymer (Scheme 4) [15a].

$$2\left[Ph_3P\text{-}Cl\right]^{\oplus} Cl^{\ominus} + HOOC\text{-}Ar^1\text{-}COOH \xrightarrow[-HCl]{} Cl^{\ominus\oplus}\left[Ph_3P\text{-}OOCAr^1\text{-}COO\text{-}PPh_3\right]^{\oplus} Cl^{\ominus}$$

$$\mathbf{6}$$

$$n \quad \mathbf{6} + n \ H_2N\ Ar\ NH_2 \xrightarrow[-Ph_3PO]{} \left[N\text{-}Ar\text{-}N\text{-}C\text{-}Ar^1\text{-}C\right]_n$$

Scheme 4

The following example has been described by Ogata et al. in the synthesis of a copolyamide [15b]: A solution was prepared from terephthalic acid (2 mmol) and triphenylphosphine (4.8 mmol) in pyridine (5 ml); this was then added to a 5 ml pyridine solution containing p-phenylenediamine (2 mmol), 4,4'-diamino-diphenylether (2 mmol) and C$_2$Cl$_6$ (6 mmol) at room temperature. The reaction was complete in one hour at room temperature. A mixture of chloro-benzene/NMP can be used to improve the molecular weight of polymer. The TPPCl$_2$ can also be prepared by dropwise addition of a solution of oxalyl chloride in chlorobenzene to triphenylphosphine oxide in chlorobenzene [15a].

Since it is well recognized that increased reactivity of carboxylic acid derivatives towards nucleophiles may be roughly correlated with greater stability of the leaving anions, Ueda et al. have exploited a series of good leaving groups in the synthesis of active esters and amides useful in the preparation of high molecular weight polyamides: (1,2-benzisoxazol-3-yl)diphenylphosphate **7** [16a], phenyl bis(2,3-dihydro-2-oxobenzothiazol-3-yl)-phosphinate **8** [16b], di-phenyl (2,3-dihydro-2-oxo-3-benzothiazolyl)-phosphonate **9** [16c] and diphenyl (2,3-dihydro-2-thioxo-3-benzoxazolyl)-phosphonate **10** [16d]. In the last case, **10** was prepared from 2-benzoxazolethiol and diphenylphosphorochloridate in the presence of triethylamine (Scheme 5). When reacted with a diacid, an active amide is formed (Scheme 6) and subsequent reaction with a diamine yields the desired polyamide (Scheme 7). An example of a polyamide prepared by this method is as follows:

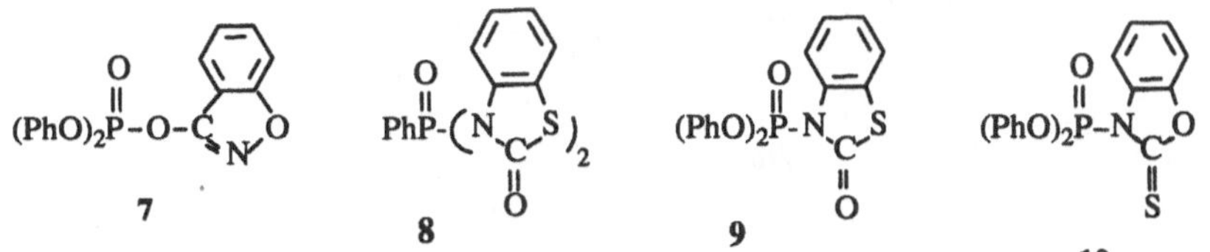

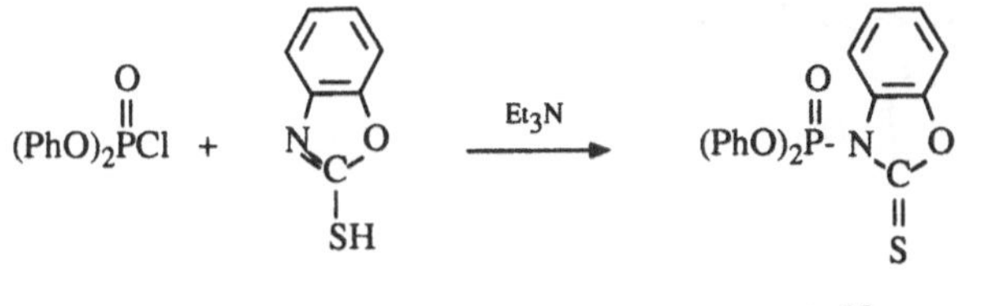

Scheme 5

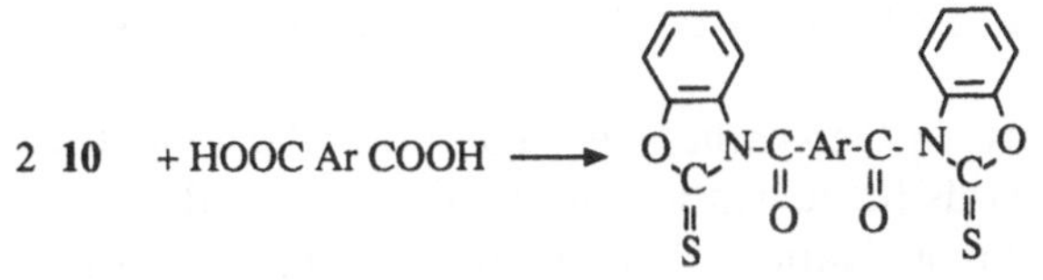

Scheme 6

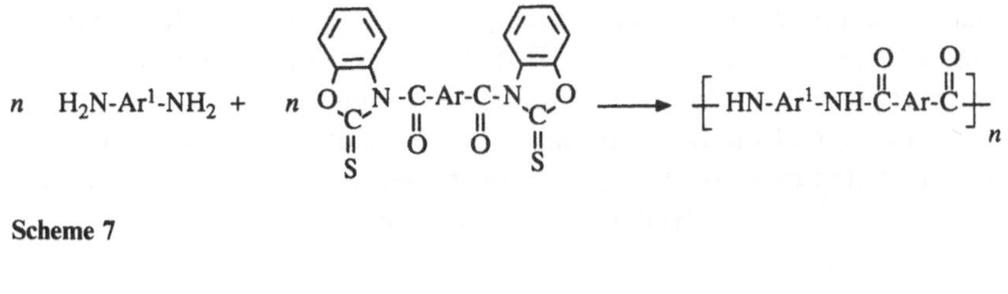

Scheme 7

Activating agent (2.2 mmol) was added to a solution of isophthalic acid (1 mmol) and triethylamine (2 mmol) in NMP (1 ml) at room temperature. After 30 min, 4,4'-diaminodiphenylether (1 mmol) was added, and the reaction was complete after 12 h.

This method is especially useful when other acylation sensitive functional groups are present in the monomers, for example, polymers **11a**, **11b**, **11c** were prepared with OH, NH_2, COOH groups unprotected by this chemoselective polycondensation [16d]:

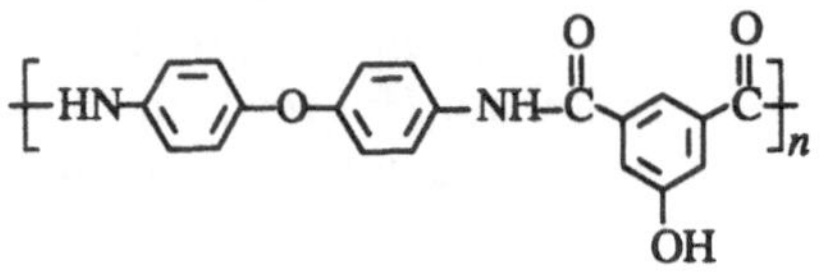

11a

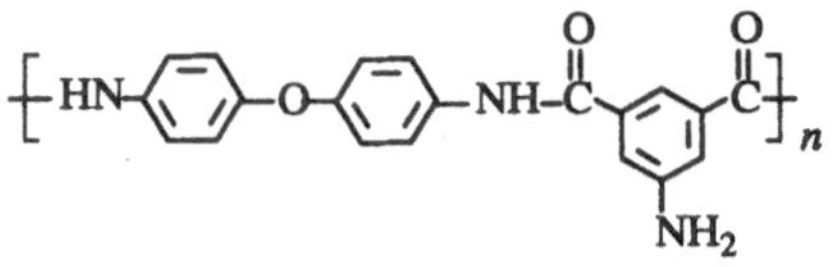

11b

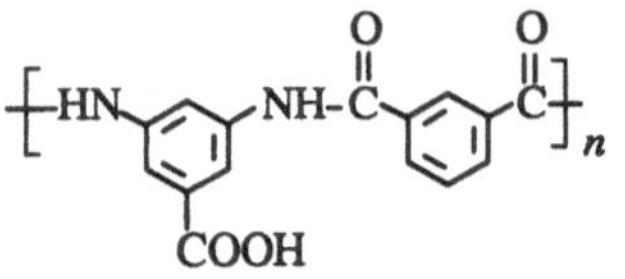

11c

2.3 Polycondensation of Silylated Diamine and Diacid Chloride

While most of the efforts in the synthesis of high molecular weight polyamides had been oriented towards the activation of the diacid, recently there has been increasing interest in the activation of the diamine component by reacting it with trimethylsilyl chloride. Indeed high molecular weight polyamides have been synthesized by low temperature polycondensation of an N-silylated aromatic diamine with an aromatic diacid chloride [17–19]. The reaction is usually carried out at − 10 °C in NMP (Scheme 8). Furthermore reaction proceeds more rapidly and affords polyamides with higher molecular weights relative to those obtained from free diamines. The following nucleophilic addition-elimination two-step mechanism has been proposed for the acyl substitution of an acid chloride with an N-silylated amine (Scheme 9).

$$n \; Me_3SiHN\text{-}Ar\text{-}NHSiMe \;+\; n \; ClOC\text{-}Ar^1\text{-}COCl \xrightarrow[-Me_3SiCl]{} \left[\begin{array}{c} H \quad H \; O \quad O \\ N\text{-}Ar\text{-}N\text{-}C\text{-}Ar^1\text{-}C \end{array} \right]_n$$

Scheme 8

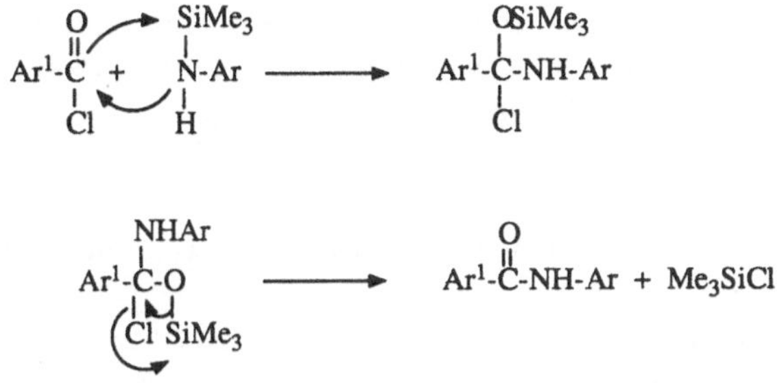

Scheme 9

In the first step, because of the strong affinity of silicon for oxygen, the carbonyl oxygen of the acid chloride is attracted to the silicon atom in the amine derivative which in turn facilitates the nucleophilic attack of the nitrogen atom of the N-silylated amine on the carbonyl carbon. In the second step, the elimination of chloride ion from the tetrahedral intermediate is enhanced by the presence of the β-silicon through the σ–π effect, affording rapidly the amide product along with trimethylsilyl chloride.

With aromatic diamines having low reactivity, it has proved difficult to produce polyamides of high molecular weight. However they can be considerably activated by conversion to their N-silylated derivatives. Using this method, for example, high molecular weight polyamides have been obtained from N,N'-diphenylaniline and tetrafluoro-m-phenylenediamine (Scheme 10).

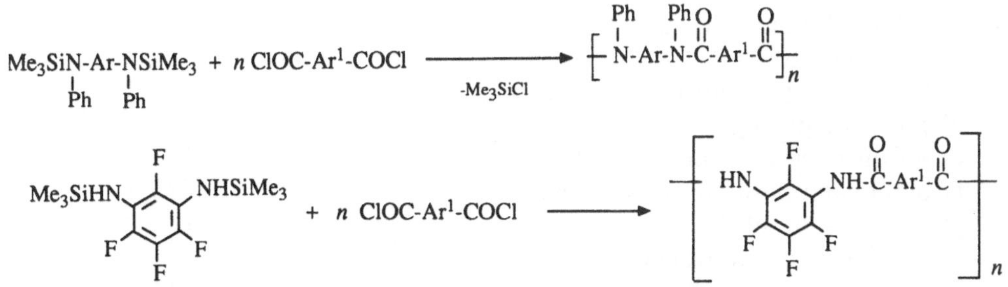

Scheme 10

The N-silylated diamine method has several advantages over conventional condensation procedures:

a) high-purity N-silylated diamines can be obtained by simple distillation

b) silylated diamines show excellent solubility in various organic solvents offering the flexibility in the polycondensation process

c) the condensation proceeds under neutral conditions with elimination of Me_3SiCl.

The synthesis of polyamide was carried out typically as follows:

To an N-silylated diamine (5 mmol) solution in NMP (10 ml) containing LiCl (0.53 g) cooled at $-10\,°C$ to $-5\,°C$ was added diacid chloride (5 mmol). The solution was stirred for 5 h, then the polymer was isolated.

2.4 Interfacial Polycondensation of Diamine and Diacid Chloride

As we have seen so far, in low temperature solution methods, the monomers are dissolved and reacted in a single solvent phase. Monomers can also be brought to react in an alternative way, e.g. at the interface of two phases. In the so-called interfacial polycondensation method, the two fast reacting intermediates are dissolved in a pair of immiscible liquids, one of which is preferably water. The water phase generally contains the diamine and usually an inorganic base to neutralise the by-product acid. The other phase contains the diacid chloride in an organic solvent such as dichloromethane, toluene or hexane (Scheme 11).

$$n\,H_2NArNH_2 \Big/_{H_2O\,+\,base} + n\,ClOCAr^1COCl \Big/_{organic\,solvent} \longrightarrow \left[\begin{array}{c} H \quad H\;O \quad\;\; O \\ | \quad\;\; | \quad\| \quad\quad \| \\ N\text{-}Ar\text{-}N\text{-}C\text{-}Ar^1\text{-}C \end{array} \right]_n$$

Scheme 11

The key factors that influence this type of polycondensation have been studied in detail and reviewed by Morgan [20, 21]. These are:

a) purity of intermediates
b) reaction rate
c) freedom from competing side reactions
d) adequate mixing
e) polymer swelling and solubility
f) nature and purity of the solvents
g) volume ratio of the phases
h) concentrations of reactants
i) addition of salts and bases
j) phase transfer agents (for some systems).

In this method where two phases are involved, the nature of the organic solvent influences the partition of monomers between the bulk solvent and the interface. In addition, a good solvent for the diamine would cause diffusion of this monomer into the organic phase, hence disturbing the balance of the two monomers and leading to low molecular weight product. The volume ratio of the two phases and concentrations of reactants influence the reaction rate of the polycondensation and also the side reactions such as hydrolysis. All of these can influence the molecular weight of the final polymer. Usually, hydrolysis is reduced by use of a weakly alkaline acid acceptor, such as sodium carbonate. Effective mixing can create a high interfacial surface area to increase the polycondensation rate. In the laboratory, a high speed blender can be used.

With the interfacial method, high molecular weight polymers can be obtained, but the molecular weight distribution, unlike that from polymers from solution polycondensation, is rather broad and this method is not suitable for the preparation of polymers for fibres and films [1c]. The solution method also has other advantages over the interfacial method. For example, it yields a solution of polymer amenable for direct fabrication of certain aromatic polyamides; some polyamides are soluble as made in solution by low temperature polycondensation but they often cannot be redissolved in solvents other than H_2SO_4 once they have been dried.

Rod-like polyamides cannot be prepared with high molecular weight by this interfacial technique because of their low solubility in organic solvents and resistance to swelling in these solvents [1c].

When the diacid chloride component has relatively high hydrolytic stability, completely or partially water-miscible solvents are particularly useful in the preparation of those aromatic polyamides which are frequently obtained only in low molecular weight when water-immiscible solvents are used. Water miscible solvents accelerate the reaction rate, enhance polymer swelling and facilitate product recovery. Typical examples are cyclohexanone, 2,4-dimethyltetramethylenesulphone, methylethylketone, tetramethylenesulphone, acetone, THF and isopropanol [20].

Amongst the different methods discussed above, the one described by Higashi involving reaction between a diamine and a diacid and using triphenylphosphite in NMP in the presence of LiCl and $CaCl_2$ as well as Imai's method using a silylated diamine and a diacid chloride are, in our opinion, the first choices for the preparation of novel aromatic polyamides. The methods are flexible, easy to use and are very successful for obtaining high molecular weight polymers.

3 Structure-Property Relationships of Aromatic Polyamides

3.1 Solubility and Thermostability

The high thermal stability and excellent mechanical properties of wholly aromatic polyamides such as poly(p-phenyleneterephthalamide) and poly-(m-phenyleneterephthalamide) have been recognised for a long time. The fabrication of most unsubstituted aromatic polyamides, however is difficult as a result of their high softening temperature and their insolubility in most organic solvents. Recently there has been an increasing requirement for new processable engineering plastics having a moderately high softening temperature and solubility in organic solvents. Attempts to increase the solubility of polyamides have been numerous and a number of strategies have been adopted: the introduction of flexible bonds in the polymer backbone; the substitution of the amide groups of the polymer chain; the introduction of bulky pendant groups onto the diamine or the diacid monomer units and the use of non-coplanar structures. The substitution of the amide groups usually decreases the overall thermal stability, although the solubility can be improved in this way. This type of modification of preformed polymers is more interesting in the preparation of liquid crystal solutions, which will be discussed in the next section, and in molecular composites [67c–f].

In the literature, efforts directed towards achieving more soluble, highly thermostable polyamides have been oriented in three main directions: a) the introduction of substituents of varying chemical nature onto the benzene ring of monomers; b) the use of monomers containing several benzene rings linked together by flexible chains; and c) the use of heterocyclic monomers.

3.1.1 Substituted Phenylene Monomers

This has been a very popular route used to modify the properties of a polyamide and is based on the direct derivatization of the precursor aromatic monomers, both the diamine as well as the diacid. In Table 1, we summarize those substituted monomers described in the literature, dividing them into several groups in order to discuss them separately. Thermogravimetric analysis (TGA) has been used widely to determine the thermal stability of polymers. The decomposition temperature (T_d) is generally taken as the point when 10% weight loss in air occurs in a TGA.

Halogen substituents

Extensive studies have been carried out separately by three research groups, Nagata et al., Diaz et al., and Pearce et al. Thus halogenated diamines and diacids have been used for the synthesis of thermostable, soluble and flame-resistant polyamides. The incorporation of halogen substituents along the polymer chains has long been recognized to lead to flame-resistant polymers. Detailed investigations have been carried out with mono, di, and tetrasubstituted diamines [22, 23b, 24] and diacids [23].

The results from the three groups indicate that:

a) The thermostability of halogen-substituted polyamides decreases in the order $H > F > Cl > Br$. This order corresponds to the bond dissociation

Table 1. Substituted diamines and diacids used to synthesize substituted polyamides

Nature of substituent	Diamines	Ref.	Diacids	Ref.
halogen	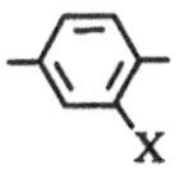 (benzene ring with X) (X = F, Br, Cl)	22a, 22b, 22c	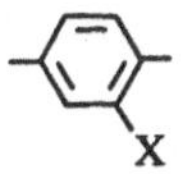 (benzene ring with X) (X = F, Br, Cl)	23a
	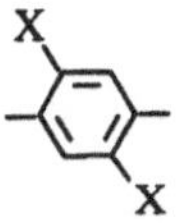 (benzene ring with X, X) (X = F, Br, Cl)	22b, 22c	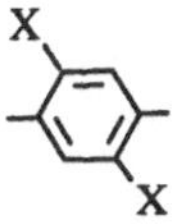 (benzene ring with X, X) (X = F, Br, Cl)	23a, 23c
	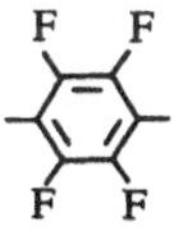 (benzene ring with F, F, F, F)	23b	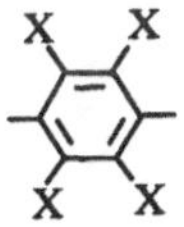 (benzene ring with X, X, X, X) (X = F, Br, Cl)	23a, 23b, 23c
	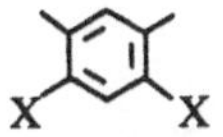 (benzene ring with X, X) (X = F, Br, Cl)	24, 23c	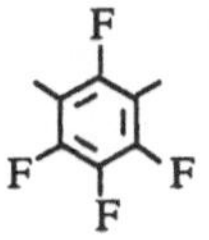 (benzene ring with F, F, F, F)	23b
	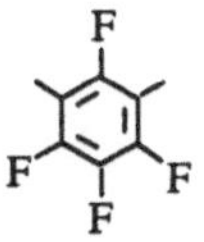 (benzene ring with F, F, F, F)	23b		
alkyl (alkoxy)	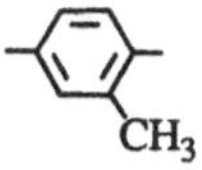 (benzene ring with CH₃)	22a, 25a	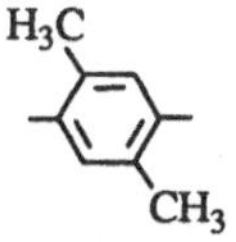 (benzene ring with H₃C, CH₃)	25b
	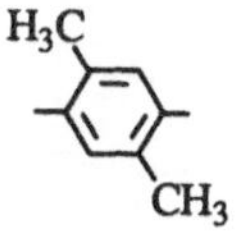 (benzene ring with H₃C, CH₃)	25a		
	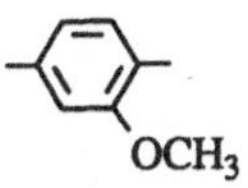 (benzene ring with OCH₃)	22a	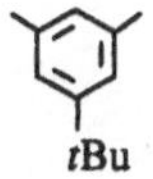 (benzene ring with tBu)	26
	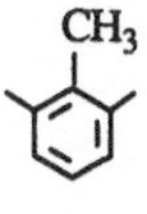 (benzene ring with CH₃)	22a, 25a		
	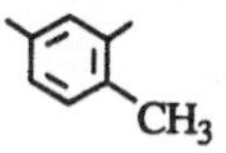 (benzene ring with CH₃)	22a, 25a		

Table 1. (Contd.)

Nature of substituent	Diamines	Ref.	Diacids	Ref.
	OCH₃ structure	22a		
	H₃C CH₃ H₃C CH₃ structure	22a		
	biphenyl structure	27		
			biphenyl structure	27, 70
	XPh structure (X = O, S)	28		
			terphenyl structure	70
			terphenyl structure	70
			OPh structure	30
			phenyl structure	29
others	NO₂ structure	22a		
			OCOPh structure	33
	OH structure	22a		
			NHCOPh structure	34
	SO₃H structure	22a		

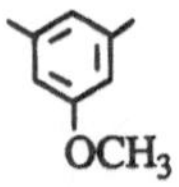

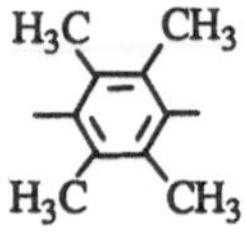

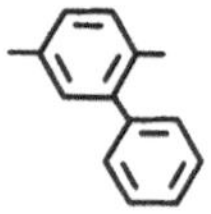

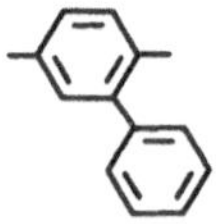

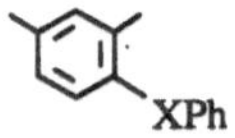

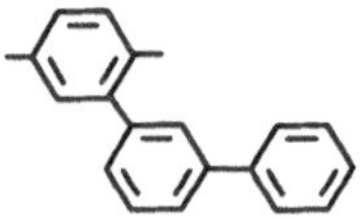

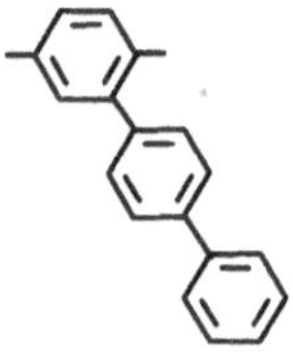

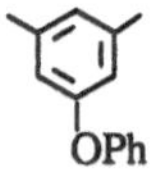

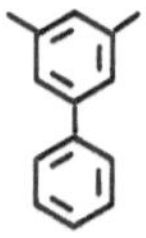

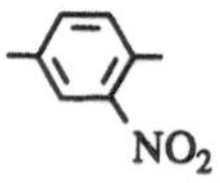

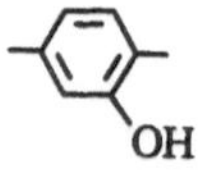

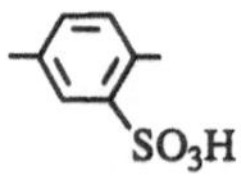

Table 1. (Contd.)

Nature of substituent	Diamines	Ref.	Diacids	Ref.
	 CONHPh	32	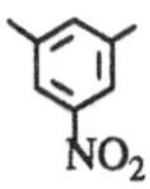 NO$_2$	31
	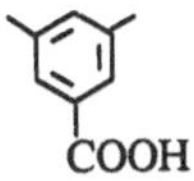 COOH	22a		

energies, i.e. C–Br: 79.9 Kcal/mol, C–Cl: 94.1 Kcal/mol, C–F: 124.8 Kcal/mol [23a]. The thermostability also generally decreases in the order: polymers made from *p*-phenylenediamine > polymers made from *m*-phenylenediamine, and polymers from terephthalic acid > polymers from isophthalic acid. This can be seen from the values of T_d of polyamides **12** (Table 2).

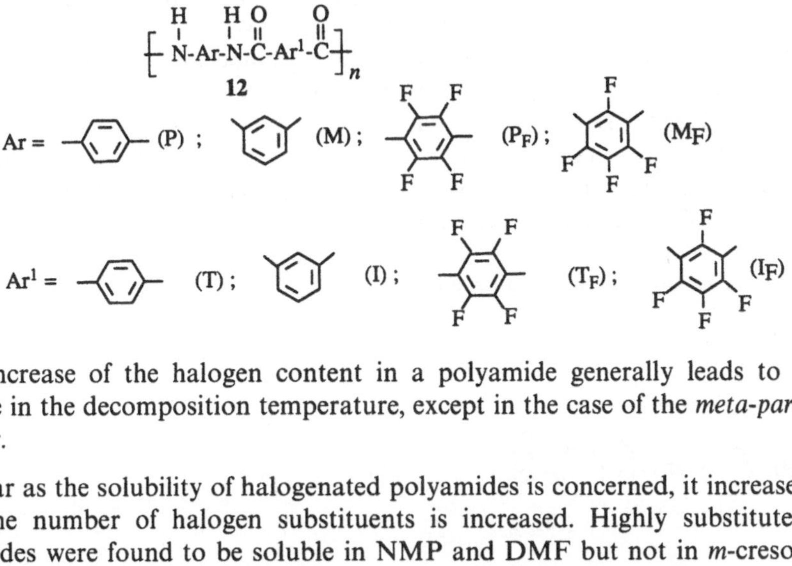

12

$$Ar = \text{(P)} ; \text{(M)} ; \text{(P}_F\text{)} ; \text{(M}_F\text{)}$$

$$Ar^1 = \text{(T)} ; \text{(I)} ; \text{(T}_F\text{)} ; \text{(I}_F\text{)}$$

b) Increase of the halogen content in a polyamide generally leads to a decrease in the decomposition temperature, except in the case of the *meta-para* polymer.

As far as the solubility of halogenated polyamides is concerned, it increases when the number of halogen substituents is increased. Highly substituted polyamides were found to be soluble in NMP and DMF but not in *m*-cresol, *o*-chlorophenol, chlorobenzene, tetrachloroethane nor DMSO [23a].

Table 2. Decomposition temperature T_d (°C) of polyamides **12**

PT (540)	PI (500)	MT (465)	MI (460)
PT$_F$ (382)	PI$_F$ (440)	MT$_F$ (366)	MI$_F$ (220)
P$_F$T (460)	P$_F$I (462)	M$_F$T (436)	M$_F$I (411)
P$_F$T$_F$ (408)	P$_F$I$_F$ (351)	M$_F$T$_F$ (392)	M$_F$I$_F$ (352)

Alkyl substituents

The methyl group has been used as a substituent to improve the solubility of polyamides by disrupting the regularity of the packing along the mainchain. Mono, di, and tetra substituted diamines and di-substituted diacids have been used as starting materials for the synthesis of substituted polyamides. The study by Takatsuka et al. [25] and Chaudhuri et al. [22a] showed that for methyl substituted polyamides of the following general structure **13**:

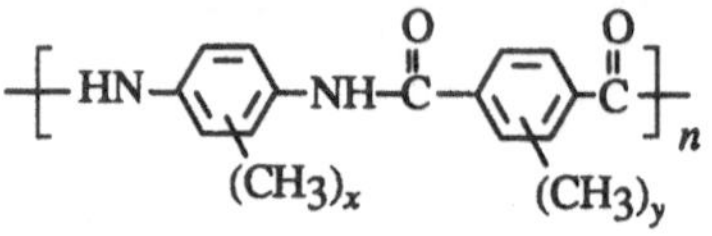

$$x = 0, 1, 2, 4 \quad y = 0, 2$$

13

   a) The solubility is increased by methylation. Those polyamides with unsymmetrical methyl substitution have higher solubility, e.g. the polymer from 2-methyl-1,4-diamine is more soluble than the one from 2,5-dimethyl-1,4-diamine, and the polymer from 6-methyl-1,3-diamine has a higher solubility than the one from 2-methyl-1,3-diamine. Greater solubility is obtained with the polymer from unsymmetrical 6-methyl-1,3-diamine and 2,5-dimethylterephthalic acid, which is not only soluble in strong acid (conc. H_2SO_4), and amide solvents (DMAc, DMF, NMP), but also in *m*-cresol and DMSO.
   b) The thermostability is generally decreased by methylation. Methylation on both diamine and diacid leads to an even more significant decrease in thermostability as can be seen in Table 3.

Table 3. Decomposition temperature of polyamides **13**

Polymer	T_d (°C)	
	$m = 0$	$m = 2$
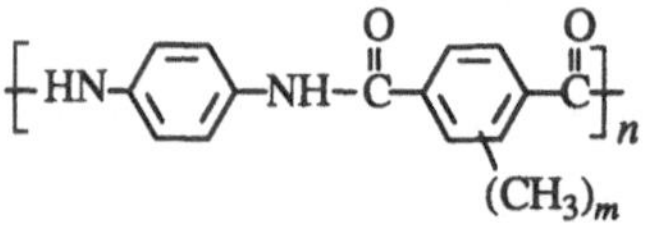	513	488
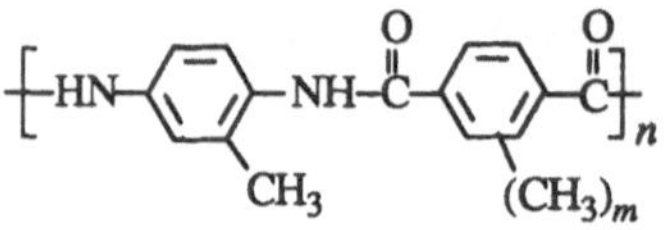	505	463
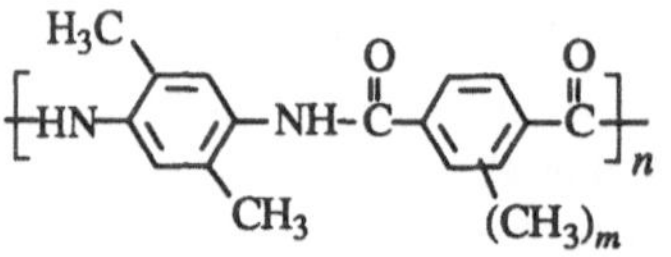	473	452

c) Methoxy substituted polyamides generally have lower stability than their methylated analogues, for example, T_d values are some 50–75 °C lower.

Yang et al. have recently reported the use of a bulky *t*-butyl group to improve the solubility of polyamides because the *t*-butyl is more thermally stable than the methyl group. However, polymer **14** has only comparable stability and solubility to the corresponding methylated one [26].

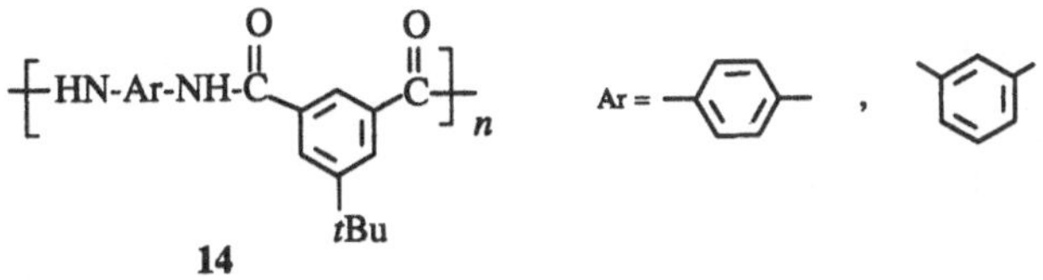

14

Aromatic substituents

The introduction of alkyl groups onto the benzene rings of aromatic polyamides has proved to impart higher solubility to the polymers. The thermostability of these however is reduced because of insufficient resistance of the alkyl substituents towards oxidation. Some studies have therefore been carried out using aromatic substituted monomers, for instance, phenyl [27], phenoxy and thiophenoxy [28] substituted diamines, and phenyl [27, 29, 70] and phenoxy [30] substituted diacids. These have been used to prepare polyamides **15**. In Table 4

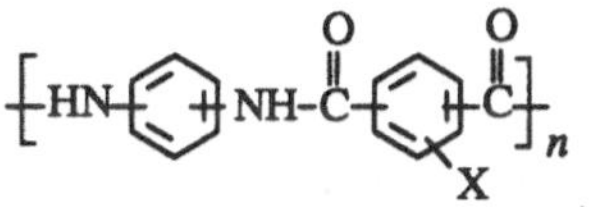

X = Ph, OPh

15a

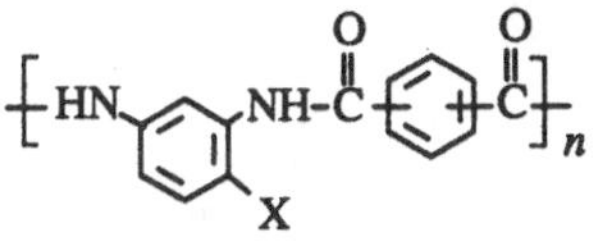

X = OPh, SPh

15b

is summarized some of the data for the polymers. It has generally been observed that:

a) While direct phenyl substitution increases T_g of a polyamide, substitution via an ether or thioether bond decreases the glass transition temperature (T_g) of the polymer. Thiophenyl is more efficient than phenyl or phenoxy group in this respect.

b) The thermostability of substituted polyamides is decreased by phenoxy, thiophenoxy substituents but not by phenyl substitution.

Although the bulky aromatic side group generally hinders free segmental motion, hence increasing the T_g of the polymer, the effect of chain separation seems to be dominant in case of OPh, SPh, and the lower hydrogen bond density probably accounting for the decrease in T_g. These polyamides with lower

Table 4. Thermal properties and solubility of polyamide **15** with or without aromatic substituent

Diamines	Diacids	T_g (°C)	T_d (°C)	Solubility[a]					Ref.
				DMF	DMAc	C[b]	Py[c]	CH[d]	
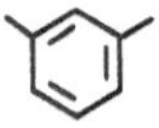 (m-phenylene)	 (p-phenylene)	295	470	−	±	−	±	−	30
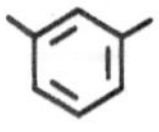 (m-phenylene)	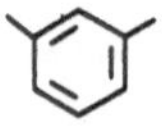 (m-phenylene)	276	455	+ +	±	−	±	−	30
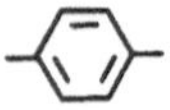 (p-phenylene)	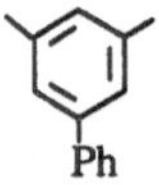 (Ph-substituted m-phenylene)	331	465	+ +	±	+	±	−	29
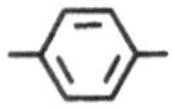 (p-phenylene)	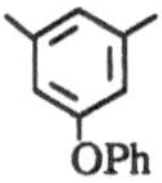 (OPh-substituted m-phenylene)	274	430	+	±	+	±	+	30
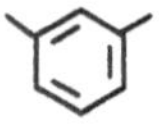 (m-phenylene)	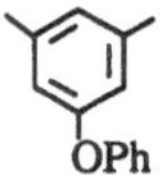 (OPh-substituted m-phenylene)	245	430	+ +	±	+	±	−	30
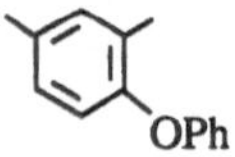 (OPh-substituted phenylene)	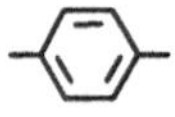 (p-phenylene)	256	400	±	+ +	+ +	+ +	±	28
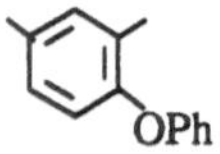 (OPh-substituted phenylene)	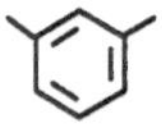 (m-phenylene)	230	435	±	+ +	+ +	+ +	±	28
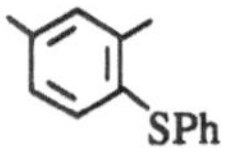 (SPh-substituted phenylene)	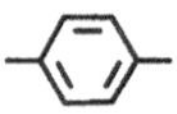 (p-phenylene)	220	400	±	+ +	+ +	+ +	±	28
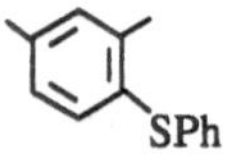 (SPh-substituted phenylene)	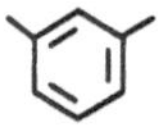 (m-phenylene)	194	425	±	+ +	+ +	+ +	±	28

[a] + + soluble at room temperature; + soluble on heating; ± partially soluble; − insoluble
[b] m-cresol
[c] pyridine
[d] cyclohexane

T_g and thermal stability are interesting candidates for melt processable high performance plastics [28].

c) Solubility is increased by these aromatic substituents especially by phenoxy or thiophenoxy.

Other substituents

Chaudhuri et al. [22a] in their study of the thermostability of wholly aromatic polyamides have introduced various polar groups onto the polymer, e.g. NO_2, COOH, OH, SO_3H, etc. Compared to methylated or chlorinated polymers, the stability of polyamides **16** is lower, and T_d decreases in the following order: NO_2 > COOH, OH > SO_3H

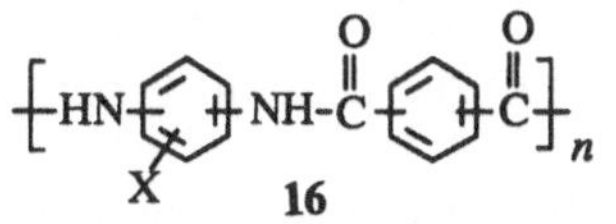

$$X = NO_2, COOH, OH, SO_3H$$

For example, the T_d of polyisophthalamides from different diamines are as follows: 402 °C, 375 °C, 364 °C and 239 °C for NO_2, OH, COOH and SO_3H containing polymers respectively. The low stability of the SO_3H substituted polymer has been attributed to the presence of the acid group which could catalyse the thermal degradation of polymer.

The last group of substituents investigated has involved an ester or an amide function [32–34] (Table 5). Polyamides containing these groups show solubility

Table 5. Properties of substituted polyamides

Polymer	Ar	T_g (°C)	T_d (°C)	Solubility[a] DMF	C^b	CH^c
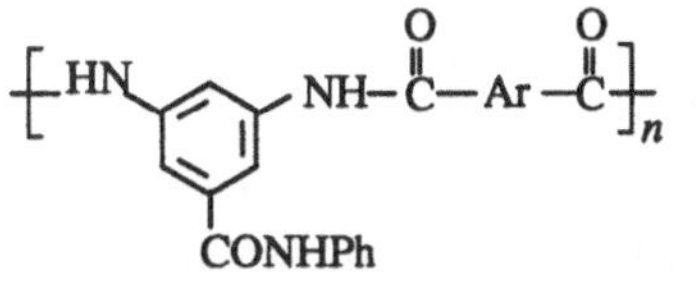 **17** (CONHPh)	 para-phenylene	280	440	+	±	±
	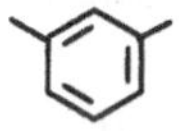 meta-phenylene	276	425	+	±	±
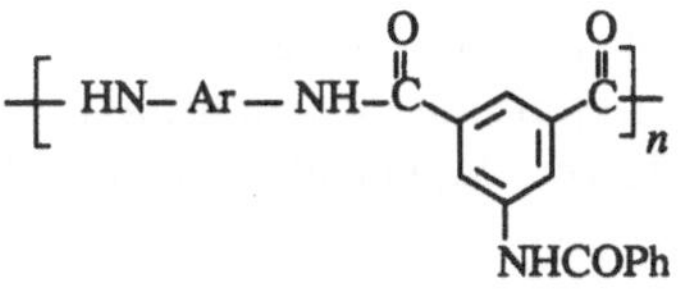 **18** (NHCOPh)	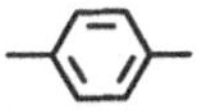 para-phenylene	317	410	+	−	−
	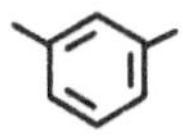 meta-phenylene	298	415	+ +	+	+
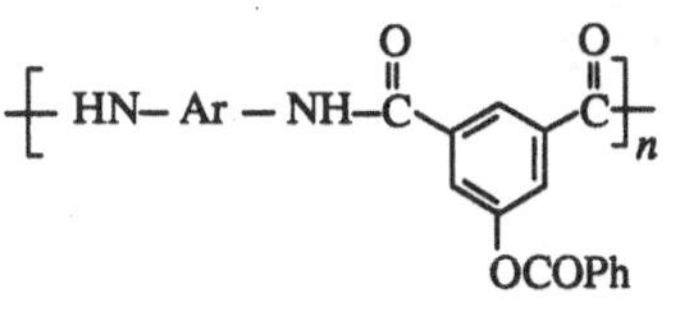 **19** (OCOPh)	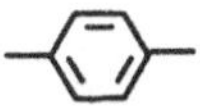 para-phenylene	290	365	+ +	+	−
	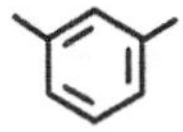 meta-phenylene	275	365	+ +	+	+

[a] + + soluble at room temperature; + soluble on heating; ± partially soluble or swollen; − insoluble
[b] *m*-cresol
[c] cyclohexane

in highly polar solvents, and some are soluble or swellable in *m*-cresol. The oxybenzoyl substituted polyamide **19** has the lowest T_d. It is quite interesting to note that while introduction of an oxybenzoyl group does not change the T_g of polymer, the T_g is nevertheless raised ca. 20 °C by introduction of an amide group onto the diacid part, and was decreased ca. 50 °C by a pendant amide group on the diamine part. De Abajo et al. explained this in terms of a double role played by the substituent. On one hand, there is the effect of asymmetry and irregularity which should give rise to a decrease of T_g, but on the other hand, there is the bulky volume and polar nature of the pendant group which might raise the T_g [33, 34].

In case of the oxybenzoyl group, these two effects seem balanced. When an iminobenzoyl group is introduced, the second effect is more important. It is interesting to notice that the introduction of a benzamide group onto a different part of the polyamide has an opposite effect on the T_g. As far as the thermostability is concerned, polyamides **17, 18** substituted with an amide group have higher stability than the oxybenzoyl substituted species.

In summary so far we can see that the introduction of phenyl, phenoxy, thiophenoxy, benzamide or iminobenzoyl groups leads to polyamides with moderately high T_g values and quite high thermal stability. They are also soluble in DMF, and some in *m*-cresol.

3.1.2 Monomers Containing Several Benzene Rings

So far we have only discussed those substituted polyamides derived from substituted phenylenediamines and isophthalic or terephthalic acids, i.e. from benzene derivatised monomers. Numerous monomers containing several aromatic rings have also been used in the synthesis of polyamides as shown in Table 6.

2,6-Naphthalenedicarboxylic acid has been employed in several studies [28, 32, 35]. Compared to polyamides derived from terephthalic or isophthalic acid, polyamides from 2,6-naphthalic acid generally have a higher glass transition temperature and a higher T_d as well.

20

$Ar = $ **a** ⟨⟩ **b** ⟨⟩ **c** ⟨⟩⟨⟩

Vernekar et al. [42] in their study of polymer **20** have shown that the T_g decreases in the order **20a > 20c > 20b** and the T_d decreases in the order **20b > 20c > 20a**

Table 6. Monomers containing several benzene rings

Diamines/diacids		Ref.
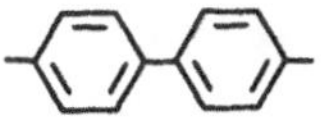		42
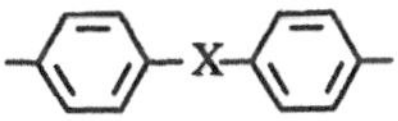	X = CH$_2$	26, 36, 42, 34
	C(CH$_3$)$_2$	34
	C(CF$_3$)$_2$	36
	O	26, 36, 42
	NPh	37
	CO	26, 36
	S	36
	SO$_2$	26, 36, 42
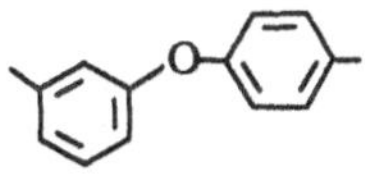		26, 36
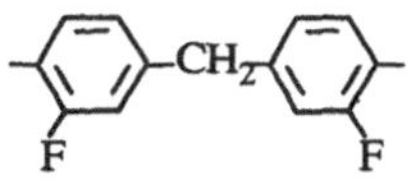		24
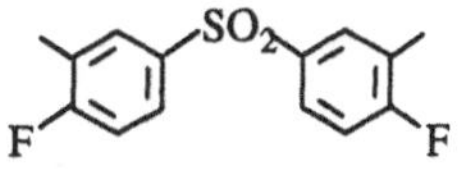		24
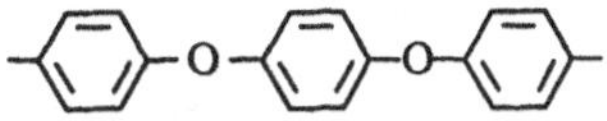		36
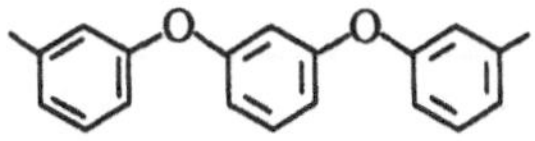		41
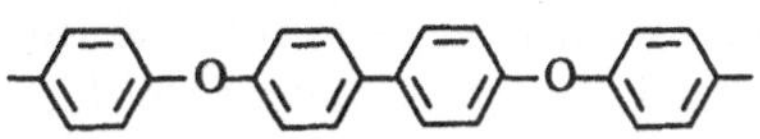		36
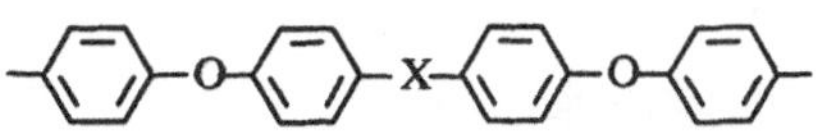	X = C(CF$_3$)$_2$	36
	SO$_2$	36, 26, 38
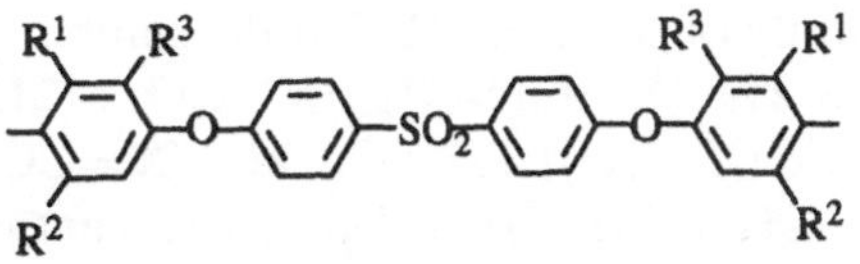 R^1 R^3 ... R^3 R^1 / R^2 ... R^2		43, 44
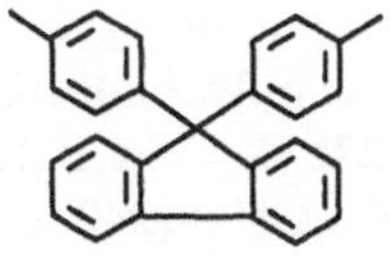		36

Table 6. (Contd.)

Diamines/diacids	Ref.
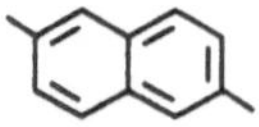	28, 32, 35
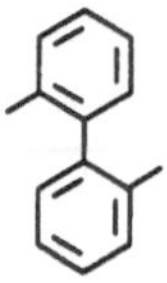	36
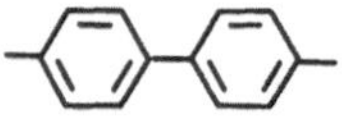	37
X = SO$_2$ / C(CF$_3$)$_2$ / SiR$_2$	37 / 37 / 38, 39
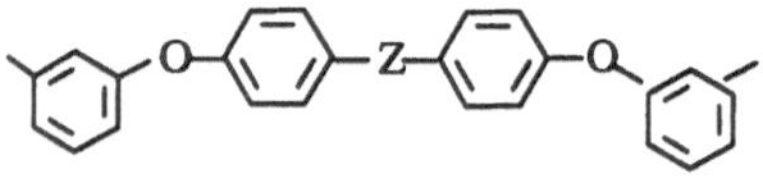 (Z = CO, SO$_2$)	41, 42
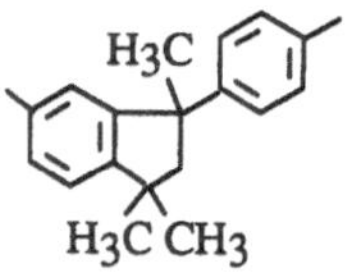	40

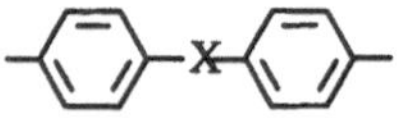

Certain monomers containing flexible links X between two aromatic rings have also been used to prepare polyamides, and the properties of the latter compared with those polymers made from monomers having only rigid links.

Ar = 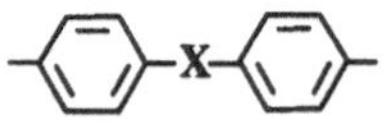   X = a CH$_2$, b C(CH$_3$)$_2$, c C(CF$_3$)$_2$, d O,

e S, f NPh, g CO, h SO$_2$, l SiR$_2$

Yang et al. [26] have shown in their study of polymers **21** that the solubility decreases in the order **21a** (X=CH$_2$) > **21d** (X=O) > **21g** (X=CO) > **21h** (X=SO$_2$); T_g decreases in the order **21h** (X=SO$_2$) (325 °C), **21g** (X=CO) (324 °C) > **21d** (X=O) (293 °C) > **21a** (X=CH$_2$) (283 °C) and T_d decreases in the order **21g** (X=CO) (470 °C) > **21d** (X=O) (460 °C) > **21a** (X=CH$_2$) (450 °C), **21h** (X=SO$_2$) (445 °C). While Imai et al. in their investigation using 2,2′-bibenzoic acid and a diverse group of diamines, observed for the polyamides **22** [36] that T_g decreases in the order **22h** (X=SO$_2$) (290 °C) > **22c** (X=C(CF$_3$)$_2$) (275 °C) > **22d** (X=O) (251 °C), **22e** (X=S) (250 °C) > **22a** (X=CH$_2$) (240 °C),

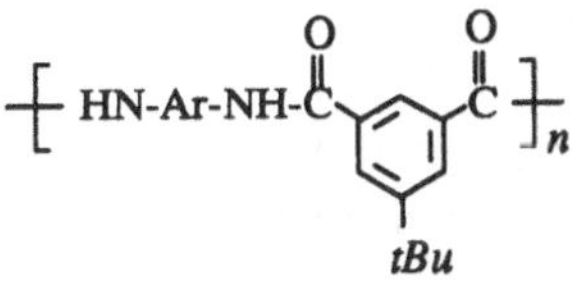

Ar with X = a, d, g, h

21

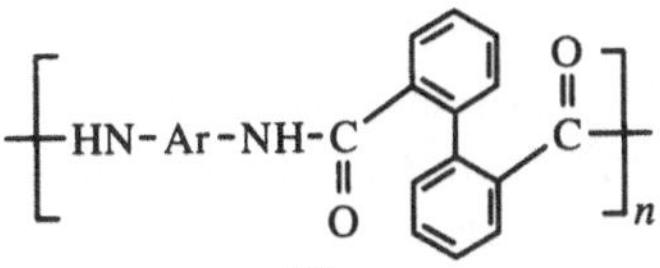

Ar with X = a, c, d, e, h

22

while T_d decreases in the order **22a** ($X=CH_2$) (440 °C) > **22e** (X = S) (420 °C), **22d** (X = O) (415 °C) > **22h** ($X=SO_2$) (400 °C) > **22c** ($X=C(CF_3)_2$) (385 °C). These polymers are also soluble in amide solvents, pyridine, *m*-cresol and even THF for some.

Hence, it is clear that the influence of different flexible links, X, can be very large, but does depend on the nature of each polyamide.

Incorporating aromatic silane functions into aromatic polyamides can also improve solubility [45], and was shown recently by Vernekar et al. [38] with polymers **23**. These polyamides from 3,3′ or 4,4′[sulfonylbis-(4,1-phenyleneoxy)] bisbenzenamine have higher solubility than those from bis(4-aminophenyl)ether. They are soluble in amide solvents, pyridine and *m*-cresol, and some in THF and nitrobenzene. They also have a high T_d in the range of 477–513 °C, and a T_g in the range of 209–246 °C.

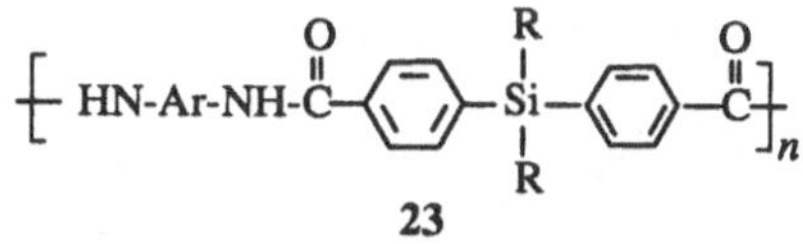

23

Ar =

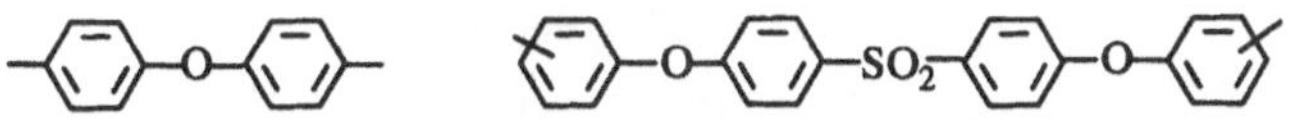

Abraham et al. [41] prepared polyamides **24**. They have very low T_g (128–144 °C) and are quite stable (T_d 413–429 °C). They offer an opportunity for facile processability while retaining high thermooxidative stability.

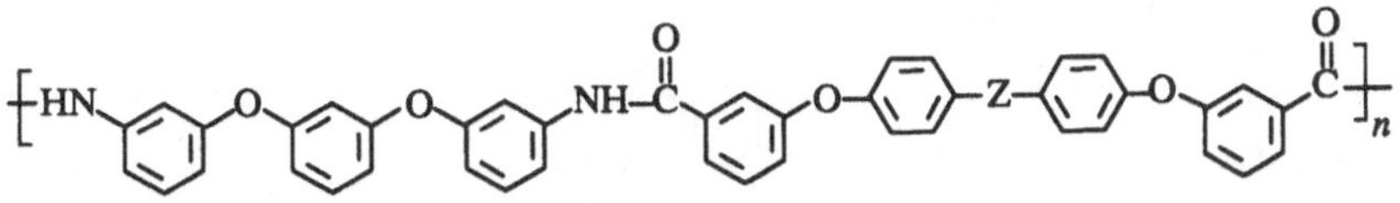

24 $Z = CO, SO_2$

3.1.3 Heterocyclic Monomers

This is the third approach used for improving the solubility, and increasing the processability and thermostability of polyamides. This method employs monomers containing heterocyclic rings (Table 7).

It is well known that a large number of polymers, in which aromatic and heterocyclic rings are linked together in chains, are resistant to high temperature [46]. In most cases, the connecting groups are less thermally stable than the rings themselves and the overall stability of the polymer is determined largely by the nature of these groups. It is also recognised that thermostability increases considerably with a decreasing number of single connecting groups in the polymer backbone, in other words, on introducing double-stranded heterocycles.

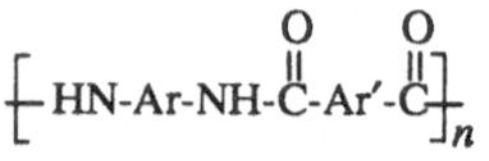

$$\left[HN\text{-}Ar\text{-}NH\text{-}\overset{\displaystyle O}{\overset{\displaystyle \|}{C}}\text{-}Ar'\text{-}\overset{\displaystyle O}{\overset{\displaystyle \|}{C}} \right]_n$$

25

Ar = a, b, c

Ar' = a′, b′ c′ = b d′ = c

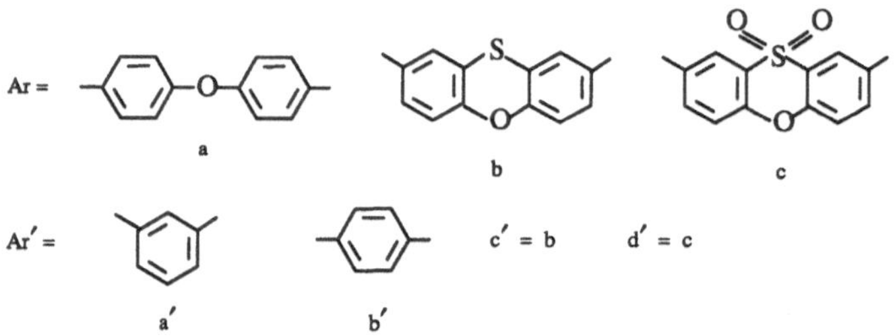

Table 7. Heterocyclic monomers used in the preparation of thermal stable polyamides

Diamines	Ref.	Diacids	Ref.
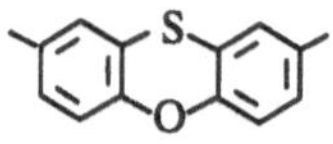	47	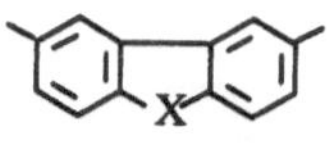 (X = S, SO₂)	57
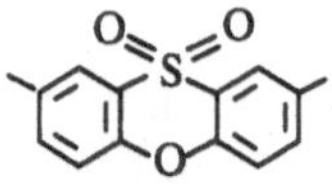	47	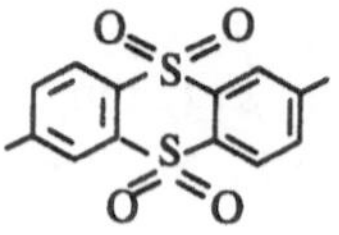	58
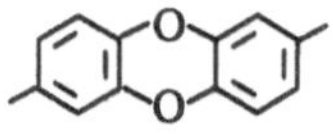	48	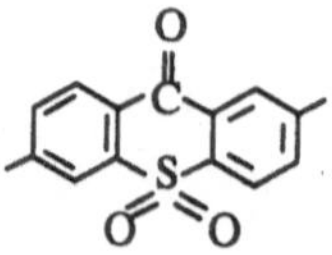	59
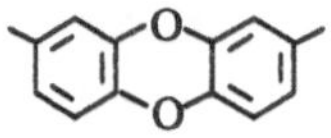	48		

Table 7. (Contd.)

Diamines	Ref.	Diacids	Ref.
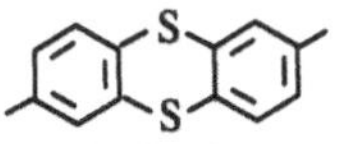	48	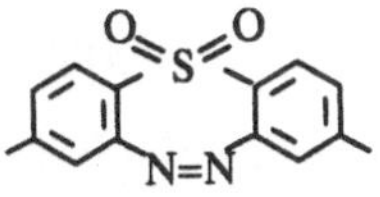	60
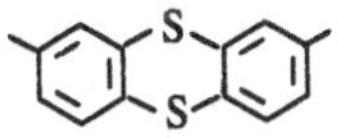	48	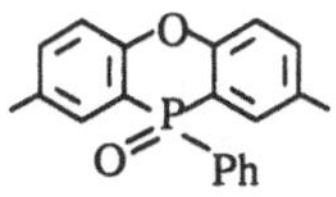	49a
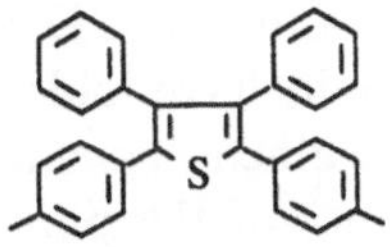	52	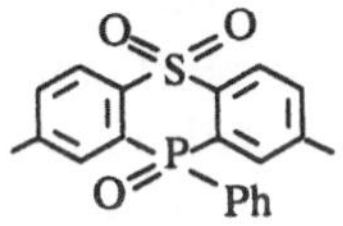	49b
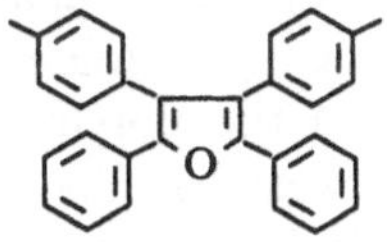	53	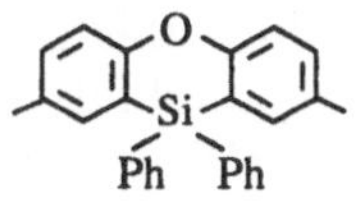	51a, b
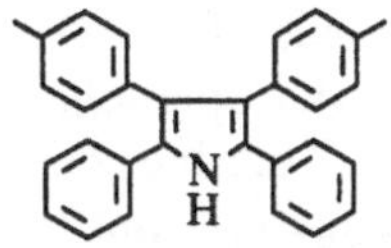	54	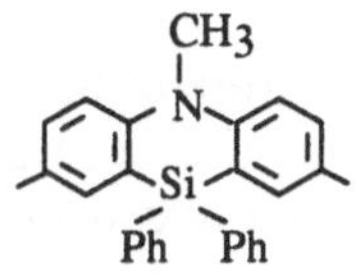	51c
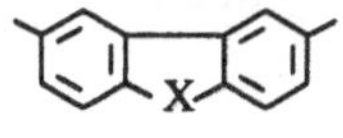 (X = S, SO₂)	57		
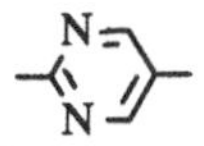	62	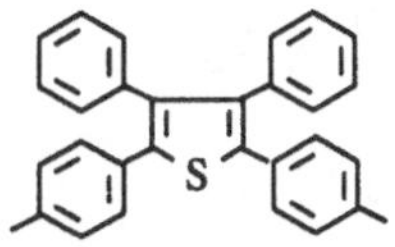	52
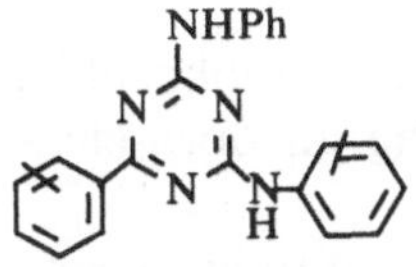	63	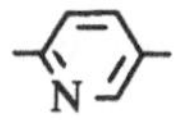	61

Imai et al. therefore started to prepare polyamides **25** with phenoxathiin rings in place of open-chain diphenyl ether linkages [47]. The thermostability follows the order **25cb″** (510 °C) > **25bc′**, **25cd′**, **25bb′** > **25ac′**, **25ad′**, **25ca′** > **25ba′**, **25aa′** (450 °C). Clearly polyamides containing the heterocycles appear to be more stable than those having open-chain diphenyl ether linkages, and the diamines DAP (2,8-diaminophenoxathiin) **26**, DAPD (2,8-diamino-phenoxathiin 10,10′-dioxide) **27** yield polymers with similarly high performance. Some DAP-containing polymers are soluble in polar aprotic solvents (NMP, DMF, DMSO) at room temperature, but polymers containing DAPD are not.

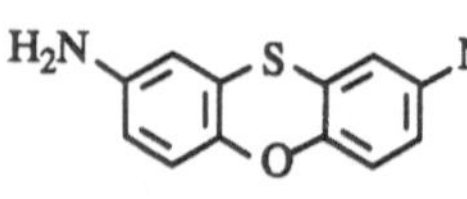

DAP, **26** 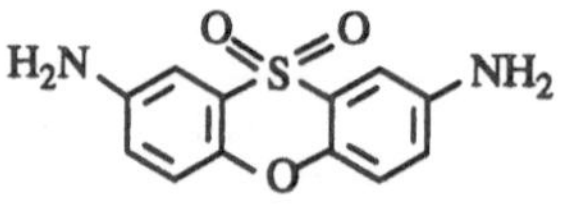   DAPD, **27**

It is well known that the properties of a polymer may not depend only on its primary structure, but also on a number of other factors such as crystallinity, intermolecular interaction and the packing situation of the polymer chains.

Niume et al. [48] hence introduced a series of tricyclic fused rings which are non-planar into the polymer backbone, for example, dibenzo-*p*-dioxins (ODP) **28**, phenoxathiin (OSP) **29** and thianthrene (SDP) **30**. These units have folded structures about the axis that combine the hetero atoms, and the dihedral angles are in the following order: ODP < OSP < SDP. The solubility of the polyamides made from these diamines with terephthalic acid or isophthalic acid follows the same order. An explanation for these results is that the highly folded structure in the mainchain interfers with the close packing of the polymer molecules and the hydrogen bonding; this make the solvation easier. These polymers show varying degrees of solubility in DMAc (LiCl). The thermostability follows a reverse order SDP < OSP < ODP. Whereas the ODP-containing polymers are more stable than the polymers with the open-chain diphenyl ether link, SDP-containing polymers are less stable. This has been attributed to the highly folded structure of SDP, and it was suggested that the effect of non-planarity on the properties of resulting polymers is influenced by the rigidity of the polymer chain and hydrogen bonding.

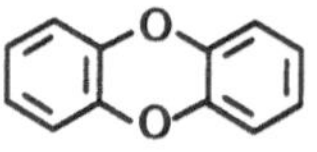

ODP, **28** 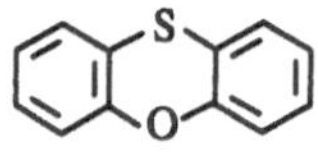   OSP, **29** 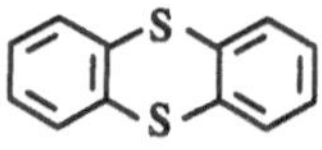   SDP, **30**

Sato et al. [49a, 49b] prepared polyamides **31** containing phenoxaphosphine rings and compared their thermostability with those having an open-chain carbon-phosphorous linkage.

These polymers have various degrees of solubility in polar aprotic solvents and also are quite soluble in *m*-cresol. T_d is in the range of 456–482 °C, and is higher than that of their open-chain analogues.

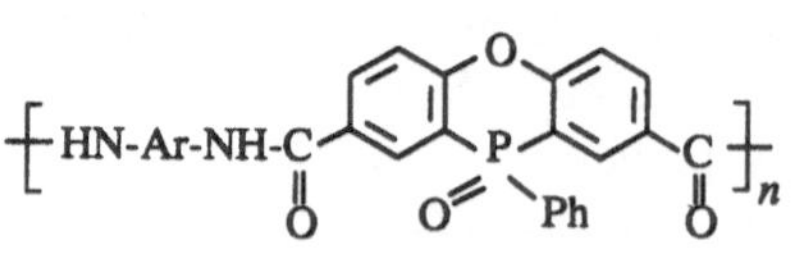

31

Polyamides **32** containing a phenothiophosphine unit, on the other hand, were shown to be more soluble in DMF, DMAc, *m*-cresol and have comparable T_d values relative to those polymers with phenoxaphosphine units.

$$\left[HN-Ar-NH-\overset{O}{\underset{\|}{C}} \underset{\underset{Ph}{\overset{O}{\|}}}{\overset{\overset{O}{\underset{\|}{S}}}{P}} \overset{O}{\underset{\|}{C}} \right]_n$$

32

In order to improve the solubility of aromatic polymers silicon-containing polymers have also been synthesized, by introduction of siloxane or silarylene linkages into polymer backbones such as polyamides, polybenzimidazoles and polyimides [50]. The resultant polymers show fair solubility in organic solvents but most of them have poorer thermal stability then their analogues without siloxane or silarylene linkages.

Since double-stranded polymers generally have higher stability than single-stranded ones, Kondo et al. [51a, b] have synthesized phenoxasilin-containing polyamides **33**. Unfortunately these proved to be hardly soluble in amide solvents at all and were even insoluble in concentrated H_2SO_4. Nevertheless, they were more thermally stable (T_d in the range 477–506 °C) compared with their corresponding open-chain polymers.

$$\left[HN-Ar-NH-\overset{O}{\underset{\|}{C}} \underset{\underset{Ph}{} \underset{Ph}{}}{\overset{\overset{O}{}}{Si}} \overset{O}{\underset{\|}{C}} \right]_n$$

33

The same authors in a continuation of their investigations of silicon-containing polyamides prepared polyamides **34** containing a phenaxasiline ring [51c]. In comparison with the phenoxasilin-based polyamides **33**, these polymers are soluble in DMF, DMAc, NMP, and even in *m*-cresol. The thermostability of phenoxasilin-containing polyamides is comparable to the wholly aromatic ones and superior to that of polyamides containing phenaxasiline as the tertiary amine linkage of the phenaxasiline ring is likely to be readily oxidized by thermal treatment in air.

$$\left[HN-Ar-NH-\overset{O}{\underset{\|}{C}} \underset{\underset{Ph}{} \underset{Ph}{}}{\overset{\overset{CH_3}{\overset{|}{N}}}{Si}} \overset{O}{\underset{\|}{C}} \right]_n$$

34

Imai et al. [52–54] used the following highly phenylated heterocyclic systems as diamine monomers: 2,5-bis(4-aminophenyl)-3,4-diphenylthiophene **35**, 3,4-bis(4-amino-phenyl)-2,5-diphenylfuran **36**, and 3,4-bis(4-aminophenyl)-2,5-diphenylpyrrole **37**.

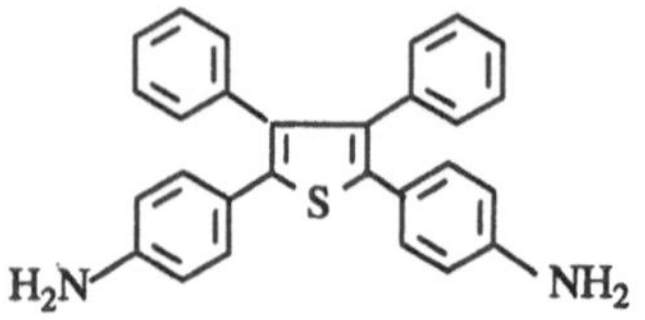

35

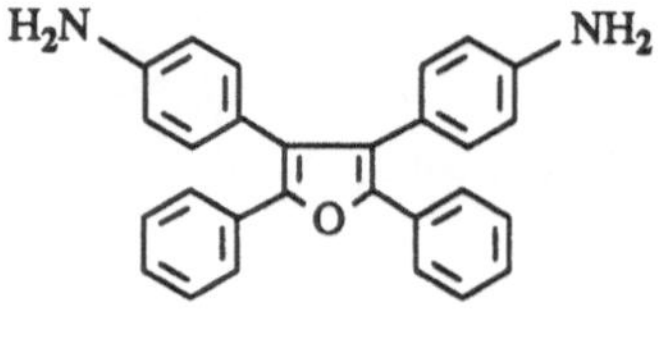

36

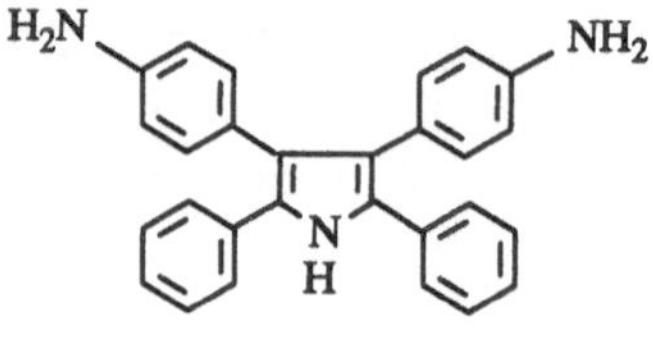

37

Polyamides **38** made from these diamines are characterised by their high thermostability and solubility in organic solvents as can be seen from the data in Table 8.

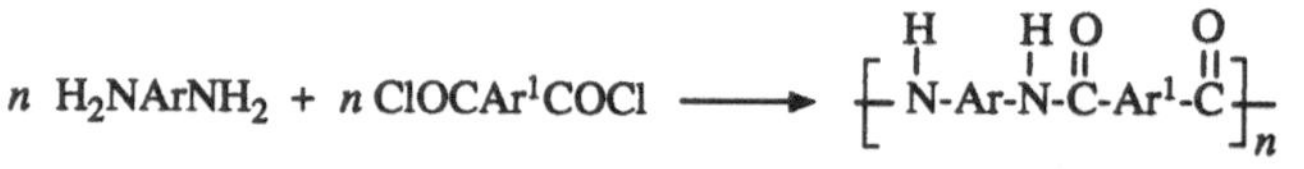

$$n\ H_2NArNH_2 + n\ ClOCAr^1COCl \longrightarrow \left[\overset{H}{\underset{|}{N}}\text{-Ar-}\overset{H}{\underset{|}{N}}\text{-}\overset{O}{\underset{||}{C}}\text{-Ar}^1\text{-}\overset{O}{\underset{||}{C}} \right]_n$$

38

$H_2NArNH_2 =$ **35, 36, 37**, etc.

In the same Table are also collected data for those polymers made from 2,5-bis(4-chloroformylphenyl)-3,4-diphenylthiophene **39** and diverse diamines.

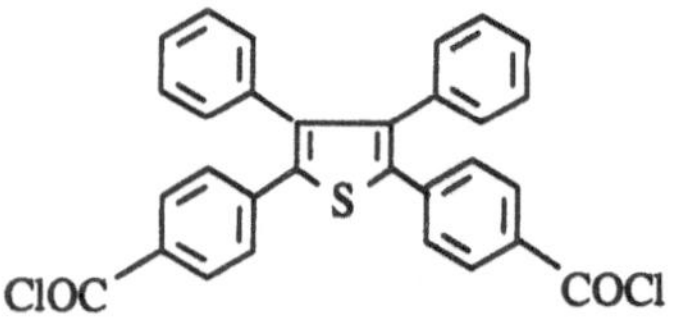

39

They are all soluble in concentrated H_2SO_4, and most are soluble in NMP, DMAc, pyridine, and m-cresol at room temperature. The order of decreasing solubility is pyrrole-containing polymers > furan-containing polymers > thiophene-containing polymers.

Polymers with the thiophene moiety in the diamine monomer and those having this unit in the diacid part have similar solubilities. Although these polymers have comparable T_g (polymers having biphenylene, naphthalene, 4,4'-diamino-phenylether have slightly higher T_g), T_d changes considerably from one polymer to another. It follows the order: thiophene-containing polymer > pyrrole-containing polymer > furan-containing polymer.

Table 8. Properties of polyamides **38** containing five member ring heterocycle unit

Diamines	Diacids/diacid chloride	T_g (°C)	T_d (°C)	Solubility[a]		
				DMAc	Py[b]	C[c]
35	(structure)	315	490	+ +	+	+
	(structure)	–	500	–	–	–
36	(structure)	308	437	+ +	±	+ +
37		306	505	+ +	+ +	+ +
36	(structure)	328	412	+ +	±	±
37		324	450	+ +	+ +	+ +
36	(structure)	342	464	+ +	±	±
37		328	430	+ +	+ +	+ +
36	(structure)	332	429	+ +	±	+ +
37		333	440	+ +	+ +	+ +
36	(structure)	302	439	+ +	+ +	+ +
37		309	485	+ +	+ +	+ +
36	(structure, –SO₂–)	312	415	+ +	+ +	+ +
37		306	465	+ +	+ +	+ +
(structure)	**39**	310	510	+ +	+ +	+ +
(structure)		–	525	–	–	–
(structure, –O–)		300	520	+ +	–	–
(structure, –CH₂–)		290	490	+ +	+ +	+ +
35		–	540	–	±	+ +

[a] + + soluble at room temperature; + soluble on heating; ± partially soluble; – insoluble
[b] pyridine
[c] m-cresol

It is generally accepted that in the case of simple heterocycles, the degree of aromaticity of the compounds determines the thermostability [55]. The general observation is that an increase in resonance energy is a quantitative measure of the increase in aromaticity, and the resonance energy increases in the order: furan < pyrrole < thiophene [56].

Diamines and diacids containing the dibenzothiophenediyl unit were used by Srinivasan et al. to synthesize aromatic polyamides 40 [57]. Those polymers with one dibenzothiophenediyl moiety in the repeating unit are soluble in concentrated H_2SO_4, amide solvents, pyridine and m-cresol when X = S and also show higher stability (395–490 °C) than the open-chain analogues. Polyamides with X = SO_2 are less soluble.

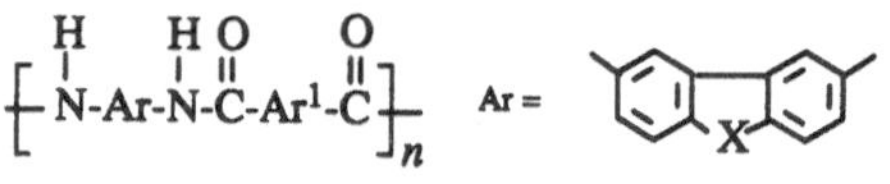

$$Ar^1 = a = a, \qquad b = b$$

$$X = S, SO_2$$

In the continuation of their study, these authors have prepared polyamides 41 containing thianthrene tetraoxide units (Table 9) [58]. They are all highly soluble in amide solvents up to 38%. Solubility follows the order: cardo diamine-containing polymers > heterocycle-containing polymers > open-chain polymers. This result may be attributed to the non-planar angular structure of thianthrene tetraoxide moiety which prevents close packing of the polymer molecules. The percentage of weight loss at 450 °C shows that thermostability follows the same order as the solubility. Polyamides from cardo amines are most stable, and this can be attributed to the bulky pendent groups about the tetrahedral carbon, giving rigidity to the macromolecular chain. In comparison, polyamides 42 containing thiaxanthone 5,5'-dioxide have comparable thermo-stability with polyamides containing thianthrene tetraoxide units, and are also quite soluble in polar aprotic solvents.

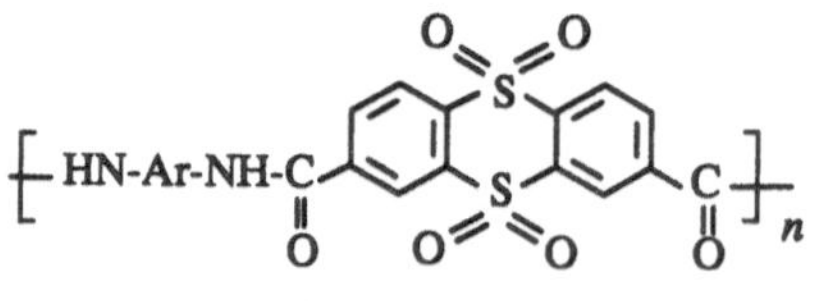

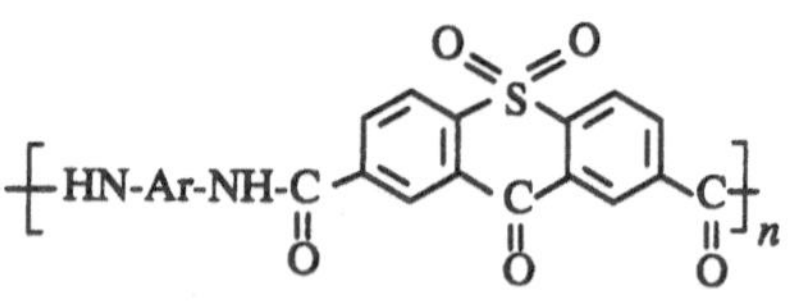

High solubility has also been achieved by using 1-thia-4,5-diaza-cyclohepta-2,4,6-triene as an heterocyclic monomer unit [60]. The polyamides 43 not only have high solubility in DMSO, HMPT, and NMP, and good solubility in DMF, and DMAc, but also have some solubility in formic acid. They decompose between 465–505 °C.

Table 9. Properties of polyamides **41**

Diacid	Diamine	Solubility[a] DMF DMSO	C[b]	wt % in DMAc	Percentage of weight loss at 450 °C
[thianthrene tetraoxide diacid structure]	[p-phenylene structure]	+	−	13	15
	[biphenylene structure]	+	−	16	12
	[diphenyl ether structure]	+	±	20	15
	[diphenyl sulfone (−SO₂−) structure]	+	−	14	6
	[fluorenylidene diphenyl structure]	+	+	33	3
	[anthrone/ketone diphenyl structure]	+	+	38	1
	[phenoxathiin structure]	+	±	26	10
	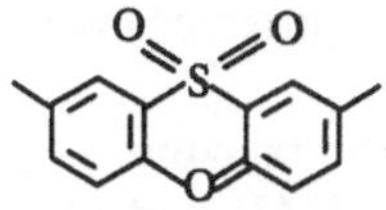 [phenoxathiin dioxide structure]	+	−	28	4

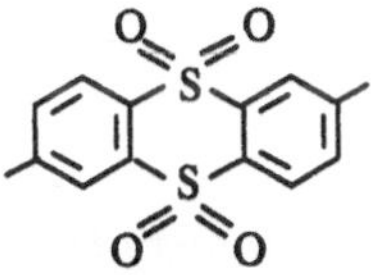

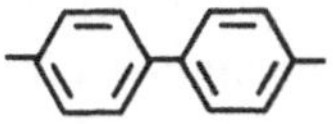

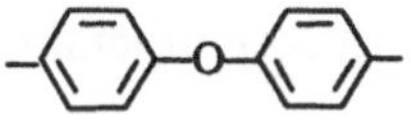

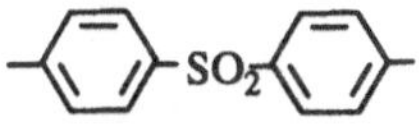

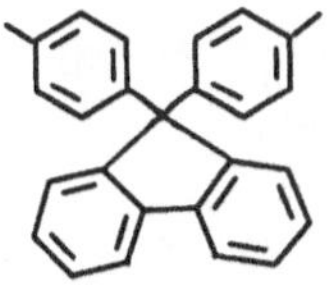

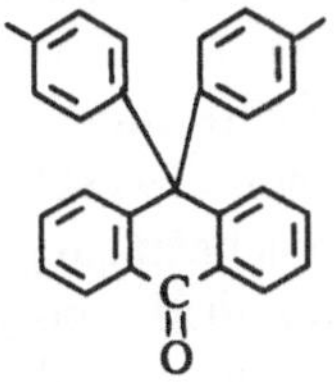

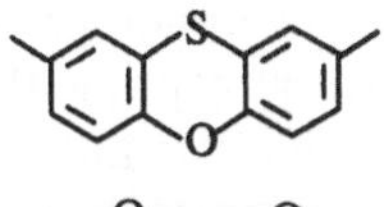

[a] + soluble at room temperature; ± partially soluble at room temperature; − insoluble
[b] C: m-cresol

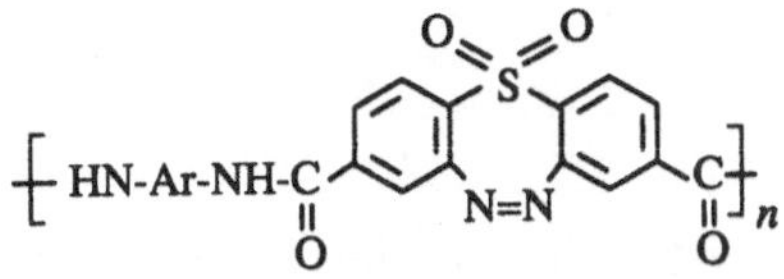

$$\left[\text{HN-Ar-NH-C} \underset{O}{\overset{\|}{}} \text{—} \underset{N=N}{} \text{—} \underset{O}{\overset{\|}{C}} \right]_n$$

[structure with O=S=O bridge, N=N azo linkage]

43

In order to modify the intermolecular interactions of polymers, and hence alter polymer properties such as solubility, lyotropism etc., six membered ring nitrogen-containing heterocycles have been used to prepare polyamides. For example, Morgan et al. [61a] and Ogata et al. [61b] prepared the following polyamide **44** containing a pyridine ring.

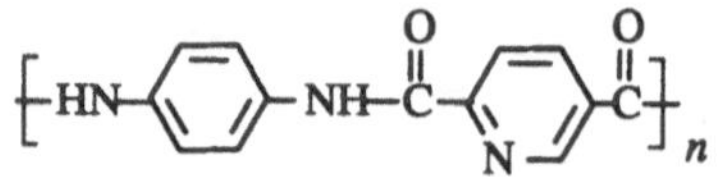

44

Suter et al. synthesized the polyamide **45** having a pyrimidine moiety [62] and polyamide **46** containing a triazine ring was made by Yuki et al. [63]. These heterocycles have polar groups imparting an additional dipole moment to the molecules, and hence they could interact with each other and alter the properties of the polymers containing these groups.

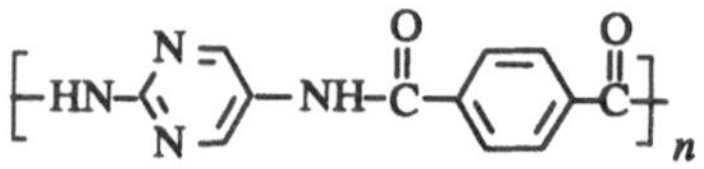

45 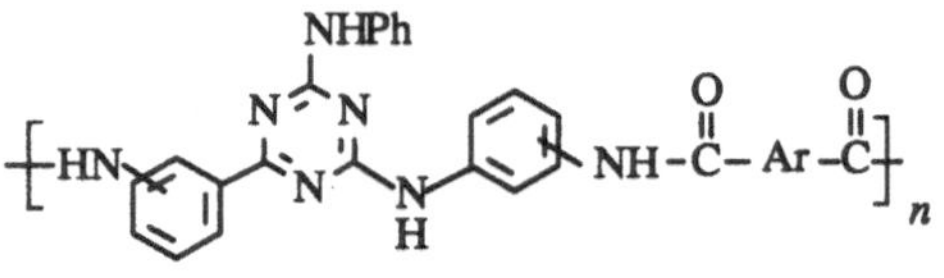   **46**

The polyamide **44** having a pyridine moiety in the chain shows much higher solubility in concentrated H_2SO_4 than does PPTA. The solubility behaviour of polyamide **45** is noteworthy. Those polymers with moderately high molecular weight are very soluble in NMP (LiCl), and conc. H_2SO_4 and also soluble in aqueous basic solution such as 5% KOH. It has been suggested that this might due to the deprotonation of the amide groups in the polymer. The increased acidity of polyamide **45** compared to PPTA is probably explained by the electronic effect of the pyrimidine moiety on the amide groups of polymer. The authors also showed that the polymer is more susceptible than PPTA to attack by strong electrophiles such as conc. H_2SO_4. This also might be explained by the presence of the basic pyrimidine moiety. The polymer does not decompose until 450 °C.

As far as those polyamides containing the triazine ring are concerned, they are soluble in DMSO, DMAc, DMF and NMP and also in formic acid and pyridine. They have T_g values in the range 232–258 °C and T_d values for 5% weight loss of 422–435 °C.

3.1.4 Summary

We have been able to see that extensive research work has been carried out over the past decades in order to obtain highly soluble, easily processable, and thermally stable polyamides with moderately high T_g values. This objective has been achieved with varying success, by for example introducing thermally stable pendant groups such as phenyl, benzamide or iminobenzoyl; also by using double-stranded fused ring structures; and finally by employing thermally stable heterocycles. In some cases, aromatic polyamides have resulted which are soluble in amide solvents, as well as in *m*-cresol, and formic acid which are common solvents for aliphatic polyamides. Polyamides containing the pyrimidine moiety are even soluble in 5% KOH. This remarkable solubility results from the polymer backbone unit structure alone, and the normal planar extended chain structure is maintained. In the case of polyamides containing a double-stranded structure, it is the non-planar configuration which gives rise to the high solubility, by virtue of the disruption of the polymer chain packing. By using pendant groups and heterocycles having polar groups, the intermolecular interactions can be altered, thus leading to greater solubility and at the same time, by using thermally stable structural units, overall polymer thermostability is maintained.

3.2 Lyotropism

Wholly aromatic rigid-rod polyamides have attracted much attention since they were first discovered because they can be spun into fibres (Aramids) exhibiting exceptional thermal and mechanical properties. Good quality fibres with a high degree of orientation can only be readily obtained from anisotropic solutions. In this liquid crystal state the polymer chains have a natural highly ordered alignment. Therefore the lyotropic behaviour of these polymers is of great interest. Processing of wholly unsubstituted aromatic polyamides such as PPDT is difficult as these polymers dissolve and form anisotropic solutions only in concentrated sulfuric acid and high boiling amide solvents (NMP, HMPA etc.) in the presence of LiCl [64, 65]. Furthermore the polymer concentration required is also high. It is therefore of interest and of potentially technological importance to have polyamides which are lyotropic in other common organic solvents and which display such properties over a large concentration range.

Formation of an anisotropic solution depends on several key factors:
a) polymer chemical primary structure and physical structure (i.e., axial ratio);
b) molecular weight; c) nature of solvent and polymer-solvent interaction;
d) solubility and concentration of polymer; and e) temperature.

Polyamide-type polymers form anisotropic solutions by virtue of their rigid-rod nature and the intermolecular hydrogen bonding of the polymer chains. One criterion for the formation of a liquid crystalline state is that the polymer chain has an axial ratio x which exceeds 6.42 [66]. The molecular weight is another important physical criterion. The higher it is, the lower the critical concentration of polymer required for anisotropy. However high molecular weight might restrict the solubility of the polymer. Solubility itself is also a critical criterion since only when a certain critical concentration is reached, can an anisotropic solution form. The solubility of a polymer, as we have seen in the last section, depends not only on the polymer primary structure, but also on its molecular weight, its crystallinity, the packing of the polymer chains, polymer-solvent interaction, the intermolecular interactions of polymer chain, and of course the temperature.

Different approaches have been described in the literature in order to obtain more soluble lyotropic polyamides, for example, substitution of the amide functions of the polymer chain [67, 68], and introduction of pendant groups on the aromatic ring of the polymer backbone [27, 69]. The introduction of pendant groups might decrease the intermolecular hydrogen bonding hence increasing the solubility. On the other hand, if an appropriate substituent is chosen, an enhancement of the interaction between the solvent and these groups might also be expected, and as a result, better solvation of the polymer chains.

3.2.1 N-Substituted Polyamides

In order to increase the solubility of the para wholly aromatic polyamide, PPTA, Katayose succeeded in producing the *N*-substituted polyamide using a metalation reaction. The polyamide was reacted with a solution of sodium methylsulfinyl carbanion in DMSO; this anion being formed from reaction of powdered sodium hydride with excess dry DMSO [67a] (Scheme 12).

The polyanion is soluble in DMSO, and different substituents were grafted onto the polyamide backbone via this anion. For instance, C_3–C_{18} alkyl chain, benzyl, 1-naphthylmethyl, 9-anthrylmethyl, carboxymethyl were successfully attached. All the substituted polyamides **47**, except the carboxymethyl substi-

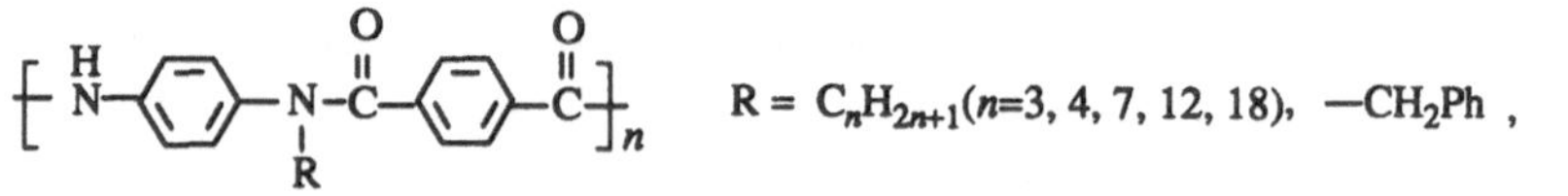

$$R = C_nH_{2n+1}(n=3, 4, 7, 12, 18), \quad -CH_2Ph ,$$

$$-CH_2COOH$$

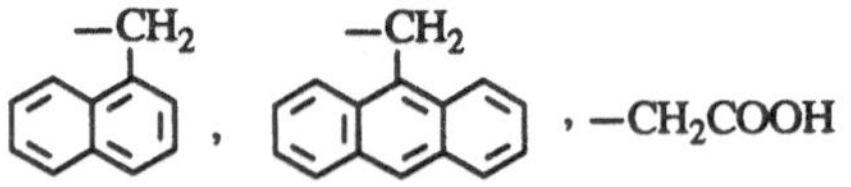

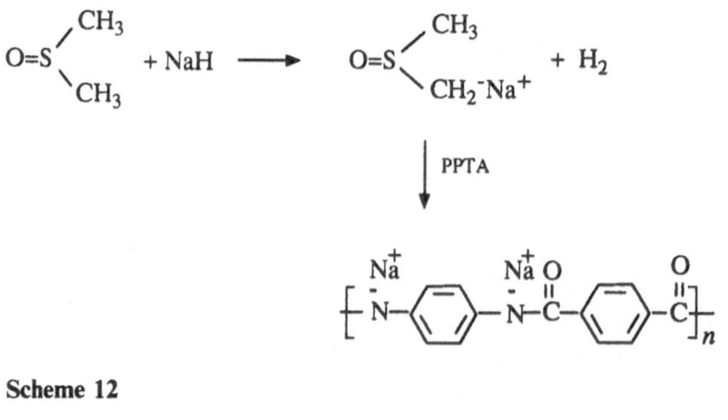

Scheme 12

tuted one, are soluble in a number of organic solvents such as phenol and tetrachloroethane. The N-alkyl substituted polymers are also soluble in THF, while polymers substituted with C_3, and C_4 chain are soluble in DMSO but not soluble in toluene. The opposite situation is true for the C_{10}, and C_{18} substituted polymers. This was explained to be a consequence of the increased hydrophobicity of longer alkyl chain.

Among these substituted polyamides, only N-(9-anthryl) methylated PPTA($M_v = 24\,700$) showed birefringence under crossed polarizers at 40% v/v concentration in bromoform. Formation of an anisotropic solution was suggested. It was explained that the bulky 9-anthrylmethyl makes the N-substituted PPTA more rigid and the group also has a tendency to stack itself, hence helping to order whole N-anthrylmethylated PPTA molecules. In a continuation of this work [67b], the same authors grafted C_{18} alkyl chains onto PPTA of different molecular weights. The polymers obtained from a PPTA of molecular weight of 36 400 showed lyotropic behaviour in THF, dichloroethane, tetrachloroethane and bromoform. The critical concentrations required are about 30%, 50%, 50% and 60% by volume respectively. No liquid crystalline behaviour was observed in benzene, even up to 60 v/v %. This was explained by the lack of the type of molecular interactions between the amide groups of PPTA and polar atoms in the polar solvents.

Recently, Reynolds [68] reported the preparation of poly(p-phenyleneterephthalamide)propanesulphonate 48 made from the PPTA polyanion and 1,3-propanesultone (Scheme 13). This polyelectrolyte can be solubilised in water to greater than 18 wt % when the amide is alkylsulphonated to 66%, but no lyotropic behaviour is observed.

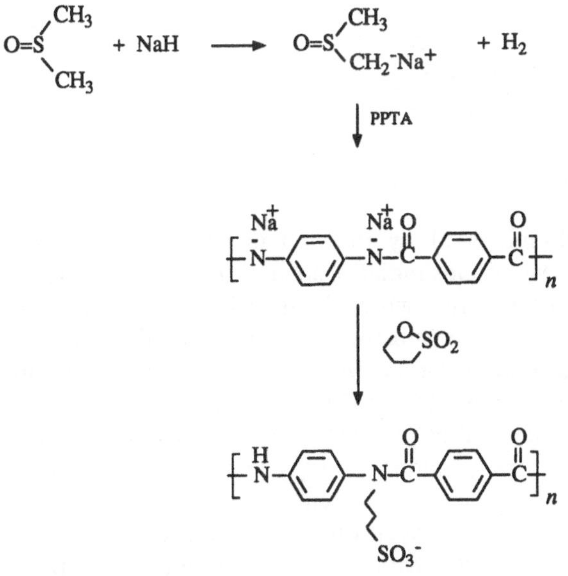

Scheme 13   **48**

3.2.2 Polyamides with Pendant Substituents

Krigbaum synthesized the phenylated *para* aromatic polyamides, **49a**, **49b**, by using a phenyl substituted diamine or diacid [27]. Their solubility in DMAc or NMP (4% LiCl) is very high, i.e. 20% and 50% by weight respectively. Polymer **49a** forms an anisotropic solution in DMAc/4% LiCl but not polymer **49b**. It has been argued that the bulky phenyl substituent *ortho* to the carbonyl group in polymer **49b**, compared to the phenyl *ortho* to the amino group in **49a**, is more effective in increasing solubility, but also creates a more flexible chain.

Vandenberg prepared the following water-soluble polyamide, poly(*N,N'*-(sulfo-phenylene)phthalamide) **50** [69]. It has a molecular weight of 5 000 and is soluble in water upon heating. On cooling the resulting solution a viscous gel is formed at concentration as low as 0.4 wt %, and this showed birefringence under crossed polarizers. It has been suggested that some type of unusual association is occuring, and indeed the concentrations involved are far below those usually required for lyotropic liquid crystal formation.

Very recently, Schmidt et al. synthesized novel polyamides **51** by using arylsubstituted terephthalic acids moieties such as *para-* or *ortho*-terphenyl-2,5-dicarboxylic acids in combination with substituted and non-coplanar diamines [70]. Those polyamides **51** from substituted diacids, diamines or non-coplanar diamines showed high solubility in DMAc, and in most cases without addition of inorganic salts (LiCl). In DMAc (LiCl), polyamide **51ca'** (η_{inh} = 1.63 dl/g) forms a lyotropic liquid crystalline phase at > 8 wt % at room temperature, and at > 5 wt % at 110 °C for polyamide **51ae'** (η_{inh} = 3.58 dl/g). On the other hand, with the copolyamides **52** the critical concentrations of liquid crystal formation are around 40–45% at room temperature. For these copolyamides, concentrated solutions in DMAc/LiCl with polymer concentration up to 50 wt % could be prepared for polymers with x > 0.6.

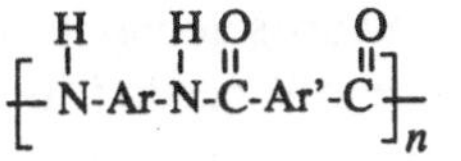

$$\left[N{-}Ar{-}N{-}\overset{O}{\underset{}{C}}{-}Ar'{-}\overset{O}{\underset{}{C}} \right]_n$$

51

Ar =

a   b   c

Ar' = a'

b'   c'   d   e' = c

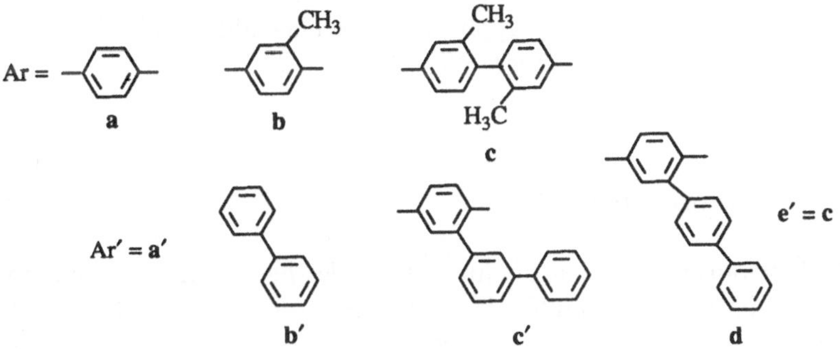

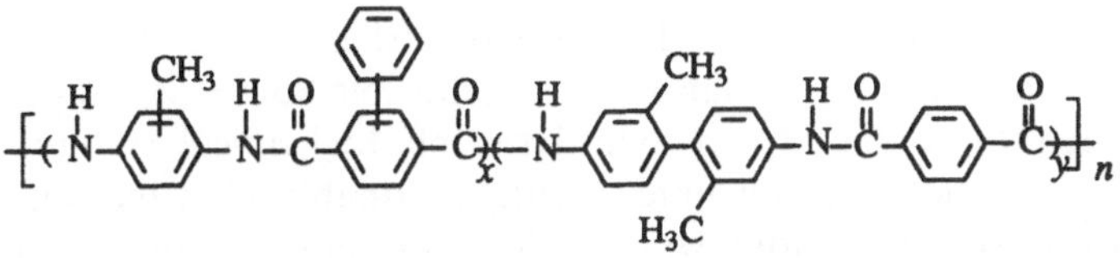

52

3.2.3 Non-Coplanar Polyamides

Gaudiana et al. in their extensive research of amorphous rigid-rod polymers have synthesized a class of high molecular weight rod-like polyamides containing all para-linked 2,2′-substituted biphenylene and/or stilbene repeat units **53** [71]. These polyamides show high solubility in common organic solvents, and some are even soluble in THF, glyme or acetone. Nevertheless, despite their high molecular weight, suitable persistence length of the polymer chain, and high solubility (prerequisites for liquid crystal formation), they did not form lyotropic liquid crystal solutions.

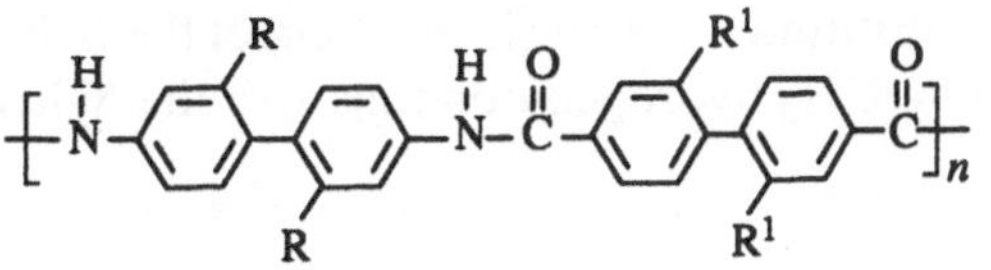

53a

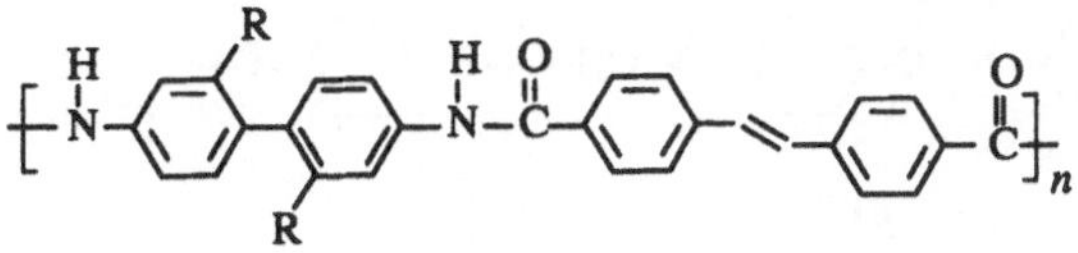

53b

In comparison with the efforts made towards achieving soluble yet thermally stable polyamides, many more studies are still needed in the field of lyotropic polyamides if anisotropic solutions in common organic solvents such as THF, chloroform etc. are to be achieved. This remains an interesting area for the fabrication of polyamides. Water soluble polyamides on the other hand, remain intriguing for some specific applications such as additives in the paper industry as described by Vandenberg [69].

3.3 Thermotropism

The lyotropism of *para* aromatic polyamides has made it possible to fabricate high strength, high modulus thermally resistant fibres, like Kevlar. These fully aromatic stiff chain polyamides usually do not melt at high temperature but decompose. The strong intermolecular hydrogen bonding of the amide linkages prevent the rigid polymer from melting. On the other hand, polyamides seem to be excellent candidates for the fabrication of durable liquid crystal polymers by virtue of their high stability to hydrolysis, photochemical stability and the possibility of synthesis of polyamides with reasonably high molecular weight [72]. Thermotropic behaviour would allow spontaneous alignment prior to melt processing, and would offer the prospect of a convenient technology for the production of very high tensile strength fibres and engineering plastics. Thermotropic polymers with relatively low transition temperatures and better solubility require structures with both rigid and flexible elements. The use of angular segments, non-symmetrically substituted units, and copolycondensation often enhance these effects [73].

One approach to achieve a decrease in the transition temperatures is to use substituents in the position *ortho* to the amide nitrogen on the aromatic ring carbons. Griffin et al. thus synthesized the following polyamides **54** [72]. Thermotropic phase transitions are indeed observed with polyamides made from 3,3'-disubstituted-4,4'-diaminophenyl. The transition temperatures decrease in the order $CH_3 > Cl > OCH_3$ as we can see from the Table 10. The authors have attributed this observation to the reduction of intermolecular hydrogen bonding caused by substitution, i.e. steric blocking of the N-hydrogen or intramolecular *ortho* Cl and OCH_3 hydrogen bonding with the N-hydrogen.

54

X = H, Cl, CH₃, OCH₃

Table 10. Thermal transition temperature of polyamides **54**

X	T_N	T_I
H	–	–
CH_3	272	332
Cl	250	275
OCH_3	226	256

T_N: solid to nematic liquid crystal transition temperature

T_I: nematic liquid crystal to isotropic melt transition temperature

Schmuchi et al. have continued in this direction [74] and have prepared the polymers **55**. Introduction of an ethylene unit into the stiff diphenylene segment not only reduces the isotropic melt temperature, but also depresses the tendency to form liquid crystals.

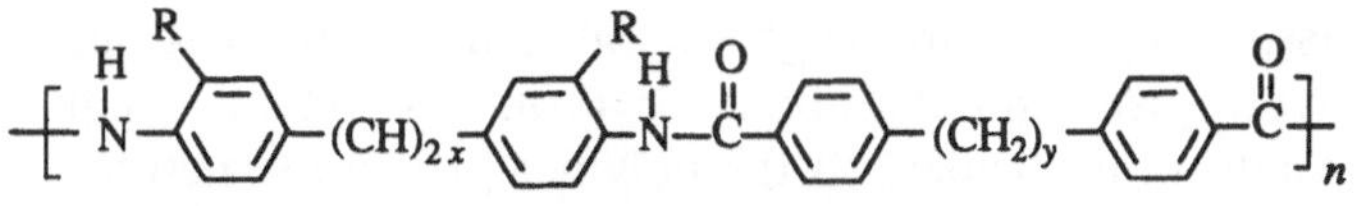

55

$x = 6, 10;\ \ y = 0, 2;\ \ R = H, CH_3$

Ringsdorf's research group have prepared novel types of rigid-rod polyesters and polyamides **56** with a disc-like mesogen in the mainchain [75]. Most polymers with six lateral substituents appear to be thermotropic liquid crystals. Polyamides with Z = H and having four substituents on the diamine component are not liquid crystalline. The two substituents on the diacid component seem to contribute to decrease further the intermolecular hydrogen bonding.

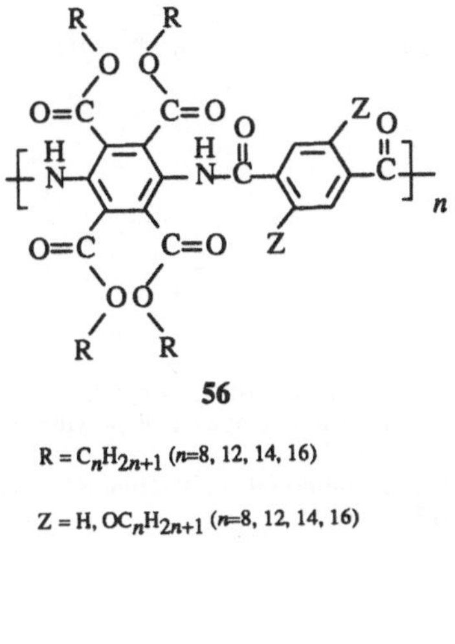

56

$R = C_nH_{2n+1}$ (n=8, 12, 14, 16)

$Z = H, OC_nH_{2n+1}$ (n=8, 12, 14, 16)

4 Conclusions

The initial commercial success of wholly aromatic polyamides (Kevlar and Nomex) has stimulated extensive further research in the following decades. Different synthetic methods have been developed in order to prepare high molecular weight polyamides under mild conditions, or with simplified procedures by starting from easily purified and manipulable monomers. Successful polycondensations have now been achieved when many other functionalities have been present on the monomer molecules. The properties of polyamides have also been studied extensively, in particular, thermal stability. Much research effort has been oriented towards obtaining more soluble, high temperature processable and thermally stable polymers. Very intriguing results have been obtained with phenyl, and benzamide substituted polyamides, and by incorporating thermally stable heterocycles or double-stranded structures into polymer backbones.

Another important property of polyamide is their lyotropism. Polyamides with higher solubility in common organic solvents are still required. Only a few examples exist in the literature. High solubility in a solvent is one of the key factors but lyotropism and the ability to form and retain a high degree of ordering depends, in particular, on the polymer primary structure.

As far as thermotropism is concerned, a limited number of papers in the literature have already shown a very promising future. Even wholly aromatic polyamides can be made thermotropic as soon as the strong intermolecular hydrogen bonding is weakened, as shown by Ringsdorf's research group [75].

By changing the primary structure of the polymer or by creating irregularities, or by introducing substituents capable of interacting efficiently with solvent, or by creating non-coplanarity, or indeed by other methods not yet known, we believe that a better understanding of the structure-property relationships can be achieved. This in turn will allow unforeseen properties to be discovered, and polyamides with novel high performance and novel applicability to be developed.

Acknowledgement. This review was prepared while J.L. was the recipient of an ICI post-doctoral fellowship which we warmly acknowledge.

5 References

1. a) Reisch MS (1987) Chem Eng News 65: 9
   b) Tanner D, Fitzgerald JA, Phillips BR (1989) Angew Chem Int Ed Engl Adv Mater 28: 649
   c) Preston J (1988) In: Mark HF, Kroschwitz JI (eds) Encyclopedia of polymer science and engineering, 2nd edn, Vol 11. Wiley-Interscience, New York, p. 381
   d) Vollbracht L (1989) In: Allen G, Bevington JC, Eastmond GC (eds) Comprehensive polymer science, 1st edn, vol 5. Pergamon, Oxford, p. 375

   e) Gaymans RJ, Sikkema DJ (1989) In: Allen G, Bevington JC, Eastmond GC (eds) Comprehensive polymer science, 1st edn, vol 5. Pergamon, Oxford, p. 357
   f) Zimmerman J (1989) In: Allen G, Bevington JC, Eastmond GC (eds) Comprehensive polymer science 315
2. Müller WT, Ringsdorf H (1990) Macromolecules 23: 2825
3. a) Du Pont de Nemours Co. (1961) Brit Patent 871581
   b) Kwolek SL, Morgan PW, Sorenson WR (1962) US Patent 3,063,966
4. a) Bair TI, Morgan PW (1972) US Patent 3,673,143
   b) Bair TI, Morgan PW (1974) US Patent 3,817,941
   c) Bair TI, Morgan PW, Killian FL (1977) Macromolecules 10: 1396
5. Herlinger H, Hoerner HP, Druschke F, Denneler W, Haiber F (1973) Appl Polym Symp 21: 201
6. Zapp Jr JA (1975) Science 190: 422
7. a) Mera H, Nakagawa Y, Yamaguchi M, Ohno M (Oct. 30, 1979) US Pat 4,172,938 (to Teijin Ltd)
   b) Vollbracht L, Veerman TJ (Dec. 29, 1981) US Pat 4,308,374 (to Akzo NV)
8. Yamazaki N, Higashi F, Kawabata J (1974) J Polym Sci Polym Chem Ed 12: 2149
9. Yamazaki N, Matsumoto M, Higashi F (1982) J Polym Sci Polym Chem Ed 20: 2081 (and references cited therein)
10. a) Krigbaum WR, Kotek R, Mihara Y, Preston J (1985) J Polym Sci Polym Chem Ed 23: 1907
    b) Krigbaum WR, Kotek R, Mihara Y, Preston J (1984) J Polym Sci Polym Chem Ed 22: 4045
11. Higashi F, Kokubo N, Goto M (1980) J Polym Sci Polym Chem Ed 18: 2879
12. Higashi F, Kubota K, Sekizuka M (1981) J Polym Sci Polym Chem Ed 19: 2681
13. Ogata N, Sanui K, Tanaka H, Yasuda S (1981) Polym J 13: 989
14. Higashi F, Hoshio A (1982) Polym Prepr Jpn 31(6): 1417
15. a) Kitayama S, Sanui K, Ogata NJ (1984) Polym Sci Polym Chem Ed 22: 2705
    b) Ogata N, Sanui K, Zao A, Watanabe M, Hanaoka T (1988) Polym J 7: 529
16. a) Ueda M, Oikawa H (1985) J Polym Sci Polym Chem Ed 23: 1607
    b) Ueda M, Mochizuki A (1985) Macromolecules 18: 2353
    c) Ueda M, Kameyama A, Ikeda C (1987) Polym J 19: 673
    d) Ueda M, Kameyama A, Hashimoto K (1988) Macromolecules 21: 19
17. Oishi Y, Kakimoto M, Imai Y (1987) Macromolecules 20: 703
18. Oishi Y, Kakimoto M, Imai Y (1988) Macromolecules 21: 547
19. Imai Y, Oishi Y (1989) Prog Polym Sci 14: 173
20. a) Morgan PW (1988) In: Mark HF, Kroschwitz JI (eds) Encyclopedia of polymer science and engineering, 2nd edn., Wiley-Interscience, New York 8: 221
    b) Morgan PW (1965) Condensation polymers: By interfacial and solution methods. Wiley-Interscience, New York
21. Morgan PM (1979) Chemtech 9: 316
22. a) Chaudhuri AK, Min BY, Pearce EM (1980) J Polym Sci Polym Chem Ed 18: 2949
    b) Kapuscinska M, Pearce EM (1984) J Polym Sci Polym Chem Ed 22: 3989
    c) Kapuscinska M, Pearce EM, Chung HFM, Ching CC, Zhou QX (1984) J Polym Sci Polym Chem Ed 22: 3999
23. a) Nagata Jr M, Tsutsumi N, Kiyotsukuri T (1988) J Polym Sci Polym Chem Ed 26: 235
    b) Kiyotsukuri T, Tsutsumi N, Okada K, Asai K, Nagata M (1988) J Polym Sci Polym Chem Ed 26: 2225
    c) Díaz FR, Tagle LH, Padilla ME (1985) J Polym Sci Polym Chem Ed 23: 2043
24. Whang WT, Pearce EM (1987) J Polym Sci Polym Chem Ed 25: 171
25. a) Takatuska R, Uno K, Toda F, Iwakura Y (1977) J Polym Sci Polym Chem Ed 15: 1905
    b) Takatuska R, Uno K, Toda F, Iwakura Y (1978) J Polym Sci Polym Chem Ed 16: 361
26. Yang CP, Oishi Y, Kakimoto M, Imai Y (1989) J Polym Sci Polym Chem Ed 27: 3895
27. a) Jadhav JY, Krigbaum WR, Preston J (1988) Macromolecules 21: 538
    b) Jadhav JY, Krigbaum WR, Preston J (1989) J Polym Sci Polym Chem Ed 27: 1175
28. Kakimoto M, Yoneyama M, Imai Y (1988) J Polym Sci Polym Chem Ed 26: 149
29. Lozano AE, de la Campa JG, de Abajo J (1990) Makromol Chem Rapid Commun 11: 471
30. Mélandez A, de la Campa J, de Abajo J (1988) Polymer 29: 1142
31. Guijarro E, de la Campa JG, de Abajo J (1984) J Polym Sci Polym Chem Ed 22: 1531
32. Kakimoto M, Padmanaban M, Yoneyama M, Imai Y (1988) J Polym Sci Polym Chem Ed 26: 2863
33. de Abajo J, Guijarro E, Serna FJ, de la Campa JG (1986) J Polym Sci Polym Chem Ed 24: 483
34. de la Campa JG, Guijarro E, Serna FJ, de Abajo J (1985) Eur Polym J 21: 1013

35. Starr L (1966) J Polym Sci A-1 4: 3041
36. Liou GS, Oishi Y, Kakimoto M, Imai Y (1991) J Polym Sci Polym Chem Ed 29: 995
37. Oishi Y, Takado N, Yoneyama M, Kakimoto M, Imai Y (1990) J Polym Sci Polym Chem Ed 28: 1763
38. Jadhav AS, Maldar NN, Shinde BM, Vernekar SP (1991) J Polym Sci Polym Chem Ed 29: 147
39. Mohite SS, Maldar NN, Marvel CS (1988) J Polym Sci Polym Chem Ed 26: 2777
40. Yoneyama M, Athula Kuruppu KD, Kakimoto M, Imai Y (1988) J Polym Sci Polym Chem Ed 26: 2917
41. Abraham T, Soloski EJ, Evers RC (1988) J Polym Sci Polym Chem Ed 26: 959
42. Idage SB, Idage BB, Vernekar SP (1989) J Appl Polym Sci 38: 2057
43. Adduci J, Chapoy LL, Jonsson G, Kops J, Shinde BM (1983) J Appl Polym Sci 23: 2069
44. Idage SB, Idage BB, Shinde BM, Vernekar SP (1989) J Polym Sci Polym Chem Ed 27: 583
45. Ghatge ND, Jadhav JY (1984) J Polym Sci Polym Chem Ed 22: 1565
46. a) Arnold Jr C (1979) J Polym Sci Macromolec Rev 14: 265
    b) Cassidy PE (1980) Thermally stable polymers, 1st edn. Dekker, New York
47. Ueda M, Aizawa T, Imai Y (1977) J Polym Sci Polym Chem Ed 15: 2739
48. Niume K, Nakamichi K, Toda F, Uno K, Hasegawa M, Iwakura Y (1980) J Polym Sci Polym Chem Ed 18: 2163
49. a) Sato M, Yokoyama M (1979) Eur Polym J 15: 733
    b) Kondo H, Sato M, Yokoyama M (1984) J Polym Sci Polym Chem Ed 22: 1055
50. a) Kovacs HN, Delman AD, Simms BB (1968) J Polym Sci A-1 6: 2103
    b) Kovacs HN, Delman AD, Simms BB (1968) J Polym Sci A-1 6: 2117
    c) Mulvaney JE, Marvel CS (1961) J Polym Sci 50: 541
    d) Breed LW, Wiley Jr JC (1976) J Polym Sci Polym Chem Ed 14: 83
    e) Kuckertz VH (1966) Makromol Chem 98: 101
51. a) Kondo H, Sato M, Yokoyama M (1982) Eur Polym J 18: 181
    b) Kondo H, Sato M, Yokoyama M (1982) Eur Polym J 18: 679
    c) Kondo H, Sato M, Yokoyama M (1983) J Polym Sci Polym Chem Ed 21: 165
52. a) Kakimoto M, Negi YS, Imai Y (1985) J Polym Sci Polym Chem Ed 23: 1787
    b) Imai Y, Maldar NN, Kakimoto M (1985) J Polym Sci Polym Chem Ed 23: 1797
53. Jeong H, Oishi Y, Kakimoto M, Imai Y (1990) J Polym Sci Polym Chem Ed 28: 3293
54. Jeong H, Kakimoto M, Imai Y (1991) J Polym Sci Polym Chem Ed 29: 767
55. Bruinsma OSC, Fromp PJJ, De S Nolting HJJ, Moulijn JA (1988) Fuel 67: 334
56. a) De Jongh HAP, Wynberg H (1965) Tetrahedron 21: 515
    b) Franklin JL (1950) J Amer Chem Soc 72: 4278
    c) Cox JD (1963) Tetrahedron 19: 1175
57. Srinivasan PR, Srinivasan M, Mahadevan V (1981) Makromol Chem 182: 1937
58. Prema S, Srinivasan M (1987) Eur Polym J 23: 897
59. Reddy TA, Srinivasan M (1988) J Polym Sci Polym Chem Ed 26: 1063
60. Reddy KA, Srinivasan M (1990) J Macromol Sci-Chem A 27: 755
61. a) US Pat 3,836,498 (1974); invs: Gulrich M, Morgan PW, Chem Abstr 82: 5299g
    b) Ogata N, Sanui K, Koyama T (1981) J Polym Sci Polym Chem Ed 19: 151
62. Oertli AG, Meyer WR, Suter UW (1991) Makromol Chem Rapid Commun 12: 57
63. Lin J, Yuki Y, Kunisada H, Kondo S (1990) J Appl Polym Sci 40: 2113
64. Schaefgen JP, Bair TI, Ballou JW, Kwolek SL, Morgan PW, Panar M, Zimmermann J (1979) In: Cifferti A, Ward IM (eds) Rigid chain polymers, properties of solution and fibers. Banking, England, p. 173
65. Kwolek SL, Memeger W, Van Trump JE (1987) In: Lewin M (ed) Liquid crystalline para aromatic polyamides. VCH, New York, p 421
66. a) Flory PJ (1956) Proc R Soc London Ser A 234: 60
    b) Flory PJ (1956) Proc R Soc London Ser A 234: 73
    c) Flory PJ, Ronca G (1979) Mol Cryst Liq Cryst 54: 289
    d) Flory PJ (1984) In: Platé NA (ed) Liquid crystal polymers I. Springer, Berlin Heidelberg New York, p 1 (Advances in Polymer Science, vol 59).
    e) Kwolek SL, Morgan PW, Schaefgen JR (1988) In: Mark HF, Kroschwitz JI (eds) Encyclopedia of polymer science and engineering, vol 9, 2nd edn. Wiley-Interscience, New York, p. 1
67. a) Takayanagi M, Katayose T (1981) J Polym Sci Polym Chem Ed 19: 1133
    b) Takayanagi M, Katayose T (1984) J Appl Polym Sci 29: 141
    c) Takayanagi M, Goto K (1984) J Appl Polym Sci 29: 2057
    d) Moore DR, Mathias LJ (1986) J Appl Polym Sci 32: 6299

   e) Takayanagi M, Goto K (1985) Polym Sci Technol 27: 247
   f) Reynolds JR, Baker CK, Gieselman M (1989) Polym Prepr (Am Chem Soc Div Polym Chem) 30: 151
68. Gieselman MB, Reynolds JR (1990) Macromolecules 23: 3118
69. a) Vandenberg EJ, Diveley WR, Filar LJ, Patel SR, Barth HG (1987) Polym Mater Sci Eng 57: 139
   b) Vandenberg EJ, Diveley WR, Filar LJ, Patel SR, Barth HG (1989) J Polym Sci Polym Chem Ed 27: 3745
70. Hatke W, Schmidt HW, Heitz W (1991) J Polym Sci Polym Chem Ed 29: 1387
71. Gaudiana RA, Minns RA, Sinta R, Weeks N, Rogers HG (1989) Prog Polym Sci 14: 47
72. Griffin AC, Britt TR, Campbell GA (1982) Mol Cryst Liq Cryst 82: 145
73. a) Blumstein A, Asrar J, Blumstein RB (1984) Liq Cryst Ordered Fluids 4: 311
   b) Al-Dujaili AH, Jenkins AD, Walton DRM (1984) Makromol Chem Rapid Commun 5: 33
74. Schmucki M, Jenkins AD (1989) Makromol Chem 190: 1303
75. Ringsdorf H, Tschiner P, Hermann-Schönherr O, Wendorff JH (1987) Makromol Chem 1: 1431

Editor: L. Ledwith
Received: April 28, 1992

Author Index Volume 101-111

Subject Index

Springer-Verlag
and the Environment

We at Springer-Verlag firmly believe that an international science publisher has a special obligation to the environment, and our corporate policies consistently reflect this conviction.

We also expect our business partners – paper mills, printers, packaging manufacturers, etc. – to commit themselves to using environmentally friendly materials and production processes.

The paper in this book is made from low- or no-chlorine pulp and is acid free, in conformance with international standards for paper permanency.